Lecture Notes in Mathematics

Volume 2314

This series reports on new developments in all areas of mathematics and their applications - quickly, informally and at a high level. Mathematical texts analysing new developments in modelling and numerical simulation are welcome. The type of material considered for publication includes:

1. Research monographs 2. Lectures on a new field or presentations of a new angle in a classical field 3. Summer schools and intensive courses on topics of current research.

Texts which are out of print but still in demand may also be considered if they fall within these categories. The timeliness of a manuscript is sometimes more important than its form, which may be preliminary or tentative.

Titles from this series are indexed by Scopus, Web of Science, Mathematical Reviews, and zbMATH.

Jialin Hong • Liying Sun

Symplectic Integration of Stochastic Hamiltonian Systems

 Springer

Jialin Hong
Academy of Mathematics and Systems
Science
Chinese Academy of Sciences
Beijing, Beijing, China

Liying Sun
Academy of Mathematics and Systems
Science
Chinese Academy of Sciences
Beijing, Beijing, China

This work was supported by National Natural Science Foundation of China No. 91530118 No. 91630312 No. 11711530017 No. 11926417 No. 11971470 No. 11871068 No. 12031020 No. 12026428 No. 12101596 No. 12171047

ISSN 0075-8434   ISSN 1617-9692   (electronic)
Lecture Notes in Mathematics
ISBN 978-981-19-7669-8   ISBN 978-981-19-7670-4   (eBook)
https://doi.org/10.1007/978-981-19-7670-4

Mathematics Subject Classification: 65P10, 37-xx, 60-xx

This Springer imprint is published by the registered company Springer Nature Singapore Pte Ltd.
The registered company address is: 152 Beach Road, #21-01/04 Gateway East, Singapore 189721, Singapore

Preface

As numerous modern challenges in scientific questions, industrial needs, and societal requirements emerge, the demand for designing numerical methods to solve tremendously complex phenomena concerning stochasticity, imprecision, and vagueness also grows. Even though the computational power increases rapidly, stochastic numerical methods still cannot compute sophisticated stochastic problems well over longer time intervals. The qualitative properties of stochastic numerical methods for such problems are of vital importance to the accuracy of the numerical simulation and the reliability of long-term predictions. Recently, stochastic structure-preserving numerical methods, which promise considerable benefits such as good numerical stability and probabilistic superiority, have been of great concern.

This monograph is devoted to stochastic structure-preserving numerical methods for stochastic differential equations. The emphasis is placed on the systematic construction and probabilistic superiority of stochastic symplectic methods, which preserve the symplectic geometric structure of the stochastic flow of finite and infinite dimensional stochastic Hamiltonian systems. The objects considered here are related to several fascinating mathematical fields: numerical analysis, stochastic analysis, symplectic geometry, rough path theory, partial differential equation, probability theory, etc.

Chapter 1 introduces the stochastic flow associated with Stratonovich stochastic differential equations, which is a cornerstone in the study of the stochastic geometric structure. It also collects long-time behaviors of stochastic (rough) Hamiltonian systems, stochastic forced Hamiltonian systems, and stochastic Poisson systems together. Compared to existing textbooks on stochastic Hamiltonian systems, this monograph presents some up-to-date mathematical tools, such as the stochastic θ-generating function with parameter $\theta \in [0, 1]$, stochastic Hamilton–Jacobi theory, which can be exploited to characterize intrinsic properties of the original systems and to construct effective stochastic symplectic methods.

The philosophy behind the preservation of the structure is to identify intrinsic properties of the continuous systems and then design numerical methods inheriting the original properties at discrete time points. At this stage, the convergence has been considered as a basic and crucial aspect for deciding the utility of stochastic numerical methods. The convergence of existing stochastic numerical methods is mainly studied in two different aspects: strong and weak convergence. In Chap. 2, we focus on the systematical construction of strongly and weakly convergent stochastic structure-preserving numerical methods proposed recently, including stochastic symplectic methods, stochastic pseudo-symplectic methods, and stochastic numerical methods preserving invariants, etc., as well as the construction of rough symplectic Runge–Kutta methods. Chapter 3 provides the methodology of constructing higher weak order and ergodic numerical methods based on the modified stochastic differential equation and stochastic generating function. Chapter 4 deals with several novel strongly convergent stochastic structure-preserving numerical methods for stochastic Maxwell equations, stochastic wave equations, and stochastic Schrödinger equations, which are infinite-dimensional stochastic Hamiltonian systems of significant value.

Different from the deterministic case, the solution of the stochastic numerical methods can be regarded as a stochastic process, namely a collection of random variables. An important shift of key ideas comes about by treating the probabilistic behaviors of stochastic numerical methods via the large deviation principle, which is concerned with the exponential decay of probabilities of very rare events. It allows a better understanding on the superiority of stochastic symplectic methods for the stochastic Hamiltonian system. One of the peculiar features in this monograph consists in the numerical merits of stochastic symplectic methods for the stochastic linear oscillator and stochastic linear Schrödinger equation, which are presented in Chaps. 2 and 4. The geometric properties of stochastic numerical methods could also be translated into the preservation of structure on the level of stochastic modified equations by the asymptotic series expansion. This characteristic is demonstrated in Chap. 3 for stochastic differential equations, especially for stochastic Hamiltonian systems.

We hope that Master or PhD students and researchers who have a prior understanding of the basic theory of stochastic processes and stochastic differential equations and are interested in the mathematical or numerical aspects of stochastic simulations will find this monograph useful. The monograph could also be used as a textbook for a semester-long graduate-level course. The topics discussed in this monograph will provide useful tools for the readers to have a good knowledge of stochastic structure-preserving integration, especially for stochastic symplectic methods of stochastic Hamiltonian systems.

Beijing, China   Jialin Hong
Beijing, China   Liying Sun
April 2022

Acknowledgments

This text would not have been possible without the invaluable support and guidance of so many people. We warmly thank Drs. Jianbo Cui, Chuchu Chen, Ziheng Chen, Tonghe Dang, Baohui Hou, Diancong Jin, Derui Sheng, Tau Zhou, and Xiaojing Zhang for helping us to track down an uncountable number of typos. Without their help, the manuscript would be much worse than it is now. We also acknowledge the National Natural Science Foundation of China (No. 91530118, No. 91630312, No. 11711530017, No. 11926417, No. 11971470, No. 11871068, No. 12031020 and No. 12026428, No. 12101596, No. 12171047) for its financial support.

Contents

Notation and Symbols

$(E, \mathbf{d})$	Metric space
$\mathbb{H}$	Hilbert space
$\langle \cdot, \cdot \rangle_{\mathbb{H}}$	Inner product in $\mathbb{H}$
$\mathbb{S}$	Action functional
$\mathbb{N}$ (resp. $\mathbb{N}_+$)	Set of all nonnegative (resp. positive) integers
$\mathbb{R}$ (resp. $\mathbb{R}_+$)	Set of all real numbers (resp. set of all positive real numbers)
$\mathbb{R}^d$ (resp. $\mathbb{T}^d$)	d-dimensional real space (resp. d-dimensional torus)
$(\Omega, \mathscr{F}, \mathbb{P})$	Probability space
$\mathscr{B}(\mathbb{H})$	Borel σ-algebra on $\mathbb{H}$
$\{\mathscr{F}_t\}_{0 \le t \le T}$	A non-decreasing family of σ-subalgebras of $\mathscr{F}$
$\mathscr{P}(\mathbb{H})$	Space of all probability measures on $(\mathbb{H}, \mathscr{B}(\mathbb{H}))$
$\mathscr{L}(\mathbb{H}, \mathbb{K})$	Space of all linear bounded operators from space $\mathbb{H}$ to $\mathbb{K}$, denoted by $\mathscr{L}(\mathbb{H})$ if $\mathbb{K} = \mathbb{H}$
$\mathscr{L}_2(\mathbb{H}, \mathbb{K})$	Space of all Hilbert–Schmidt operators from space $\mathbb{H}$ to $\mathbb{K}$, denoted by $\mathscr{L}_2(\mathbb{H})$ if $\mathbb{K} = \mathbb{H}$
$\mathbf{L}^p(\Omega, \mathbb{K})$	Space of all $\mathbb{K}$-valued functions which are pth integrable with respect to $\mathbb{P}$
$\mathbf{L}^p(\mathbb{R}^d, \mathbb{K})$	Space of all $\mathbb{K}$-valued functions defined on $\mathbb{R}^d$ which are pth integrable
$\mathbf{L}^p(\mathbb{H}, \mu)$	Space of all functions defined on $\mathbb{H}$ which are pth integrable with respect to measure μ
$\mathbf{L}^\infty(\mathbb{H}, \mu)$	Space of all essentially bounded measurable function in $\mathbb{H}$ with respect to measure μ
$\mathbf{B}_b(\mathbb{H})$	Space of all Borel bounded functions on $\mathbb{H}$
$\mathbf{C}(\mathbb{H})$	Space of continuous functions on $\mathbb{H}$
$\mathbf{C}_b(\mathbb{H})$	Space of all uniformly continuous and bounded functions on $\mathbb{H}$
$\mathbf{C}_b^q(\mathbb{H})$	Space of all q-times differential continuous functions with bounded derivatives up to order q on $\mathbb{H}$

$\mathbf{C}_b^\infty(\mathbb{H})$	Space of all smooth and bounded functions with bounded derivatives of any order
$\mathbf{C}_P^\infty(\mathbb{H})$	Space of all smooth functions with polynomial growth
$\mathbf{C}_P^q(\mathbb{H})$	Space of all functions together with their partial derivatives of order up to q growing polynomially
$\mathbf{C}^k(U, \mathbb{R}^n)$	Space of k times continuously differentiable functions from $U \subset \mathbb{R}^n$ to $\mathbb{R}^n$
$\mathbf{C}^p(\mathbb{H})$	Space of continuous functions in $\mathbf{C}(\mathbb{H})$ with continuous derivatives up to order p
$\mathbf{C}_0^p(\mathbb{H})$	Space of functions in $\mathbf{C}^p(\mathbb{H})$ with compact support
$\mathbf{W}^{k,p}(\mathbb{H})$	Sobolev space consisting of all $\mathbb{R}$-valued functions on $\mathbb{H}$ whose first k weak derivatives are functions in $\mathbf{L}^p(\mathbb{H})$
$\mathbf{W}^{k,\infty}(\mathbb{H})$	Sobolev space of all functions on $\mathbb{H}$ whose derivatives up to order k have finite $\mathbf{L}^\infty(\mathbb{H}, \mu)$-norm
$\mathbf{H}^p(\mathbb{R}^d)$	Sobolev spaces of all functions on $\mathbb{R}^d$ whose derivatives up to order p are square integrable
$\mathbf{S}^d$	d-dimensional unit sphere
$\mathbb{E}$	Expectation operator
$\mathbb{VAR}$	Variance operator
$\mathbb{COR}(X, Y)$	Correlation operator of X and Y
$(\mathbb{M}, \mathscr{G}, \mu)$	Measure space
$\pi_t(x, G)$	Transition probability for a stochastic process starting from x to hit set G at time
$\mathscr{N}(0, 1)$	Standard normal distribution
$\lvert \cdot \rvert$	Modulus of a number in $\mathbb{R}$ or $\mathbb{C}$
$\lVert \cdot \rVert$	Euclidean norm in the finite dimensional real space
$\lVert \cdot \rVert_{\mathbb{F}}$	Frobenius norm of matrices
$\langle M \rangle$	Quardratic variation process of stochastic process M
$(\cdot, \cdot)$	Inner product in a finite dimensional complex-valued space
$\{\cdot, \cdot\}$	Poisson bracket
$s \wedge t \,(\text{resp.}\, s \vee t)$	Minimum (resp. maximum) of $s, t \in \mathbb{R}$
$\mathfrak{R}(u)(\text{resp.}\,\mathfrak{J}(u))$	Real (resp. imaginary) part of u
$\mathrm{Dom}(A)$	Domain of operator A
I_n	$n \times n$ identity matrix
J_{2n}	$2n \times 2n$ standard symplectic matrix
$\mathbf{I}$	Rate function
I_α	Itô integral with index α
J_α	Stratononvich integral with index α
$\mathbb{I}$	Invariant quantity
$\mathbb{L}$	Generator operator with respect to the backward Kolmogrov equation
∇^2	Hessian matrix operator

Chapter 1
Stochastic Hamiltonian Systems

Let $(\Omega, \mathscr{F}, \{\mathscr{F}_t\}_{t\geq 0}, \mathbb{P})$ be a complete filtered probability space, and $\mathbb{E}$ be the expectation operator. To simplify the notation, we will suppress the sample $\omega \in \Omega$ and the notation '$\mathbb{P}$-a.s.' or 'a.s.' unless it is necessary to avoid confusion. Moreover, throughout the manuscript, we will use C or K as a generic constant, which may be different from line to line.

1.1 Stratonovich Stochastic Differential Equations

The time evolution of physical systems is usually characterized by a differential equation. Besides such a deterministic motion, it is essential to consider a random motion, which turns out to be a stochastic process. Many important stochastic processes have been expressed as exact solutions of stochastic differential equations developed in various different disciplines like physics, biology or mathematical finance since 1942. An significant one concerns the equation itself, considering it as a dynamical system perturbed by certain noise.

A d-dimensional Stratonovich stochastic differential equation can be written as

$$dX(t) = \sigma_0(X(t))dt + \sum_{r=1}^{m} \sigma_r(X(t)) \circ dW_r(t), \quad t \in (0, T], \tag{1.1}$$

where $X(0) = x \in \mathbb{R}^d$, $W_r(\cdot)$, $r = 1, \ldots, m$, are independent standard Wiener processes, and "$\circ$" represents the Stratonovich product. The last term in (1.1) can be regarded as a random disturbance adjoined to the deterministic ordinary differential equation $dX(t) = \sigma_0(X(t))dt$. Assume that coefficients $\sigma_r \colon \mathbb{R}^d \to \mathbb{R}^d$, $r = 0, 1, \ldots, m$, and vectors $\frac{\partial \sigma_l}{\partial x}\sigma_l$, $l = 1, \ldots, m$, are globally Lipschitz continuous,

© The Author(s), under exclusive license to Springer Nature Singapore Pte Ltd. 2022
J. Hong, L. Sun, *Symplectic Integration of Stochastic Hamiltonian Systems*,
Lecture Notes in Mathematics 2314, https://doi.org/10.1007/978-981-19-7670-4_1

i.e., there exists $C > 0$ such that for any $y_1, y_2 \in \mathbb{R}^d$,

$$\sum_{r=0}^{m} \|\sigma_r(y_1) - \sigma_r(y_2)\| + \sum_{r=1}^{m} \left\| \left(\frac{\partial \sigma_r}{\partial x}\sigma_r\right)(y_1) - \left(\frac{\partial \sigma_r}{\partial x}\sigma_r\right)(y_2) \right\| \le C\|y_1 - y_2\|$$

(1.2)

with $\frac{\partial \sigma_l(x)}{\partial x}$ being the Jacobian matrix of $\sigma_l(x)$. Then (1.1) admits a unique strong solution which is $\mathscr{F}_t$-measurable (see [151]), and the integral form is

$$X(t) = x + \int_0^t \sigma_0\left(X(s)\right) ds + \sum_{r=1}^{m} \int_0^t \sigma_r\left(X(s)\right) \circ dW_r(s), \quad t \in [0, T],$$

where $\int_0^t \sigma_r\left(X(s)\right) \circ dW_r(s)$ is a Stratonovich integral for $r \in \{1, \ldots, m\}$. One of the advantages of the Stratonovich integral is that the chain rule of ordinary calculus holds. It makes the Stratonovich integral natural to utilize for instance in connection with stochastic differential equations on manifolds. However, the Itô integral possesses the martingale property, while the Stratonovich integral does not (see e.g., [16, 151] and references therein). This gives the Itô integral a significant computational advantage, and it is often straightforward to convert a Stratonovich integral to an Itô integral. It turns out that (1.1) coincides with the Itô stochastic differential equation

$$dX(t) = \left(\sigma_0(X(t)) + \frac{1}{2}\sum_{r=1}^{m} \left(\frac{\partial \sigma_r}{\partial x}\sigma_r\right)(X(t))\right) dt + \sum_{r=1}^{m} \sigma_r(X(t))dW_r(t)$$

with $X(0) = x$ (see e.g., [151, 197] and references therein).

The stochastic differential equation (1.1) possesses ν *invariants* $\mathbb{I}_i \in \mathbf{C}^1(\mathbb{R}^d, \mathbb{R})$, $i = 1, \ldots, \nu$, if

$$(\nabla \mathbb{I}_i(y))^\top \sigma_r(y) = 0, \quad r = 0, 1, \ldots, m, \quad i = 1, \ldots, \nu \qquad (1.3)$$

for any $y \in \mathbb{R}^d$. In other words, if we define vector $\mathbb{I}(y) := (\mathbb{I}_1(y), \ldots, \mathbb{I}_\nu(y))^\top$ with $y \in \mathbb{R}^d$, then

$$\frac{\partial \mathbb{I}(y)}{\partial y}\sigma_r(y) = 0, \quad r = 0, 1, \ldots, m,$$

where $\frac{\partial \mathbb{I}(y)}{\partial y}$ is the Jacobian matrix of $\mathbb{I}(y)$. According to the definition of the invariant, it follows from the Stratonovich chain rule that $d\mathbb{I}(X(t)) = 0$ with $X(t)$ being the exact solution of (1.1). This implies that $\mathbb{I}$ along the exact solution $X(t)$ is invariant almost surely. Moreover, if the stochastic differential equation (1.1) possesses v invariants $\mathbb{I}_i$, $i = 1, \ldots, v$, we have

$$X(t) \in \mathcal{M}_x := \left\{ y \in \mathbb{R}^d \mid \mathbb{I}_i(y) = \mathbb{I}_i(x), \ i = 1, \ldots, v \right\}, \quad t \in [0, T],$$

which indicates that $X(t)$ will be confined to the invariant submanifold $\mathcal{M}_x$ generated by $\mathbb{I}_i$, $i = 1, \ldots, v$. This is a direct geometric property for stochastic systems possessing invariants.

Example 1.1.1 Consider the following system of stochastic differential equations with commutative noises

$$\begin{cases} dP_1(t) = Q_1(t)dt + c_1 Q_1(t) \circ dW_1(t) + c_2 Q_1(t) \circ dW_2(t), \\ dP_2(t) = Q_2(t)dt + c_1 Q_2(t) \circ dW_1(t) + c_2 Q_2(t) \circ dW_2(t), \\ dQ_1(t) = -P_1(t)dt - c_1 P_1(t) \circ dW_1(t) - c_2 P_1(t) \circ dW_2(t), \\ dQ_2(t) = -P_2(t)dt - c_1 P_2(t) \circ dW_1(t) - c_2 P_2(t) \circ dW_2(t), \end{cases}$$

where c_1, c_2 are constants, and $W_i(\cdot)$, $i = 1, 2$, are independent standard Wiener processes. This system has three invariants

$$\mathbb{I}_1(p_1, p_2, q_1, q_2) = p_1 q_2 - p_2 q_1,$$

$$\mathbb{I}_2(p_1, p_2, q_1, q_2) = \frac{1}{2} \left(p_1^2 - p_2^2 + q_1^2 - q_2^2 \right),$$

$$\mathbb{I}_3(p_1, p_2, q_1, q_2) = p_1 p_2 + q_1 q_2.$$

As in the theory of deterministic ordinary differential equations, it is fruitful to look at the solution of (1.1) as a function of the initial value x in the stochastic case. Let $\hat{X}(x, t)$ be the solution of (1.1) at time $t \in [0, T]$ starting from x. For any $p \geq 2$ and $T > 0$, there exists $C > 0$ such that

$$\mathbb{E}\left[\sup_{0 \leq t \leq T} \|\hat{X}(y_1, t) - \hat{X}(y_2, t)\|^p \right] \leq C\|y_1 - y_2\|^p, \quad y_1, y_2 \in \mathbb{R}^d,$$

and

$$\mathbb{E}\left[\|\hat{X}(y_1, s) - \hat{X}(y_2, t)\|^p \right] \leq C(\|y_1 - y_2\|^p + |t - s|^{\frac{p}{2}}), \quad y_1, y_2 \in \mathbb{R}^d, \ s, t \in [0, T],$$

which implies that the solution $\hat{X}$ of (1.1) has a continuous modification by the Kolmogorov's continuity criterion of random fields in [163, Theorem 1.4.1].

One of the main topics discussed in this chapter is the relationship between the stochastic flow and the stochastic differential equation (1.1). To begin with, we shall give some definitions of the stochastic flow of homeomorphisms (see [163, Chapter

4]). Let $\phi_{s,t}(y, \cdot)$, $s, t \in [0, T]$, $y \in \mathbb{R}^d$, be a continuous $\mathbb{R}^d$-valued random field defined on $(\Omega, \mathscr{F}, \mathbb{P})$. Then for almost all $\omega \in \Omega$, $\phi_{s,t}(\cdot, \omega)$ defines a continuous mapping from $\mathbb{R}^d$ into itself for any $s, t \in [0, T]$. It is called a *stochastic flow of homeomorphisms* if for almost all ω, the family of continuous mappings $\{\phi_{s,t}(\cdot, \omega) : s, t \in [0, T]\}$ possesses the following properties:

(1) $\phi_{t,u}(\phi_{s,t}(y, \omega), \omega) = \phi_{s,u}(y, \omega)$ for any $s, t, u \in [0, T]$ and any $y \in \mathbb{R}^d$.
(2) $\phi_{s,s}(\cdot, \omega)$ is an identity mapping for any $s \in [0, T]$.
(3) the mapping $\phi_{s,t}(\cdot, \omega) \colon \mathbb{R}^d \to \mathbb{R}^d$ is a homeomorphism for any $s, t \in [0, T]$.

Further $\phi_{s,t}(\cdot, \omega)$, $s, t \in [0, T]$, is called a *stochastic flow of* $\mathbf{C}^k$-*diffeomorphisms* if for almost all ω, $\{\phi_{s,t}(\cdot, \omega) : s, t \in [0, T]\}$ satisfies (1)–(3) and the following condition

(4) $\phi_{s,t}(y, \omega)$ is k-times differentiable with respect to y for any $s, t \in [0, T]$, and
      the derivatives are continuous in $(s, t, y) \in [0, T] \times [0, T] \times \mathbb{R}^d$.

A stochastic flow of homeomorphisms can be regarded as a continuous random field with values in $\mathbf{G}$ satisfying the flow properties (1) and (2), where $\mathbf{G}$ is a complete topological group consisting of all homeomorphisms of $\mathbb{R}^d$. It is also called a *stochastic flow* with values in $\mathbf{G}$. For the stochastic flow $\phi_{s,t}$ with two parameters $s, t \in [0, T]$, the first parameter s stands for the initial time, and $\phi_{s,t}$ represents the state of the stochastic flow at time t. In the analysis of stochastic flows, it is of convenience to divide the stochastic flow into the forward flow $\phi_{s,t}$, $0 \le s \le t \le T$, and the backward flow $\phi_{s,t}$, $0 \le t \le s \le T$. Here a continuous random field $\phi_{s,t}$, $0 \le s \le t \le T$, with values in $\mathbf{G}$ satisfying (1) and (2) is called a *forward stochastic flow*. Given a forward stochastic flow $\phi_{s,t}$, $0 \le s \le t \le T$, with a suitable regularity condition, there exists a unique continuous semimartingale $\mathscr{M}(\cdot)$ such that for every fixed $s \in [0, T]$ and $y \in \mathbb{R}^d$, $\phi_{s,t}(y, \cdot)$ satisfies the Itô stochastic differential equation based on $\mathscr{M}(\cdot)$ (see e.g., [22, 106, 170] and references therein). Conversely, the solution of the Itô stochastic differential equation driven by Wiener processes defines a forward stochastic flow of homeomorphisms, provided that coefficients of the equation are globally Lipschitz continuous (see e.g., [97, 142, 163, 243] and references therein). Thereby, a forward stochastic flow of homeomorphisms can be constructed by (1.1), which is stated in the following theorem. For the sake of simplicity, we denote $\phi_{s,t}(y) := \phi_{s,t}(y, \cdot)$ for any $0 \le s \le t \le T$ and any $y \in \mathbb{R}^d$.

Theorem 1.1.1 *The solution of the Stratonovich stochastic differential equation* (1.1) *defines a forward stochastic flow of homeomorphisms, i.e.,*

$$\phi_{s,t}(y) = y + \int_s^t \sigma_0(\phi_{s,u}(y))du + \sum_{r=1}^m \int_s^t \sigma_r(\phi_{s,u}(y)) \circ dW_r(u), \qquad (1.4)$$

where $0 \le s \le t \le T$ *and* $y \in \mathbb{R}^d$.

It indicates that $\phi_{0,t}(x) = \hat{X}(x, t)$ for any $t \in [0, T]$ and $x \in \mathbb{R}^d$. More precisely, $\phi_{0,t}(x, \omega) = \hat{X}(x, t, \omega)$, where $t \in [0, T]$ and $\omega \in \Omega$. To prove Theorem 1.1.1, it suffices to consider the homeomorphic property of $\phi_{s,t}(\cdot, \omega)$, since the semi-flow property can be deduced directly by the existence and uniqueness of (1.1). To this end, we quote some lemmas on L^p-estimates for the random field $\phi_{s,t}(y)$, which has a continuous modification still denoted by $\phi_{s,t}(y)$ for $0 \leq s \leq t \leq T$ and $y \in \mathbb{R}^d$ (see e.g., [138, 163] and references therein).

Lemma 1.1.1 (See [163]) *For any $p \in \mathbb{R}$, there exists $C := C(p) > 0$ such that*

$$\mathbb{E}[(1 + \|\phi_{s,t}(y)\|^2)^p] \leq C(1 + \|y\|^2)^p \tag{1.5}$$

for any $y \in \mathbb{R}^d$ and $0 \leq s \leq t \leq T$.

Lemma 1.1.2 (See [163]) *For any $p \in \mathbb{R}$, there exists $C := C(p) > 0$ such that*

$$\mathbb{E}[(\epsilon + \|\phi_{s,t}(y_1) - \phi_{s,t}(y_2)\|^2)^p] \leq C(\epsilon + \|y_1 - y_2\|^2)^p \tag{1.6}$$

for any $y_1, y_2 \in \mathbb{R}^d$, $0 \leq s \leq t \leq T$ and any $\epsilon > 0$.

The proof of the above two lemmas could be carried out by means of the Itô formula and the Grönwall inequality. Lemma 1.1.2 implies that if $y_1 \neq y_2$, then the set $\{\omega : \phi_{s,t}(y_1, \omega) = \phi_{s,t}(y_2, \omega)\}$ has measure zero but may depend on $y_1 \in \mathbb{R}^d$ and $y_2 \in \mathbb{R}^d$ for any $0 \leq s \leq t \leq T$. In order to prove the 'one to one' property, one requires a set of measure zero independent of both $y_1 \in \mathbb{R}^d$ and $y_2 \in \mathbb{R}^d$. Thus the following lemma is used.

Lemma 1.1.3 (See [163]) *For $0 \leq s \leq t \leq T$ and $x \neq y$, $x, y \in \mathbb{R}^d$, set*

$$\eta_{s,t}(x, y) := \frac{1}{\|\phi_{s,t}(x) - \phi_{s,t}(y)\|}.$$

Then for any $p > 1$, there exists $C := C(p) > 0$ such that for any $\delta > 0$,

$$\mathbb{E}[|\eta_{s,t}(x, y) - \eta_{s',t'}(x', y')|^{2p}]$$
$$\leq C\delta^{-4p}\big(\|x - x'\|^{2p} + \|y - y'\|^{2p}$$
$$+ (1 + \|x\| + \|x'\| + \|y\| + \|y'\|)^{2p}(|t - t'|^p + |s - s'|^p)\big)$$

for any $s < t$, $s' < t'$, $s, t, s', t' \in [0, T]$ and $x, y, x', y' \in \mathbb{R}^d$ such that $\|x-y\| \geq \delta$ and $\|x' - y'\| \geq \delta$.

Based on the Hölder inequality and (1.6), the proof of Lemma 1.1.3 can be given. The Kolmogorov's continuity criterion of random fields indicates that η has a modification which is continuous in the domain $\{(s, t, x, y) : s < t, \|x - y\| \geq \delta\}$. According to the arbitrariness of δ, for almost all ω, the modification is continuous in the domain $\{(s, t, x, y) : s < t, x \neq y\}$. As a result, the mapping $\phi_{s,t}(\cdot, \omega)$ is one to one, i.e., for almost all ω, if $x \neq y$, $\phi_{s,t}(x, \omega) \neq \phi_{s,t}(y, \omega)$ for all $0 \leq s < t \leq T$.

In order to arrive at the 'onto' property of $\phi_{s,t}(\cdot, \omega)$, $0 \leq s \leq t \leq T$, we first recall some concepts in the algebraic topology. Let $\mathscr{E}$ and $\tilde{\mathscr{E}}$ be two topological spaces. If f and g are continuous mappings from $\mathscr{E}$ to $\tilde{\mathscr{E}}$, we call that f is *homotopic* to g if there exists a continuous mapping $F \colon \mathscr{E} \times [0, 1] \to \tilde{\mathscr{E}}$ such that

$$F(y, 0) = f(y) \quad \text{and} \quad F(y, 1) = g(y)$$

for each $y \in \mathscr{E}$. It can be verified that if a mapping is homotopic to the identity mapping from $\mathbf{S}^d$ to $\mathbf{S}^d$, where $\mathbf{S}^d$ is the unit sphere with $d \in \mathbb{N}_+$, then it is a surjection. As the one-point compactification $\bar{\mathbb{R}}^d = \mathbb{R}^d \cup \{\infty\}$ is homeomorphic to $\mathbf{S}^d$, a mapping homotopic to the identity mapping from $\bar{\mathbb{R}}^d$ to $\bar{\mathbb{R}}^d$ must be a surjection (see [138]). Now we turn to showing that the extension of $\phi_{s,t}(y)$ with $0 \leq s \leq t \leq T$ as follows

$$\tilde{\phi}_{s,t}(y) = \begin{cases} \phi_{s,t}(y), & \text{if} \quad y \in \mathbb{R}^d; \\ \infty, & \text{if} \quad y = \infty, \end{cases}$$

is homotopic to the identity mapping from $\bar{\mathbb{R}}^d$ to $\bar{\mathbb{R}}^d$. To this end, the following lemma is used.

Lemma 1.1.4 (See [163]) *Set $\hat{y} = \|y\|^{-2} y$ if $y \neq 0$ and define*

$$\eta_{s,t}(\hat{y}) = \begin{cases} \dfrac{1}{1 + \|\phi_{s,t}(y)\|}, & \text{if} \quad \hat{y} \neq 0; \\ 0, & \text{if} \quad \hat{y} = 0, \end{cases} \tag{1.7}$$

where $0 \leq s \leq t \leq T$ and $y, \hat{y} \in \mathbb{R}^d$. Then for each $p > 1$, there exists a positive constant $C := C(p)$ such that

$$\mathbb{E}[|\eta_{s,t}(\hat{y}) - \eta_{s',t'}(\hat{y}')|^{2p}] \leq C(\|\hat{y} - \hat{y}'\|^{2p} + |t - t'|^p + |s - s'|^p)$$

holds for any $s < t$, $s' < t'$, $s, t, s', t' \in [0, T]$ and $\hat{y}, \hat{y}' \in \mathbb{R}^d$.

Taking advantage of the Kolmogorov's continuity criterion of random fields, we derive that for almost all ω, $\eta_{s,t}(\hat{y})$ is continuous at $\hat{y} = 0$ for any $s \leq t$. Then the extension $\tilde{\phi}_{s,t}(y)$ is continuous in $(s, t, y) \in [0, T] \times [0, T] \times \mathbb{R}^d$. The "onto" property of $\phi_{s,t}(\cdot, \omega)$ follows from removing the restriction $\tilde{\phi}_{s,t}(\infty) = \infty$ for any $0 \leq s \leq t \leq T$. Since the continuous mapping $\phi_{s,t}(\cdot, \omega) \colon \mathbb{R}^d \to \mathbb{R}^d$ is a bijection for any $0 \leq s < t \leq T$, it has the homeomorphic property, which means that the inverse mapping $\phi_{s,t}(\cdot, \omega)^{-1} = \phi_{t,s}(\cdot, \omega)$ is also one to one, onto and continuous for any $0 \leq s < t \leq T$. Therefore, (1.1) defines a forward stochastic flow of homeomorphisms on $\mathbb{R}^d$.

Remark 1.1.1 The forward stochastic flow of homeomorphisms $\phi_{s,t}(y, \cdot)$, $0 \leq s \leq t \leq T$, $y \in \mathbb{R}^d$, in (1.4) can also be expressed by the exact solution of (1.1). In detail, let $\phi_{s,t}(y,\omega) := \hat{X}(\hat{X}(x,s,\omega), t-s, \theta_s(\omega))$ and $y = \hat{X}(x,s,\omega)$, where θ_s, $s \geq 0$, is the Wiener shift. It can be verified that properties *(1)-(3)* are satisfied.

Remark 1.1.2 (See [163]) If ϕ_t is a continuous process with values in **G** such that ϕ_0 is an identity mapping almost surely, then $\phi_{s,t} = \phi_t \circ \phi_s^{-1}$ with ϕ_s^{-1} being the inverse of ϕ_s is a stochastic flow with values in **G**, where $s, t \in [0,T]$. Conversely, if $\phi_{s,t}$, $s,t \in [0,T]$, is a stochastic flow with values in **G**, then $\phi_t := \phi_{0,t}$ is a continuous process with values in **G**, and satisfies $\phi_{s,t} = \phi_t \circ \phi_s^{-1}$. Thus, a continuous stochastic flow with values in **G** is equivalent to a continuous process ϕ_t, $t \in [0,T]$, with values in **G** such that ϕ_0 is an identity mapping almost surely.

Remark 1.1.3 (See [21]) If $\sigma_0(\cdot) = 0$ and $\sigma_r \in C^\infty(\mathbb{R}^d, \mathbb{R}^d)$ for any $r \in \{1,\ldots,m\}$, then the pull-back action ϕ_t^* on functions of $\phi_t := \phi_{0,t}$, $t \in [0,T]$, associated with the stochastic differential equation (1.1) can formally solve the stochastic differential equation

$$\phi_t^* = I_d + \sum_{i=1}^m \int_0^t \phi_s^* \sigma_i \circ dW_i(s),$$

so that ϕ_t^*, $t \geq 0$, can formally be seen as a lift of $W(t)$ in the formal object $\exp(\mathfrak{L}(\sigma_1,\ldots,\sigma_m))$ with $\mathfrak{L}(\sigma_1,\ldots,\sigma_m)$ being the Lie algebra generated by $\sigma_1,\ldots,\sigma_m$. In detail, the pull-back action ϕ_t^*, $t \in [0,T]$, is given by

$$(\phi_t^* f)(x) = (f \circ \phi_t)(x) = f(X(t)),$$

where $f : \mathbb{R}^d \to \mathbb{R}$ is a smooth function. By the Itô formula, we have

$$f(X(t)) = f(x) + \sum_{i=1}^m \int_0^t (\sigma_i f)(X(s)) \circ dW_i(s), \quad t \in [0,T].$$

Here, the vector field σ_i with $i \in \{1,\ldots,m\}$ defines a differential operator acting on the smooth function f, i.e., $(\sigma_i f)(x) = \sum_{j=1}^d \sigma_i^j(x)\frac{\partial f(x)}{\partial x_j}$, where σ_i^j is the jth component of σ_i. Applying the Itô formula to $\sigma_i f(X(s))$ leads to

$$f(X(t)) = f(x) + \sum_{i=1}^m (\sigma_i f)(x)W_i(t) + \sum_{i,j=1}^m \int_0^t \int_0^s (\sigma_j\sigma_i f)(X(u)) \circ dW_j(u) \circ dW_i(s).$$

Continuing this procedure by N steps, we achieve

$$f(X(t)) = f(x) + \sum_{k=1}^N \sum_{I=(i_1,\ldots i_k)} (\sigma_{i_1}\cdots\sigma_{i_k} f)(x)\int_{\Delta^k[0,t]} \circ dW^I + R_N(t).$$

for some remainder term R_N, where

$$\int_{\Delta^k[0,t]} \circ dW^I = \int_{0 \le t_1 \le \cdots \le t_k \le t} \circ dW_{i_1}(t_1) \circ \cdots \circ dW_{i_k}(t_k)$$

with $\Delta^k[0,t] = \{(t_1, \ldots, t_k) \in [0,t]^k, t_1 \le \cdots \le t_k\}$ and $I = (i_1, \ldots, i_k) \in \{1, \ldots, m\}^k$, $k \in \mathbb{N}_+$. Letting $N \to +\infty$ and assuming that $R_N \to 0$ (convergence questions are discussed in [18]), we obtain a nice formal formula

$$f(X(t)) = f(x) + \sum_{k=1}^{+\infty} \sum_{I=(i_1,\ldots i_k)} (\sigma_{i_1} \cdots \sigma_{i_k} f)(x) \int_{\Delta^k[0,t]} \circ dW^I,$$

which yields

$$\phi_t^* = I_d + \sum_{k=1}^{+\infty} \sum_{I=(i_1,\ldots i_k)} \sigma_{i_1} \cdots \sigma_{i_k} \int_{\Delta^k[0,t]} \circ dW^I.$$

Though this formula does not make sense from an analytical point of view, at least, the probabilistic information contained in the stochastic flow associated with the stochastic differential equation (1.1) can be given by the set of Stratonovich chaos.

Assume in addition that σ_0, σ_r, and $\frac{\partial \sigma_r}{\partial x} \sigma_r$ belong to $\mathbf{C}^k(\mathbb{R}^d, \mathbb{R}^d)$ for $r = \{1, \ldots, m\}$, and $k \ge 2$. Then the solution of (1.1) defines a forward stochastic flow of $\mathbf{C}^{k-1}$-diffeomorphisms (see e.g., [138, 142, 162, 163] and references therein). Denote by $Y(t) = \frac{\partial X(t)}{\partial x}$ the Jacobian matrix of $X(t)$ for any $t \in [0, T]$. It can be proved that

$$Y(t) = I_d + \int_0^t \frac{\partial \sigma_0(X(u))}{\partial x} Y(u) du + \sum_{r=1}^m \int_0^t \frac{\partial \sigma_r(X(u))}{\partial x} Y(u) \circ dW_r(u).$$

$$(1.8)$$

Consider another matrix $Z(t)$ which is the solution of the equation

$$Z(t) = I_d - \int_0^t Z(u) \frac{\partial \sigma_0(X(u))}{\partial x} du - \sum_{r=1}^m \int_0^t Z(u) \frac{\partial \sigma_r(X(u))}{\partial x} \circ dW_r(u),$$

where $t \in [0, T]$. By using the chain rule, one can check that for any $t \in [0, T]$,

$$Z(t)Y(t) = Y(t)Z(t) = I_d.$$

As a consequence, the matrix $Y(t)$ is invertible for any $t \in [0, T]$ and $Y(t)^{-1} = Z(t)$. Moreover, (1.8) plays an important role in studying the stochastic symplectic structure of the phase flow for the stochastic Hamiltonian system in Sect. 1.2.

1.2 Stochastic Hamiltonian Systems

The stochastic Hamiltonian system, as an extension of Nelson's stochastic mechanics, is ubiquitous in applications of physics, chemistry and so on (see e.g., [29, 199, 200, 203, 216, 233] and references therein). It can be recognized as an 'open' Hamiltonian system perturbed by random fluctuations from an external world. From the perspective of finance, the stochastic Hamiltonian system may be considered as dynamical systems describing some risky assets (see [203]). In accelerator physics, the synchrotron oscillations of particles in storage rings under the influence of external fluctuating electromagnetic fields can be characterized with the aid of the stochastic Hamiltonian system (see [233]). The stochastic Hamiltonian system is also introduced in the theory of stochastic optimal control as a necessary condition of an optimal control, known as a stochastic version of the maximum principle of Pontryagin's type (see [216]).

In this section, we are interested in the following $2n$-dimensional stochastic Hamiltonian system

$$
\begin{cases}
dP(t) = -\dfrac{\partial H_0(P(t), Q(t))}{\partial Q}dt - \displaystyle\sum_{r=1}^{m} \dfrac{\partial H_r(P(t), Q(t))}{\partial Q} \circ dW_r(t), \ P(0) = p, \\[3mm]
dQ(t) = \dfrac{\partial H_0(P(t), Q(t))}{\partial P}dt + \displaystyle\sum_{r=1}^{m} \dfrac{\partial H_r(P(t), Q(t))}{\partial P} \circ dW_r(t), \quad Q(0) = q,
\end{cases}
$$

$$(1.9)$$

where $t \in (0, T]$, p, q are n-dimensional column vectors, $H_j \in \mathbf{C}^{\infty}(\mathbb{R}^{2n}, \mathbb{R})$, $j = 0, 1, \ldots, m$, are Hamiltonians, and $W_r(\cdot)$, $r = 1, \ldots, m$, are independent standard Wiener processes. By denoting $X = (P^{\top}, Q^{\top})^{\top} \in \mathbb{R}^{2n}$, (1.9) has an equivalent compact form

$$
dX(t) = J_{2n}^{-1}\nabla H_0(X(t))dt + \sum_{r=1}^{m} J_{2n}^{-1}\nabla H_r(X(t)) \circ dW_r(t), \tag{1.10}
$$

where $X(0) = (p^{\top}, q^{\top})^{\top}$ and $J_{2n} = \begin{bmatrix} 0 & I_n \\ -I_n & 0 \end{bmatrix}$ with I_n being the $n \times n$ identity matrix. Suppose that $\sigma_r(X) := J_{2n}^{-1}\nabla H_r(X)$, $r = 0, 1, \ldots, m$, satisfy the condition (1.2), and then the stochastic Hamiltonian system (1.9) admits a forward stochastic flow of diffeomorphisms $\phi_{s,t}(\cdot, \omega) \colon \mathbb{R}^{2n} \to \mathbb{R}^{2n}$, where $0 \leq s \leq t \leq T$ (see e.g., [122, 142, 163] and references therein). Moreover, it follows that $(P, Q) \in \mathbf{C}([0, T])$, where

$$
\mathbf{C}([0, T]) := \{(P, Q) \colon \Omega \times [0, T] \to \mathbb{R}^n \times \mathbb{R}^n \mid P, Q \text{ are almost surely}
$$

$$
\text{continuous } \mathscr{F}_t\text{-adapted semimartingales}\}.
$$

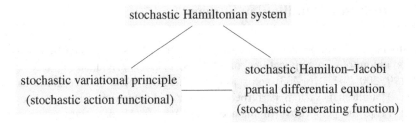

An outstanding property of (1.9) is the symplecticity of the stochastic flow. Moreover, as shown in the diagram, the stochastic Hamiltonian system operates in interrelated domains including stochastic variational principle and stochastic Hamilton–Jacobi partial differential equation, which will be studied one after the other.

1.2.1   Stochastic Symplectic Structure

We now introduce the stochastic symplectic structure, which is a geometric property of stochastic flow of the stochastic Hamiltonian system (1.9). We refer to [122, 169, 197] and references therein for more details.

Consider the symplectic form $\omega_2 = \mathrm{d}p \wedge \mathrm{d}q = \sum\limits_{i=1}^{n} \mathrm{d}p_i \wedge \mathrm{d}q_i$ on $\mathbb{R}^{2n}$, where 'd' denotes the exterior derivative operator. One can think of this either as a differential form on $\mathbb{R}^{2n}$ with constant coefficients, or as a nondegenerate skew-symmetric bilinear form $\omega_2 : \mathbb{R}^{2n} \times \mathbb{R}^{2n} \to \mathbb{R}$. These two notions coincide when the vector space $\mathbb{R}^{2n}$ is considered as the tangent space of the manifold $\mathbb{R}^{2n}$ at a point (see [258]). Given a pair of vectors $\zeta = (\xi^\top, \eta^\top)^\top$ and $\zeta' = (\xi'^\top, \eta'^\top)^\top$ with $\xi, \eta, \xi', \eta' \in \mathbb{R}^n$,

$$\omega_2\left(\zeta, \zeta'\right) = \mathrm{d}p \wedge \mathrm{d}q\left(\zeta, \zeta'\right) = \sum_{i=1}^{n}\left(\xi_i \eta_i' - \eta_i \xi_i'\right) = \zeta^\top J_{2n} \zeta',$$

which represents the sum of the oriented areas of projections onto the coordinate planes $\{p_i, q_i\}$ with $i = 1, \ldots, n$. One of the most significant features of differential forms is the way they behave under differentiable mappings. Let $\tilde{\omega}$ be a k-form in $\mathbb{R}^{n_2}$, and $f : \mathbb{R}^{n_1} \to \mathbb{R}^{n_2}$ be a differentiable mapping, where $k \in \mathbb{N}$ and $n_1, n_2 \in \mathbb{N}_+$. The *pull-back $f^*\tilde{\omega}$ of $\tilde{\omega}$ under f* is a k-form in $\mathbb{R}^{n_1}$ (see [258]) and given by

$$\left(f^*\tilde{\omega}\right)(x)(v_1, \ldots, v_k) = \tilde{\omega}(f(x))\left(df_x(v_1), \ldots, df_x(v_k)\right).$$

Here, $x \in \mathbb{R}^{n_1}$, $v_1, \ldots, v_k \in \mathbb{R}_x^{n_1}$, and $df_x : \mathbb{R}_x^{n_1} \to \mathbb{R}_{f(x)}^{n_2}$ is the differential of the mapping f at x, where $\mathbb{R}_x^{n_1}$ is the tangent space of $\mathbb{R}^{n_1}$ at x and $\mathbb{R}_{f(x)}^{n_2}$ is the tangent space of $\mathbb{R}^{n_2}$ at $f(x)$.

Actually, the stochastic flow $\phi_t := \phi_{0,t}(\cdot, \omega)$, $t \in [0, T]$, of (1.9) preserves the symplectic form ω_2, i.e., for every $x \in \mathbb{R}^{2n}$,

$$\phi_t^* \omega_2(\xi, \eta) = \omega_2\left(\frac{\partial \phi_t(x)}{\partial x}\xi, \frac{\partial \phi_t(x)}{\partial x}\eta\right) = \omega_2(\xi, \eta) \quad \forall \, \xi, \eta \in \mathbb{R}_x^{2n} \tag{1.11}$$

with ϕ_t^* being the pull-back by the stochastic flow ϕ_t and $\frac{\partial \phi_t(x)}{\partial x}$ being the Jacobian matrix of $\phi_t(x)$. Since (1.11) holds, we say that ϕ_t, $t \in [0, T]$, of (1.9) preserves the stochastic symplectic structure, which implies the invariance of oriented area under ϕ_t on $\mathbb{R}^{2n}$. Denote by $Sp(\mathbb{R}^{2n}) := \{g \in \mathbf{C}^1(\mathbb{R}^{2n}, \mathbb{R}^{2n}) \mid g^* \omega_2 = \omega_2\}$ and $Sp(2n)$ the space of $2n \times 2n$ symplectic matrices, respectively (see [100]). Here, a matrix M of order $2n$ is called a *symplectic matrix* if it satisfies $M^\top J_{2n} M = J_{2n}$. A diffeomorphism $g: \mathbb{R}^{2n} \to \mathbb{R}^{2n}$, $\widehat{x} = g(x)$ is called a *canonical transformation* on $\mathbb{R}^{2n}$, if for any $x \in \mathbb{R}^{2n}$, $M = \frac{\partial \widehat{x}}{\partial x} \in Sp(2n)$.

g is a canonical transformation

$$\Longleftrightarrow M^\top J_{2n} M = J_{2n} \qquad\qquad \forall \, x \in \mathbb{R}^{2n}$$

$$\Longleftrightarrow \xi^\top M^\top J_{2n} M \eta = \xi^\top J_{2n} \eta \qquad \forall \, \xi, \eta \in \mathbb{R}_x^{2n}, \, x \in \mathbb{R}^{2n}$$

$$\Longleftrightarrow g^* \omega_2(\xi, \eta) = \omega_2(\xi, \eta) \qquad \forall \, \xi, \eta \in \mathbb{R}_x^{2n}$$

$$\Longleftrightarrow g^* \omega_2 = \omega_2.$$

It can be verified that for almost all ω and any $t \in [0, T]$, $\phi_t \in Sp(\mathbb{R}^{2n})$ if and only if $\frac{\partial \phi_t(x)}{\partial x}$ satisfies

$$\left(\frac{\partial \phi_t(x)}{\partial x}\right)^\top J_{2n} \frac{\partial \phi_t(x)}{\partial x} = J_{2n} \tag{1.12}$$

for any $x \in \mathbb{R}^{2n}$. By means of differential 2-form, the stochastic symplectic structure of $\phi_t: (p^\top, q^\top)^\top \mapsto ((P(t))^\top, (Q(t))^\top)^\top$ can also be characterized as

$$dP(t) \wedge dQ(t) = dp \wedge dq, \quad a.s. \tag{1.13}$$

for any $t \in [0, T]$, which is shown in the following theorem.

Theorem 1.2.1 (See [13, 93, 200]) *The phase flow* $\phi_t: (p^\top, q^\top)^\top \mapsto ((P(t))^\top, (Q(t))^\top)^\top$ *of* (1.9) *preserves the stochastic symplectic structure, that is,* (1.13) *holds for any* $t \in [0, T]$.

Proof It follows from (1.8) that both P and Q are differentiable with respect to the initial value p and q. To avoid notational complexity, we let $P_p^{jk} := \frac{\partial P_j}{\partial p_k}$, $Q_p^{jk} := \frac{\partial Q_j}{\partial p_k}$, $P_q^{jk} := \frac{\partial P_j}{\partial q_k}$ and $Q_q^{jk} := \frac{\partial Q_j}{\partial q_k}$, $j, k \in \{1, \ldots, n\}$, where p_j, q_j, P_j, Q_j, are the jth component of p, q, P, Q, respectively. Since

$$dP_j = \sum_{k=1}^{n} P_p^{jk} dp_k + \sum_{l=1}^{n} P_q^{jl} dq_l, \quad dQ_j = \sum_{k=1}^{n} Q_p^{jk} dp_k + \sum_{l=1}^{n} Q_q^{jl} dq_l$$

for $j \in \{1, \ldots, n\}$, we obtain

$$dP \wedge dQ = \sum_{j=1}^{n} \sum_{k=1}^{n} \sum_{l=1}^{n} \left(P_p^{jk} Q_q^{jl} - P_q^{jl} Q_p^{jk} \right) dp_k \wedge dq_l$$

$$+ \sum_{j=1}^{n} \sum_{k=1}^{n} \sum_{l=1}^{k-1} \left(P_p^{jk} Q_p^{jl} - P_p^{jl} Q_p^{jk} \right) dp_k \wedge dp_l$$

$$+ \sum_{j=1}^{n} \sum_{k=1}^{n} \sum_{l=1}^{k-1} \left(P_q^{jk} Q_q^{jl} - P_q^{jl} Q_q^{jk} \right) dq_k \wedge dq_l.$$

Based on Stratonovich chain rule, applying (1.8) and (1.9), one can get

$$d \left(\sum_{j=1}^{n} \left(P_p^{jk} Q_q^{jl} - P_q^{jl} Q_p^{jk} \right) \right) = 0 \quad \forall k, l \in \{1, \ldots, n\},$$

$$d \left(\sum_{j=1}^{n} \left(P_p^{jk} Q_p^{jl} - P_p^{jl} Q_p^{jk} \right) \right) = 0 \quad \forall k \neq l,$$

$$d \left(\sum_{j=1}^{n} \left(P_q^{jk} Q_q^{jl} - P_q^{jl} Q_q^{jk} \right) \right) = 0 \quad \forall k \neq l.$$

A straight computation yields $\sum_{j=1}^{n} dP_j \wedge dQ_j = \sum_{j=1}^{n} dp_j \wedge dq_j$, which completes the proof.   □

Remark 1.2.1 We can also arrive at the symplecticity of ϕ_t, $t \in [0, T]$, by verifying (1.12). In detail, the Jacobian matrix $Y(t) = \frac{\partial X(t)}{\partial x}$ associated with (1.10) satisfies

$$dY(t) = J_{2n}^{-1} \nabla^2 H_0(X(t)) Y(t) dt + \sum_{r=1}^{m} J_{2n}^{-1} \nabla^2 H_r(X(t)) Y(t) \circ dW_r(t),$$

where $\nabla^2 H_r$ is the Hessian matrix of H_r for $r \in \{0, 1, \ldots, m\}$. As a result, $Z(t) = Y(t)^{-1}$ satisfies

$$dZ(t) = -Z(t)J_{2n}^{-1}\nabla^2 H_0(X(t))dt - \sum_{r=1}^{m} Z(t)J_{2n}^{-1}\nabla^2 H_r(X(t)) \circ dW_r(t).$$

Due to $J_{2n}(Y(0))^\top = Z(0)J_{2n} = J_{2n}$ and the fact that

$$dJ_{2n}(Y(t))^\top = J_{2n}(Y(t))^\top \nabla^2 H_0(X(t))J_{2n}dt + \sum_{r=1}^{m} J_{2n}(Y(t))^\top \nabla^2 H_r(X(t))J_{2n} \circ dW_r(t),$$

$$dZ(t)J_{2n} = Z(t)J_{2n}\nabla^2 H_0(X(t))J_{2n}dt + \sum_{r=1}^{m} Z(t)J_{2n}\nabla^2 H_r(X(t))J_{2n} \circ dW_r(t),$$

we deduce $J_{2n}(Y(t))^\top = Z(t)J_{2n}$ for any $t \in [0, T]$. Thus,

$$(Y(t))^\top J_{2n}Y(t)$$
$$= J_{2n}Y(t) - \int_0^t (Y(s))^\top \nabla^2 H_0(X(s))ds Y(t) - \sum_{r=1}^{m} \int_0^t (Y(s))^\top \nabla^2 H_r(X(s)) \circ dW_r(s)Y(t)$$
$$= J_{2n}\left(I_{2n} - \int_0^t Z(s)J_{2n}^{-1}\nabla^2 H_0(X(s))ds - \sum_{r=1}^{m} \int_0^t Z(s)J_{2n}^{-1}\nabla^2 H_r(X(s)) \circ dW_r(s)\right)Y(t)$$
$$= J_{2n}Z(t)Y(t),$$

which implies $(Y(t))^\top J_{2n}Y(t) = J_{2n}$ for any $t \in [0, T]$.

The symplecticity of ϕ_t, $t \in [0, T]$, can be also obtained by utilizing the stochastic Liouville's theorem, and we refer to [169] for more details. Now we present two linear stochastic Hamiltonian systems to illustrate the stochastic symplectic structure.

Example 1.2.1 (Kubo Oscillator) One of the simplest stochastic Hamiltonian systems is the following 2-dimensional Kubo oscillator

$$\begin{cases} dP(t) = -Q(t)dt - Q(t) \circ dW(t), & P(0) = p, \\ dQ(t) = P(t)dt + P(t) \circ dW(t), & Q(0) = q, \end{cases} \tag{1.14}$$

where $W(\cdot)$ is a standard Wiener process. The solution of (1.14) can be expressed explicitly

$$P(t) = p\cos(t + W(t)) - q\sin(t + W(t)),$$
$$Q(t) = p\sin(t + W(t)) + q\cos(t + W(t)). \tag{1.15}$$

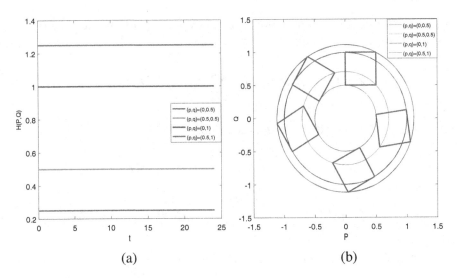

Fig. 1.1 (**a**) Evolution of $H(P, Q)$, (**b**) Images under the phase flow with initial image being a square determined by (0,0.5), (0.5,0.5), (0,1), (0.5,1)

It is known that $H(P, Q) = P^2 + Q^2$ is invariant, which can be also seen from Fig. 1.1a. This conservation law means that the phase trajectory of (1.14) forms a circle centered at the origin. For the symplecticity of the phase flow, we choose the original square which is produced by four initial points $(0, 0.5)$, $(0.5, 0.5)$, $(0, 1)$ and $(0.5, 1)$. As shown in Fig. 1.1b, the area keeps invariant under the phase flow of (1.14).

Remark 1.2.2 The Kubo oscillator is a typical stochastic Hamiltonian system with a conserved quantity. In fact, not all stochastic Hamiltonian systems preserve Hamiltonians $H_i, i = 0, 1, \ldots, m$. By the Stratonovich chain rule, we have

$$
dH_i = \sum_{k=1}^{n} \left(\frac{\partial H_i}{\partial P_k} dP_k + \frac{\partial H_i}{\partial Q_k} dQ_k \right)
$$

$$
= \sum_{k=1}^{n} \left(\frac{\partial H_i}{\partial Q_k} \frac{\partial H_0}{\partial P_k} - \frac{\partial H_i}{\partial P_k} \frac{\partial H_0}{\partial Q_k} \right) dt + \sum_{r=1}^{m} \sum_{k=1}^{n} \left(\frac{\partial H_i}{\partial Q_k} \frac{\partial H_r}{\partial P_k} - \frac{\partial H_i}{\partial P_k} \frac{\partial H_r}{\partial Q_k} \right) \circ dW_r.
$$

Thus, Hamiltonians $H_i, i = 0, 1, \ldots, m$, are invariants for the flow of (1.9), i.e., $dH_i = 0$, if and only if $\{H_i, H_j\} := \sum_{k=1}^{n} \left(\frac{\partial H_j}{\partial Q_k} \frac{\partial H_i}{\partial P_k} - \frac{\partial H_i}{\partial Q_k} \frac{\partial H_j}{\partial P_k} \right) = 0$ for $i, j = 0, 1, \ldots, m$.

Example 1.2.2 (Linear Stochastic Oscillator) Consider the linear stochastic oscillator

$$\begin{cases} \ddot{X} + X = \alpha \dot{W}, \\ \dot{X}(0) = p, \ X(0) = q \end{cases}$$

with $\alpha \in \mathbb{R}$ and $W(\cdot)$ being a 1-dimensional standard Wiener process. Denote $P :=$ $\dot{X}$ and $Q := X$, respectively. The linear stochastic oscillator can be rewritten as a 2-dimensional stochastic Hamiltonian system

$$\begin{cases} dP(t) = -Q(t)dt + \alpha dW(t), \quad P(0) = p, \\ dQ(t) = P(t)dt, \quad Q(0) = q \end{cases} \tag{1.16}$$

with Hamiltonians $H_0(P, Q) = \frac{1}{2}P^2 + \frac{1}{2}Q^2$ and $H_1(P, Q) = -\alpha Q$. The exact solution $(P(t), Q(t))$ (see e.g., [185, 260] and references therein) of (1.16)

$$P(t) = p\cos(t) - q\sin(t) + \alpha \int_0^t \cos(t-s)dW(s),$$

$$Q(t) = p\sin(t) + q\cos(t) + \alpha \int_0^t \sin(t-s)dW(s),$$

possesses the following two properties:

(1) the second moment grows linearly, i.e., $\mathbb{E}[P(t)^2 + Q(t)^2] = p^2 + q^2 + \alpha^2 t$.
(2) almost surely, $P(t)$ has infinitely many zeros, all simple, on each half line $t_0 < t < \infty$ with $t_0 \geq 0$.

In the numerical experiments, we choose $\alpha = 1$. Figure 1.2 shows that the projection area keeps invariant, and there are infinitely many zeros on the positive real axis. These illustrate the symplecticity of the stochastic flow and the oscillation of the exact solution.

1.2.2   Stochastic Variational Principle

The variational principles underlie the Lagrangian and Hamiltonian analytical formulations of classical mechanics. The variational methods, which were developed for classical mechanics during the eighteenth and nineteenth centuries, have become the preeminent formalisms for classical dynamics and many other branches of modern science and engineering.

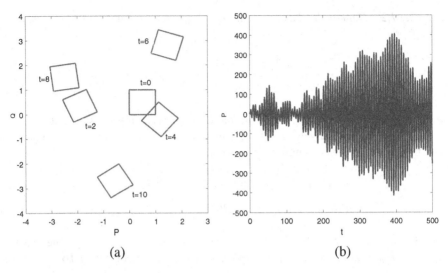

Fig. 1.2 (**a**) Images under the phase flow with initial image being a square determined by (0, 0), (0, 1), (1, 0), (1, 1), (**b**) Oscillation of P

Denote the position of a mechanical system with n degrees of freedom by $q = (q_1, \ldots, q_n)^\top$. Suppose that $U := U(q)$ and $T := T(q, \dot{q}) = \frac{1}{2}\dot{q}^\top M(q)\dot{q}$ with $M(q)$ being a positive definite matrix represent the potential and kinetic energy, respectively. Then the position q of the system satisfies the following Lagrange equation

$$\frac{d}{dt}\left(\frac{\partial L}{\partial \dot{q}}\right) = \frac{\partial L}{\partial q}, \tag{1.17}$$

where $L = T - U$ is the Lagrangian of the system. Through the Legendre transform $p = \left(\frac{\partial L}{\partial \dot{q}}\right)^\top = M(q)\dot{q}$, we have the Hamiltonian H given by

$$H(p, q) = p^\top \dot{q} - L(q, \dot{q}) = \frac{1}{2}p^\top (M(q))^{-1}p + U(q) = T(q, \dot{q}) + U(q),$$

which is the total energy of the mechanical system. The invariance of the Hamiltonian coincides with the conservation law of the total energy of the mechanical system. Further, the Lagrange equation is equivalent to the deterministic Hamiltonian system

$$dp(t) = -\frac{\partial H(p(t), q(t))}{\partial q}dt, \quad dq(t) = \frac{\partial H(p(t), q(t))}{\partial p}dt.$$

The classical Hamilton's principle in [111] indicates that the Lagrange equation extremizes the action functional $\mathbb{S}_{det}(q) := \int_0^T L(q(t), \dot{q}(t))dt$ on the space of curves joining $q(0) = q_0$ and $q(T) = q_1$. Moreover, it is always exploited to solve conservative problems.

In general, almost all nonconservative dynamical systems are not amenable to solution by means of variational methods in the sense of Hamilton's principle. The usual exceptions are systems with one degree of freedom subject to a linear viscous force, and the so-called rheonomic systems whose Lagrangians are explicitly dependent on time but where the total energy is not conserved (see [259]). As shown in Example 1.2.2, the Hamiltonian H_0 for the linear stochastic oscillator (1.16) is not conserved, i.e.,

$$\mathbb{E}[H_0(P(t), Q(t))] = \frac{1}{2}(P(0)^2 + Q(0)^2 + \alpha^2 t), \quad t \in [0, T].$$

The linear growth of H_0 with respect to the time t indicates that the stochastic Hamiltonian system is disturbed by certain nonconservative force, which may add the energy to the system (see [260]). A natural association of the force is caused by the driving process $W(\cdot)$, since $W(\cdot)$ is the resource of the random disturbance of the system and does not depend on Q. In the presence of nonconservative forces, a different sort of the action functional and the variational principle should be considered.

Now we define the stochastic action functional $\mathbb{S}$ on $\mathbf{C}([0, T])$ by

$\mathbb{S}(P(\cdot), Q(\cdot))$

$$= \int_0^T (P(t))^\top \circ dQ(t) - \int_0^T H_0(P(t), Q(t))dt - \sum_{r=1}^m \int_0^T H_r(P(t), Q(t)) \circ dW_r(t),$$

and we have omitted writing $\omega \in \Omega$ as arguments of functions. Denote the variations of P and Q by δP and δQ, respectively. Based on the above notations, we state the stochastic variational principle for (1.9) below, and refer to [35, 122, 169, 260] for more details.

Theorem 1.2.2 *Under the condition* (1.2), *if* $(P(t), Q(t))$ *satisfies the stochastic Hamiltonian system* (1.9) *for* $t \in [0, T]$, *then the pair* $(P(\cdot), Q(\cdot))$ *is a critical point of the stochastic action functional* $\mathbb{S}$, *i.e.,*

$$\delta\mathbb{S}(P(\cdot), Q(\cdot)) = \frac{d}{d\epsilon}\bigg|_{\epsilon=0} \mathbb{S}(P(\cdot) + \epsilon\delta P(\cdot), Q(\cdot) + \epsilon\delta Q(\cdot)) = 0, \quad a.s.,$$

for all $(\delta P(\cdot), \delta Q(\cdot))$ *with* $\delta Q(0) = \delta Q(T) = 0$, *a.s.*

Proof Since the Hamiltonian $H_j \in \mathbb{C}^\infty(\mathbb{R}^{2n}, \mathbb{R})$ for $j \in \{0, 1, \ldots, m\}$, the variation becomes

$$
\delta\mathbb{S}(P(\cdot), Q(\cdot))
$$

$$
= \int_0^T (\delta P(t))^\top \circ dQ(t) + \int_0^T (P(t))^\top \circ d\delta Q(t)
$$

$$
- \int_0^T \left(\left(\frac{\partial H_0}{\partial P}(P(t), Q(t))\right)^\top \delta P(t) + \left(\frac{\partial H_0}{\partial Q}(P(t), Q(t))\right)^\top \delta Q(t) \right) dt
$$

$$
- \sum_{r=1}^m \int_0^T \left(\left(\frac{\partial H_r}{\partial P}(P(t), Q(t))\right)^\top \delta P(t) + \left(\frac{\partial H_r}{\partial Q}(P(t), Q(t))\right)^\top \delta Q(t) \right) \circ dW_r(t).
$$

By means of the Stratonovich integration by parts formula in [218, Corollary 2],

$$
\int_0^T (P(t))^\top \circ d\delta Q(t) = (P(T))^\top \delta Q(T) - (P(0))^\top \delta Q(0) - \int_0^T (\delta Q(t))^\top \circ dP(t)
$$

$$
= - \int_0^T (\delta Q(t))^\top \circ dP(t),
$$

which yields

$$
\delta\mathbb{S}(P(\cdot), Q(\cdot))
$$

$$
= \int_0^T (\delta P(t))^\top \left(\circ dQ(t) - \frac{\partial H_0}{\partial P}(P(t), Q(t))dt - \sum_{r=1}^m \frac{\partial H_r}{\partial P}(P(t), Q(t)) \circ dW_r(t) \right)
$$

$$
- \int_0^T (\delta Q(t))^\top \left(\circ dP(t) + \frac{\partial H_0}{\partial Q}(P(t), Q(t))dt + \sum_{r=1}^m \frac{\partial H_r}{\partial Q}(P(t), Q(t)) \circ dW_r(t) \right).
$$

Substituting the formulae

$$
Q(t) = q + \underbrace{\int_0^t \frac{\partial H_0}{\partial P}(P(s), Q(s))ds}_{M_1(t)} + \underbrace{\sum_{r=1}^m \int_0^t \frac{\partial H_r}{\partial P}(P(s), Q(s)) \circ dW_r(s)}_{M_2(t)},
$$

$$
P(t) = p - \underbrace{\int_0^t \frac{\partial H_0}{\partial Q}(P(s), Q(s))ds}_{M_3(t)} - \underbrace{\sum_{r=1}^m \int_0^t \frac{\partial H_r}{\partial Q}(P(s), Q(s)) \circ dW_r(s)}_{M_4(t)}
$$

into the variation $\delta\mathbb{S}(P(\cdot), Q(\cdot))$, we obtain

$$\delta\mathbb{S}(P(\cdot), Q(\cdot))$$

$$= \int_0^T (\delta P(t))^\top \left(dM_1(t) + \circ dM_2(t) - \frac{\partial H_0}{\partial P}(P(t), Q(t))dt \right.$$

$$\left. - \sum_{r=1}^m \frac{\partial H_r}{\partial P}(P(t), Q(t)) \circ dW_r(t) \right)$$

$$+ \int_0^T (\delta Q(t))^\top \left(dM_3(t) + \circ dM_4(t) \right.$$

$$\left. - \frac{\partial H_0}{\partial Q}(P(t), Q(t))dt - \sum_{r=1}^m \frac{\partial H_r}{\partial Q}(P(t), Q(t)) \circ dW_r(t) \right)$$

$$= 0.$$

$\square$

The stochastic action functional $\mathbb{S}$ can be applied to constructing the stochastic generating function for the stochastic flow $\phi_t(\cdot, \omega)\colon \mathbb{R}^{2n} \to \mathbb{R}^{2n}$, $t \in [0, T]$, of (1.9) for almost all $\omega \in \Omega$. Fix time $t \in [0, T]$ and suppose that for almost all $\omega \in \Omega$, there are open neighborhoods $\mathscr{U}(\omega) \subset \mathbb{R}^n$ of q, $\mathscr{V}(\omega) \subset \mathbb{R}^n$ of $Q(\omega, t)$, and $\mathscr{W}(\omega) \subset \mathbb{R}^n \times \mathbb{R}^n$ of $(q, Q(\omega, t))$ such that for all $q_a \in \mathscr{U}(\omega)$ and $q_b \in \mathscr{V}(\omega)$, the exact solution $(P(\omega, t; q_a, q_b), Q(\omega, t; q_a, q_b))$ of (1.9) satisfying $Q(\omega, 0; q_a, q_b) = q_a$ and $Q(\omega, t; q_a, q_b) = q_b$ belongs to $\mathscr{W}(\omega)$. Define the function

$$S(q, Q(t)) := \mathbb{S}(P(\cdot; q, Q(t)), Q(\cdot; q, Q(t)))$$

$$= \int_0^t (P(s))^\top \circ dQ(s) - \int_0^t H_0(P(s), Q(s))ds$$

$$- \sum_{r=1}^m \int_0^t H_r(P(s), Q(s)) \circ dW_r(s),$$

where for notational convenience we have omitted writing $(q, Q(t))$ explicitly as arguments of $P(s)$ and $Q(s)$ for $s \in (0, t]$. Now we prove that $S(q, Q(t))$ generates the stochastic flow ϕ_t of (1.9).

Theorem 1.2.3 *The mapping $\phi_t\colon (p^\top, q^\top)^\top \mapsto ((P(t))^\top, (Q(t))^\top)^\top$, $t \in [0, T]$, is implicitly given by equations*

$$P(t) = \frac{\partial S}{\partial Q}(q, Q(t)), \quad p = -\frac{\partial S}{\partial q}(q, Q(t)). \tag{1.18}$$

Proof Fix time $t \in [0, T]$. With the help of $S(q, Q(t))$, we obtain

$$\frac{\partial S}{\partial q}(q, Q(t))$$

$$= \int_0^t \left(\frac{\partial P(s)}{\partial q}\right)^\top \circ dQ(s) + \left(\int_0^t (P(s))^\top \circ d\frac{\partial Q(s)}{\partial q}\right)^\top$$

$$- \int_0^t \left(\left(\frac{\partial P(s)}{\partial q}\right)^\top \frac{\partial H_0}{\partial P}(P(s), Q(s)) + \left(\frac{\partial Q(s)}{\partial q}\right)^\top \frac{\partial H_0}{\partial Q}(P(s), Q(s))\right) ds$$

$$- \sum_{r=1}^m \int_0^t \left(\left(\frac{\partial P(s)}{\partial q}\right)^\top \frac{\partial H_r}{\partial P}(P(s), Q(s)) + \left(\frac{\partial Q(s)}{\partial q}\right)^\top \frac{\partial H_r}{\partial Q}(P(s), Q(s))\right) \circ dW_r(s).$$

Since $Q(s)$ is mean-square differentiable with respect to the parameters q for $s \in [0, t]$ and $\frac{\partial Q}{\partial q}$ is a semimartingale (see [122]), by applying the Stratonovich integration by parts formula in [218, Corollary 2], we have

$$\left(\int_0^t (P(s))^\top \circ d\frac{\partial Q(s)}{\partial q}\right)^\top = -P(0) - \int_0^t \left(\frac{\partial Q(s)}{\partial q}\right)^\top \circ dP(s).$$

Based on the above equation,

$$\frac{\partial S}{\partial q}(q, Q(t)) + p = \frac{\partial S}{\partial q}(q, Q(t)) + P(0)$$

$$= \int_0^t \left(\frac{\partial P(s)}{\partial q}\right)^\top \left(\circ dQ(s) - \frac{\partial H_0}{\partial P}(P(s), Q(s))ds - \sum_{r=1}^m \frac{\partial H_r}{\partial P}(P(s), Q(s)) \circ dW_r(s)\right)$$

$$- \int_0^t \left(\frac{\partial Q(s)}{\partial q}\right)^\top \left(\circ dP(s) + \frac{\partial H_0}{\partial Q}(P(s), Q(s))ds + \sum_{r=1}^m \frac{\partial H_r}{\partial Q}(P(s), Q(s)) \circ dW_r(s)\right)$$

$$= 0,$$

which implies that $p = -\frac{\partial S}{\partial q}(q, Q(t))$. Similar arguments lead to the first equation $P(t) = \frac{\partial S}{\partial Q}(q, Q(t))$. By means of the definition of stochastic flow, we obtain $\phi_t(X(0)) = X(t)$, where $X(0) = (p^\top, q^\top)^\top$ and $X(t) = ((P(t))^\top, (Q(t))^\top)^\top$.   □

Remark 1.2.3 If there exists locally a smooth random mapping S such that $\frac{\partial^2 S}{\partial Q \partial q}$ is invertible almost surely and

$$P = \frac{\partial S}{\partial Q}(q, Q), \quad p = -\frac{\partial S}{\partial q}(q, Q),$$

which is equivalent to

$$dS = P^{\top}dQ - p^{\top}dq, \tag{1.19}$$

then the mapping $\psi: (p^{\top}, q^{\top})^{\top} \mapsto (P^{\top}, Q^{\top})^{\top}$ is symplectic (see [93]).

Now we consider the point $(P(t), Q(t))$ moving in the phase flow of the stochastic Hamiltonian system (1.9) for $t \in [0, T]$. A natural problem is whether there exists a function $S(q, Q(t), t)$ depending on $t \in [0, T]$ such that it can generate the symplectic mapping $(p, q) \mapsto (P(t), Q(t))$ associated with (1.9). In accordance with (1.18), $S(q, Q(t), t)$ has to satisfy

$$P(t) = \frac{\partial S}{\partial Q}(q, Q(t), t), \quad p = -\frac{\partial S}{\partial q}(q, Q(t), t), \quad t \in (0, T]. \tag{1.20}$$

Consider the total stochastic differential $\overline{d}S(q, Q(t), t)$ with respect to time t. On one hand, similar arguments in [122, Proposition 2.1] lead to

$$\overline{d}S(q, Q(t), t) = dS(q, Q(t), t) + \left(\frac{\partial S}{\partial Q}(q, Q(t), t)\right)^{\top} \circ dQ(t),$$

where $dS(q, Q(t), t)$ denotes the partial stochastic differential of S with respect to t. On the other hand, we have

$$\overline{d}S(q, Q(t), t) = (P(t))^{\top} \circ dQ(t) - H_0(P(t), Q(t))dt - \sum_{r=1}^{m} H_r(P(t), Q(t)) \circ dW_r(t).$$

Using $P(t) = \frac{\partial S}{\partial Q}(q, Q(t), t)$ for $t \in [0, T]$, we deduce

$$dS = -H_0\left(\frac{\partial S}{\partial Q}, Q\right)dt - \sum_{r=1}^{m} H_r\left(\frac{\partial S}{\partial Q}, Q\right) \circ dW_r(t). \tag{1.21}$$

In fact, $S(q, Q(t), t)$ is the stochastic generating function and (1.21) is the so-called *stochastic Hamilton–Jacobi partial differential equation*. The stochastic generating function associated with stochastic Hamiltonian system (1.9) is rigorously introduced in [29] as the solution of the stochastic Hamilton–Jacobi partial differential equation (1.21). It is revealed that the phase flow of the stochastic Hamiltonian system can be generated by the solution of the stochastic Hamilton–Jacobi partial differential equation. Now we state this result in the following proposition (see e.g., [29, 93] and references therein).

Proposition 1.2.1 *Let $S(q, Q, t)$ be a local solution of (1.21) with $S(q, Q, 0) =$ $\mathscr{S}(q, Q)$ satisfying*

$$\frac{\partial \mathscr{S}}{\partial q_i}(q, q) + \frac{\partial \mathscr{S}}{\partial Q_i}(q, q) = 0, \quad i = 1, \ldots, n.$$

If there exists a stopping time $\tau > 0$ almost surely such that

(1) $S(q, Q, t)$, $\frac{\partial S(q,Q,t)}{\partial Q}$ *and* $\frac{\partial S(q,Q,t)}{\partial q}$, $t \in [0, \tau)$, *are semimartingales* ;

(2) $S(q, Q, t)$, $\frac{\partial S(q,Q,t)}{\partial Q}$ *and* $\frac{\partial S(q,Q,t)}{\partial q}$ *are continuous at (q, Q, t) and $\mathbf{C}^\infty$ functions of (q, Q) for $t \in [0, \tau)$;*

(3) *the matrix $\frac{\partial^2 S(q,Q,t)}{\partial q \partial Q}$ is almost surely invertible for $t \in [0, \tau)$,*

then the mapping $\phi_t : (p^\top, q^\top)^\top \mapsto ((P(t))^\top, (Q(t))^\top)^\top$, $t \in [0, \tau)$, by (1.20) is the stochastic flow of stochastic Hamiltonian system (1.9).

Proof The invertibility of the matrix $\frac{\partial^2 S}{\partial q \partial Q}$ guarantees that the mapping $(p^\top, q^\top)^\top \mapsto ((P(t))^\top, (Q(t))^\top)^\top$ is well defined for $0 \leq t < \tau$ due to the implicit function theorem. Based on the second equation of (1.20), it can be derived that

$$d\left(\frac{\partial S}{\partial q_i}\right) + \sum_{j=1}^{n} \frac{\partial^2 S}{\partial q_i \partial Q_j} \circ dQ_j = 0$$

for $i = 1, \ldots, n$ (see [163, Theorem 3.3.2]). According to (1.21),

$$d\left(\frac{\partial S}{\partial q_i}\right) + \sum_{j=1}^{n} \frac{\partial H_0}{\partial P_j} \frac{\partial^2 S}{\partial q_i \partial Q_j} dt + \sum_{r=1}^{m} \sum_{j=1}^{n} \frac{\partial H_r}{\partial P_j} \frac{\partial^2 S}{\partial q_i \partial Q_j} \circ dW_r(t) = 0$$

for $i = 1, \ldots, n$ (see [29, Corollary of Theorem 6.14]). Comparing the above two equations and using the invertibility of the matrix $\frac{\partial^2 S}{\partial q \partial Q}$, the second equation of (1.9) holds. The first equation of (1.9) can be deduced likewise. $\qquad\square$

Remark 1.2.4 The stochastic generating function $\bar{S}(P(t), q, t)$, $t \in [0, T]$, considered in [122] and the associated stochastic Hamilton–Jacobi partial differential equation can be constructed via the stochastic action functional

$$\bar{S}(P(\cdot), Q(\cdot)) = (P(t))^\top Q(t) - \int_0^t (P(s))^\top \circ dQ(s)$$

$$+ \int_0^t H_0(P(s), Q(s)) ds + \sum_{r=1}^{m} \int_0^t H_r(P(s), Q(s)) \circ dW_r(s),$$

where $(P(\cdot), Q(\cdot)) \in \mathbf{C}([0, t])$.

1.2.3 Stochastic θ-Generating Function

In addition to the aforementioned stochastic generating function $S(q, Q(t), t)$, there are the other three kinds of generating functions S^i, $i = 1, 2, 3$, under different coordinates (see e.g., [12, 13, 93, 122, 260, 263] and references therein). Through different stochastic action functionals, these stochastic generating functions and stochastic Hamilton–Jacobi partial differential equations can be given analogously. In this subsection, we first give a brief introduction to S^i, $i = 1, 2, 3$, and the associated stochastic Hamilton–Jacobi partial differential equations. Then we unify and extend them to the stochastic θ-generating function with parameter $\theta \in [0, 1]$.

Now we start with the stochastic type-I generating function $S^1(P(t), q, t)$. Fix time $t \in (0, T]$ and assume that the exact solution $(P(\cdot), Q(\cdot))$ of (1.9) can be rewritten as $(P(\cdot; q, P(t)), Q(\cdot; q, P(t))) \in \mathbb{R}^n \times \mathbb{R}^n$ satisfying $P(t; q, P(t)) = P(t)$ and $Q(0; q, P(t)) = q$. Define

$$S^1(P(t), q) := (P(t))^\top (Q(t) - q) - S(q, Q(t)).$$

One can check that

$$d\left((P(t))^\top q + S^1(q, P(t))\right) = d\left((P(t))^\top Q(t) - S(q, Q(t))\right)$$
$$= (P(t))^\top dQ(t) + (Q(t))^\top dP(t) + p^\top dq - (P(t))^\top dQ(t)$$
$$= (Q(t))^\top dP(t) + p^\top dq.$$

Consequently, by comparing coefficients of dP and dq on both sides of the equality and letting time t vary, the stochastic flow $\phi_t : (p^\top, q^\top)^\top \mapsto ((P(t))^\top, (Q(t))^\top)^\top$ is implicitly given by

$$p = P(t) + \frac{\partial S^1}{\partial q}(P(t), q, t), \quad Q(t) = q + \frac{\partial S^1}{\partial P}(P(t), q, t), \quad t \in (0, T],$$

where

$$S^1(P(t), q, t) := (P(t))^\top (Q(t) - q) - S(q, Q(t), t).$$

This formula yields

$$dS^1 = -\left(\frac{\partial S^1}{\partial P}\right)^\top \circ dP + (Q-q)^\top \circ dP + P^\top \circ dQ - \left(\frac{\partial S}{\partial Q}\right)^\top \circ dQ - dS = -dS.$$

Further, the stochastic Hamilton–Jacobi partial differential equation with respect to S^1 reads

$$dS^1 = H_0\left(P, q + \frac{\partial S^1}{\partial P}\right) dt + \sum_{r=1}^{m} H_r\left(P, q + \frac{\partial S^1}{\partial P}\right) \circ dW_r(t), \quad S^1(P, q, 0) = 0.$$

Remark 1.2.5 If we define the stochastic action functional as

$$\mathbb{S}^1(P(\cdot), Q(\cdot)) = (P(t))^\top (Q(t) - q) - \int_0^t (P(s))^\top \circ dQ(s) \tag{1.22}$$

$$+ \int_0^t H_0(P(s), Q(s))ds + \sum_{r=1}^{m} \int_0^t H_r(P(s), Q(s)) \circ dW_r(s)$$

with $(P(\cdot), Q(\cdot)) \in C([0, t])$, which is a slight variant of the action functional $\tilde{\mathbb{S}}(P(\cdot), Q(\cdot))$, we can obtain the stochastic type-I generating function $S^1(P(t), q, t)$ that equals the one given in [93, 260], where $t \in [0, T]$.

It is straightforward to obtain the stochastic Hamilton–Jacobi partial differential equation for the stochastic generating function $S^2(p, Q(t), t)$, which is just the adjoint case of $S^1(P(t), q, t)$. Similarly, let $t \in (0, T]$ be fixed and the exact solution be rewritten as $(P(\cdot; p, Q(t)), Q(\cdot; p, Q(t))) \in \mathbb{R}^n \times \mathbb{R}^n$ satisfying $P(0; p, Q(t)) = p$ and $Q(t; p, Q(t)) = Q(t)$. Consider

$$S^2(p, Q(t)) = p^\top (Q(t) - q) - S(q, Q(t)).$$

It can be verified that for any $t \in [0, T]$, the stochastic flow $\phi_t : (p^\top, q^\top)^\top \mapsto ((P(t))^\top, (Q(t))^\top)^\top$ of (1.9) can be defined by

$$P(t) = p - \frac{\partial S^2}{\partial Q}(p, Q(t), t), \quad Q(t) = q + \frac{\partial S^2}{\partial p}(p, Q(t), t),$$

and the associated stochastic Hamilton–Jacobi partial differential equation is

$$dS^2 = H_0\left(p - \frac{\partial S^2}{\partial Q}, Q\right) dt + \sum_{r=1}^{m} H_r\left(p - \frac{\partial S^2}{\partial Q}, Q\right) \circ dW_r(t), \quad S^2(p, Q, 0) = 0.$$

Analogously, under appropriate conditions, consider the stochastic generating function

$$S^3\left(\frac{P(t) + p}{2}, \frac{Q(t) + q}{2}\right) := \frac{1}{2}(P(t) + p)^\top (Q(t) - q) - S(q, Q(t), t)$$

for $t \in (0, T]$. It can be deduced that its associated stochastic Hamilton–Jacobi partial differential equation is

$$dS^3 = H_0 \left(u - \frac{1}{2}\frac{\partial S^3}{\partial v}, v + \frac{1}{2}\frac{\partial S^3}{\partial u} \right) dt + \sum_{r=1}^{m} H_r \left(u - \frac{1}{2}\frac{\partial S^3}{\partial v}, v + \frac{1}{2}\frac{\partial S^3}{\partial u} \right) \circ dW_r(t),$$

$$S^3(u, v, 0) = 0$$

with $u = \frac{P+p}{2}$ and $v = \frac{Q+q}{2}$, and the mapping $(p^\top, q^\top)^\top \mapsto ((P(t))^\top, (Q(t))^\top)^\top$ $t \in [0, T]$, defined by

$$P(t) = p - \frac{\partial S^3}{\partial v} \left(\frac{P(t) + p}{2}, \frac{Q(t) + q}{2}, t \right),$$

$$Q(t) = q + \frac{\partial S^3}{\partial u} \left(\frac{P(t) + p}{2}, \frac{Q(t) + q}{2}, t \right)$$

is the phase flow of the stochastic Hamiltonian system (1.9).

Now we turn to giving a systematic treatment of stochastic generating functions S^i, $i = 1, 2, 3$. Specifically, in the following, we shall unify and extend S^i, $i = 1, 2, 3$. to the *stochastic θ-generating function*

$$S_\theta(\hat{P}, \hat{Q}) := \hat{P}^\top (Q - q) - S(q, Q), \quad \theta \in [0, 1], \tag{1.23}$$

where $\hat{P} = (1 - \theta)p + \theta P$ and $\hat{Q} = (1 - \theta)Q + \theta q$, respectively.

Lemma 1.2.1 *If there exists a smooth function $S_\theta(\hat{P}, \hat{Q})$ with $\theta \in [0, 1]$ such that*

$$p^\top d\hat{Q} + Q^\top d\hat{P} = d\left(\hat{P}^\top \hat{Q} \right) + \theta dS_\theta, \tag{1.24}$$

then the mapping $\psi : (p^\top, q^\top)^\top \mapsto (P^\top, Q^\top)^\top$ is symplectic.

Proof   The formula (1.24) leads to

$$p^\top d\hat{Q} + Q^\top d\hat{P} = d\left(\hat{P}^\top Q \right) - \theta dS.$$

Fix $\theta \in (0, 1)$. Multiplying both sides of the above equation by $(1 - \theta)$ yields

$$\theta(1 - \theta)dS + (1 - \theta)(p^\top d\hat{Q} + Q^\top d\hat{P}) = (1 - \theta)d\left(\hat{P}^\top Q \right).$$

Based on the fact that

$$\theta P^\top d\hat{Q} + \hat{Q}^\top d(\theta P) + (1 - \theta)^2 d\left(Q^\top p \right) - \theta^2 d\left(P^\top q \right) = (1 - \theta)d\left(\hat{P}^\top Q \right),$$

we obtain

$$d\hat{Q}^\top(\theta P) = \theta P^\top d\hat{Q} + \hat{Q}^\top d(\theta P)$$

$$= \theta(1-\theta)dS + (1-\theta)(p^\top d\hat{Q} + Q^\top d\hat{P}) - (1-\theta)^2 d\left(Q^\top p\right) + \theta^2 d\left(P^\top q\right)$$

$$= \theta(1-\theta)dS + \theta P^\top d(\theta q) - (1-\theta)p^\top d((1-\theta)Q) + (1-\theta)p^\top d\hat{Q}$$

$$+ (1-\theta)Q^\top d\hat{P} - (1-\theta)Q^\top d((1-\theta)p) + \theta q^\top d(\theta P).$$

It implies that

$$\theta P^\top d\hat{Q} - (1-\theta)p^\top d\hat{Q} = \theta(1-\theta)dS + \theta P^\top d(\theta q) - (1-\theta)p^\top d((1-\theta)Q).$$

Subtracting the term $\theta P^\top d(\theta q) - (1-\theta)p^\top d((1-\theta)Q)$ from both sides of the equation above, we get

$$\theta P^\top d((1-\theta)Q) - (1-\theta)p^\top d(\theta q) = \theta(1-\theta)dS,$$

which means (1.19). Thus, $dP \wedge dQ = dp \wedge dq$. For the cases of $\theta = 1$ or $\theta = 0$, the function $S_\theta(\hat{P}, \hat{Q})$ becomes $S^1(P, q)$ or $S^2(p, Q)$, respectively, which implies the result immediately (see e.g., [13, 93, 260] and references therein).   □

Remark 1.2.6 When $\theta = \frac{1}{2}$, $S_{\frac{1}{2}}(\hat{P}, \hat{Q})$ equals $S^3(\frac{P+p}{2}, \frac{Q+q}{2})$ (see [93]).

Similarly, the stochastic θ-generating function $S_\theta(\hat{P}(t), \hat{Q}(t), t)$ with $\theta \in [0, 1]$ is also associated with a stochastic Hamilton–Jacobi partial differential equation, which is assumed to satisfy similar conditions as in [163] assuring the existence of a local solution S_θ.

Theorem 1.2.4 (See [132]) *Let $S_\theta(\hat{P}, \hat{Q}, t)$, $\theta \in [0, 1]$, be a local solution of the stochastic Hamilton–Jacobi partial differential equation*

$$dS_\theta = \sum_{r=0}^m H_r\left(\hat{P} - (1-\theta)\frac{\partial S_\theta}{\partial \hat{Q}}, \hat{Q} + \theta\frac{\partial S_\theta}{\partial \hat{P}}\right) \circ dW_r(t) \qquad (1.25)$$

with $S_\theta(\hat{P}, \hat{Q}, 0) = 0$, $dW_0(t) := dt$. If there exists a stopping time $\tau > 0$ almost surely such that

(1) *$S_\theta(\hat{P}, \hat{Q}, t)$, $\frac{\partial S_\theta(\hat{P}, \hat{Q}, t)}{\partial \hat{P}}$, and $\frac{\partial S_\theta(\hat{P}, \hat{Q}, t)}{\partial \hat{Q}}$, $t \in [0, \tau)$, are semimartingales;*

(2) *$S_\theta(\hat{P}, \hat{Q}, t)$, $\frac{\partial S_\theta(\hat{P}, \hat{Q}, t)}{\partial \hat{P}}$, and $(\hat{P}, \hat{Q}, t)$, and $\mathbf{C}^\infty$ functions of $(\hat{P}, \hat{Q})$ for $t \in [0, \tau)$;*

(3) *the following matrices are almost surely invertible for $t \in [0, \tau)$,*

(i) $\dfrac{1}{\theta - 1}\left(I_n - (1 - \theta)\dfrac{\partial^2 S_\theta}{\partial \hat{P} \partial \hat{Q}}\right)\left(\dfrac{\partial^2 S_\theta}{\partial \hat{Q}^2}\right)^{-1}\left(I_n + \theta\dfrac{\partial^2 S_\theta}{\partial \hat{Q} \partial \hat{P}}\right) - \theta\dfrac{\partial^2 S_\theta}{\partial \hat{P}^2},$

$\dfrac{1}{\theta}\left(I_n + \theta\dfrac{\partial^2 S_\theta}{\partial \hat{Q} \partial \hat{P}}\right)\left(\dfrac{\partial^2 S_\theta}{\partial \hat{P}^2}\right)^{-1}\left(I_n - (1 - \theta)\dfrac{\partial^2 S_\theta}{\partial \hat{P} \partial \hat{Q}}\right) + (1 - \theta)\dfrac{\partial^2 S_\theta}{\partial \hat{Q}^2},$

$\dfrac{\partial^2 S_\theta}{\partial \hat{Q}^2},\ \dfrac{\partial^2 S_\theta}{\partial \hat{P}^2}$ *for $\theta \in (0, 1)$;*

(ii) $I_n - \dfrac{\partial^2 S_0}{\partial p \partial Q}$ *for $\theta = 0$;*

(iii) $I_n + \dfrac{\partial^2 S_1}{\partial P \partial q}$ *for $\theta = 1$,*   (1.26)

then the mapping $(p^\top, q^\top)^\top \mapsto ((P(t))^\top, (Q(t))^\top)^\top, 0 \le t < \tau$, defined by

$$P(t) = p - \dfrac{\partial S_\theta(\hat{P}(t), \hat{Q}(t), t)}{\partial \hat{Q}}, \quad Q(t) = q + \dfrac{\partial S_\theta(\hat{P}(t), \hat{Q}(t), t)}{\partial \hat{P}} \tag{1.27}$$

is the stochastic flow of the stochastic Hamiltonian system (1.9).

Proof Based on the invertibility of the matrices in (1.26), it can be verified that the mapping $(p^\top, q^\top)^\top \mapsto ((P(t))^\top, (Q(t))^\top)^\top, t \in [0, \tau)$, is well-defined by (1.27) via the implicit function theorem. Since $\frac{\partial S_\theta(\hat{P},\hat{Q},t)}{\partial \hat{Q}}$ and $\frac{\partial S_\theta(\hat{P},\hat{Q},t)}{\partial \hat{P}}$ with $t \in [0, \tau)$ are semimartingales, so are $P(t)$ and $Q(t)$ by (1.27). According to [163, Theorem 3.3.2], we derive

$$d\left(\dfrac{\partial S_\theta}{\partial \hat{Q}}\right) = -\left(I_n + \theta\dfrac{\partial^2 S_\theta}{\partial \hat{Q} \partial \hat{P}}\right) \circ dP - (1 - \theta)\dfrac{\partial^2 S_\theta}{\partial \hat{Q}^2} \circ dQ, \tag{1.28}$$

$$d\left(\dfrac{\partial S_\theta}{\partial \hat{P}}\right) = -\theta\dfrac{\partial^2 S_\theta}{\partial \hat{P}^2} \circ dP + \left(I_n - (1 - \theta)\dfrac{\partial^2 S_\theta}{\partial \hat{P} \partial \hat{Q}}\right) \circ dQ, \tag{1.29}$$

via $d\hat{P} = \theta dP$ and $d\hat{Q} = (1 - \theta)dQ$. By making use of the stochastic Hamilton–Jacobi partial differential equation (1.25) and similar arguments as in [29, Corollary of Theorem 6.14], we obtain

$$d\left(\dfrac{\partial S_\theta}{\partial \hat{Q}}\right) = \sum_{r=0}^{m}\left((\theta - 1)\dfrac{\partial^2 S_\theta}{\partial \hat{Q}^2}\dfrac{\partial H_r}{\partial P}(P, Q) + \left(I_n + \theta\dfrac{\partial^2 S_\theta}{\partial \hat{P} \partial \hat{Q}}\right)\dfrac{\partial H_r}{\partial Q}(P, Q)\right) \circ dW_r(t),$$

$$\tag{1.30}$$

$$d\left(\frac{\partial S_\theta}{\partial \hat{P}}\right) = \sum_{r=0}^{m}\left((I_n - (1-\theta)\frac{\partial^2 S_\theta}{\partial \hat{Q}\partial \hat{P}})\frac{\partial H_r}{\partial P}(P, Q) + \theta\frac{\partial^2 S_\theta}{\partial \hat{P}^2}\frac{\partial H_r}{\partial Q}(P, Q)\right)\circ dW_r(t).$$

$$(1.31)$$

Comparing (1.28) with (1.30) and (1.29) with (1.31) yields (1.9) for $t \in [0, \tau)$.   □

The integral form of the stochastic Hamilton–Jacobi partial differential equation (1.25) under the initial condition is

$$S_\theta(\hat{P}, \hat{Q}, t) = \sum_{r=0}^{m}\int_0^t H_r\left(\hat{P} - (1-\theta)\frac{\partial S_\theta}{\partial \hat{Q}}, \hat{Q} + \theta\frac{\partial S_\theta}{\partial \hat{P}}\right)\circ dW_r(s).$$

Following the idea for the deterministic case in [116], since $H_r \in \mathbf{C}^\infty(\mathbb{R}^{2n}, \mathbb{R})$ for $r = 0, 1, \ldots, m$, we perform a Stratonovich–Taylor expansion of the integrands

$$H_r\left(\hat{P} - (1-\theta)\frac{\partial S_\theta}{\partial \hat{Q}}, \hat{Q} + \theta\frac{\partial S_\theta}{\partial \hat{P}}\right)$$

at $(\hat{P}, \hat{Q})$, which leads to the following formal series expansion of

$$S_\theta(\hat{P}, \hat{Q}, t) = \sum_\alpha G_\alpha^\theta(\hat{P}, \hat{Q})J_\alpha(t),$$

$$(1.32)$$

where

$$J_\alpha(t) = \int_0^t \int_0^{s_l} \cdots \int_0^{s_2} \circ dW_{j_1}(s_1) \circ dW_{j_2}(s_2) \circ \cdots \circ dW_{j_l}(s_l)$$

is a multiple Stratonovich stochastic integral with multi-index $\alpha = (j_1, j_2, \ldots, j_l)$, $j_i \in \{0, 1, \ldots, m\}, i = 1, \ldots, l, l \geq 1$, and $dW_0(s) := ds$.

One may wonder what the coefficient $G_\alpha^\theta(\hat{P}, \hat{Q})$ is in (1.32). To answer it, we first specify some notations (see [93]). Let $l(\alpha)$ and $\alpha-$ denote the length of α and the multi-index resulted from discarding the last index of α, respectively. For any multi-index $\alpha = (j_1, j_2, \ldots, j_{l(\alpha)})$ without duplicated elements, i.e., $j_m \neq j_n$ if $m \neq n$ with $m, n \in \{1, \ldots, l(\alpha)\}$,

$$R(\alpha) = \begin{cases} \emptyset, & \text{if } l(\alpha) = 1; \\ \{(j_m, j_n) \mid m < n, m, n = 1, \ldots, l(\alpha)\}, & \text{if } l(\alpha) \geq 2. \end{cases}$$

Given $\alpha = (j_1, \ldots, j_l)$ and $\alpha' = (j'_1, \ldots, j'_{l'})$, we define the concatenation operator '$*$' as $\alpha * \alpha' := (j_1, \ldots, j_l, j'_1, \ldots, j'_{l'})$. Furthermore, set

$$
\Lambda_{\alpha, \alpha'} =
\begin{cases}
\{(j_1, j'_1), (j'_1, j_1)\}, & \text{if } l = l' = 1; \\[2mm]
\{\Lambda_{(j_1), \alpha'-} * (j'_{l'}), \alpha' * (j_1)\}, & \text{if } l = 1, l' \neq 1; \\[2mm]
\{\Lambda_{\alpha-, (j'_1)} * (j_l), \alpha * (j'_1)\}, & \text{if } l \neq 1, l' = 1; \\[2mm]
\{\Lambda_{\alpha-, \alpha'} * (j_l), \Lambda_{\alpha, \alpha'-} * (j'_{l'})\}, & \text{if } l \neq 1, l' \neq 1,
\end{cases}
$$

where the concatenation '$*$' between a set of multi-indices Λ and α is $\Lambda * \alpha = \{\beta * \alpha \mid \beta \in \Lambda\}$. For $k > 2$, let $\Lambda_{\alpha, \ldots, \alpha_k} = \{\Lambda_{\beta, \alpha_k} \mid \beta \in \Lambda_{\alpha, \ldots, \alpha_{k-1}}\}$.

Lemma 1.2.2 (See [93]) *If there are no duplicated elements in or between any of the multi-indices $\alpha_1 = \left(j_1^{(1)}, j_2^{(1)}, \ldots, j_{l_1}^{(1)}\right), \ldots, \alpha_n = \left(j_1^{(n)}, j_2^{(n)}, \ldots, j_{l_n}^{(n)}\right)$, then*

$$
\Lambda_{\alpha_1, \ldots, \alpha_n} = \left\{ \beta \in \mathcal{M} \mid l(\beta) = \sum_{k=1}^{n} l(\alpha_k) \text{ and } \cup_{k=1}^{n} R(\alpha_k) \subseteq R(\beta) \right.
$$

and there are no duplicated elements in β $\left. \right\}$,

where $\mathcal{M} = \left\{ (\hat{j}_1, \hat{j}_2, \ldots, \hat{j}_{\hat{l}}) \mid \hat{j}_i \in \{j_1^{(1)}, j_2^{(1)}, \ldots, j_{l_1}^{(1)}, \ldots, j_1^{(n)}, j_2^{(n)}, \ldots, j_{l_n}^{(n)}\} \right.$, $i = 1, \ldots, \hat{l}, \hat{l} = l_1 + \cdots + l_n \left. \right\}$.

Employing similar techniques as in [93], we substitute the series expansion (1.32) into the stochastic Hamilton–Jacobi partial differential equation (1.25) and take the Taylor series expansion of H_r, $r = 0, 1, \ldots, m$, at $(\hat{P}, \hat{Q})$ to obtain the expression of $S_\theta(\hat{P}, \hat{Q}, t)$ as follows, i.e.,

$$
S_\theta(\hat{P}, \hat{Q}, t)
$$

$$
= \sum_{r=0}^{m} \int_0^t H_r \left(\hat{P} + (\theta - 1) \sum_\alpha \frac{\partial G_\alpha^\theta}{\partial \hat{Q}} J_\alpha, \, \hat{Q} + \theta \sum_\alpha \frac{\partial G_\alpha^\theta}{\partial \hat{P}} J_\alpha \right) \circ dW_r(s)
$$

$$
= \sum_{r=0}^{m} \int_0^t \sum_{i=0}^{\infty} \frac{1}{i!} \sum_{k_1, \ldots, k_i = 1}^{n} \sum_{j=0}^{i} \frac{\partial^i H_r(\hat{P}, \hat{Q})}{\partial \hat{P}_{k_1} \cdots \partial \hat{P}_{k_j} \partial \hat{Q}_{k_{j+1}} \cdots \partial \hat{Q}_{k_i}} C_i^j (\theta - 1)^j \theta^{i-j}
$$

$$
\times \left(\sum_\alpha \frac{\partial G_\alpha^\theta}{\partial \hat{Q}} J_\alpha \right)_{k_1} \cdots \left(\sum_\alpha \frac{\partial G_\alpha^\theta}{\partial \hat{Q}} J_\alpha \right)_{k_j} \left(\sum_\alpha \frac{\partial G_\alpha^\theta}{\partial \hat{P}} J_\alpha \right)_{k_{j+1}} \cdots \left(\sum_\alpha \frac{\partial G_\alpha^\theta}{\partial \hat{P}} J_\alpha \right)_{k_i} \circ dW_r(s)
$$

$$= \sum_{r=0}^{m} \int_0^t H_r(\hat{P}, \hat{Q}) \circ dW_r(s) + \sum_{r=0}^{m} \sum_{i=1}^{\infty} \frac{1}{i!} \sum_{k_1,\ldots,k_i=1}^{n} \sum_{j=0}^{i} \frac{\partial^i H_r(\hat{P}, \hat{Q})}{\partial \hat{P}_{k_1} \cdots \partial \hat{P}_{k_j} \partial \hat{Q}_{k_{j+1}} \cdots \partial \hat{Q}_{k_i}}$$

$$\times C_i^j (\theta - 1)^j \theta^{i-j} \sum_{\alpha_1,\ldots,\alpha_i} \frac{\partial G_{\alpha_1}^\theta}{\partial \hat{Q}_{k_1}} \cdots \frac{\partial G_{\alpha_j}^\theta}{\partial \hat{Q}_{k_j}} \frac{\partial G_{\alpha_{j+1}}^\theta}{\partial \hat{P}_{k_{j+1}}} \cdots \frac{\partial G_{\alpha_i}^\theta}{\partial \hat{P}_{k_i}} \int_0^t \prod_{k=1}^{i} J_{\alpha_k} \circ dW_r(s),$$

$$(1.33)$$

where $\left(\sum_\alpha \frac{\partial G_\alpha^\theta}{\partial \hat{P}} \right)_{k_i}$ and $\left(\sum_\alpha \frac{\partial G_\alpha^\theta}{\partial \hat{Q}} \right)_{k_i}$ are the k_ith component of column vectors $\left(\sum_\alpha \frac{\partial G_\alpha^\theta}{\partial \hat{P}} \right)$ and $\left(\sum_\alpha \frac{\partial G_\alpha^\theta}{\partial \hat{Q}} \right)$, respectively. By means of the relation (see [93, 151])

$$\prod_{k=1}^{i} J_{\alpha_k} = \sum_{\beta \in \Lambda_{\alpha_1,\ldots,\alpha_i}} J_\beta,$$

and after equating coefficients on both sides of Eq. (1.33), we obtain that $G_\alpha^\theta(\hat{P}, \hat{Q}) = H_r(\hat{P}, \hat{Q})$ for $\alpha = (r)$ with $l(\alpha) = 1$, and

$$G_\alpha^\theta(\hat{P}, \hat{Q})$$

$$= \sum_{i=1}^{l(\alpha)-1} \frac{1}{i!} \sum_{k_1,\ldots,k_i=1}^{n} \sum_{j=0}^{i} \frac{\partial^i H_r(\hat{P}, \hat{Q})}{\partial \hat{P}_{k_1} \cdots \partial \hat{P}_{k_j} \partial \hat{Q}_{k_{j+1}} \cdots \partial \hat{Q}_{k_i}} C_i^j (\theta - 1)^j \theta^{i-j}$$

$$\times \sum_{\substack{l(\alpha_1)+\cdots+l(\alpha_i)=l(\alpha)-1 \\ \alpha - \in \Lambda_{\alpha_1,\ldots,\alpha_i}}} \frac{\partial G_{\alpha_1}^\theta}{\partial \hat{Q}_{k_1}} \cdots \frac{\partial G_{\alpha_j}^\theta}{\partial \hat{Q}_{k_j}} \frac{\partial G_{\alpha_{j+1}}^\theta}{\partial \hat{P}_{k_{j+1}}} \cdots \frac{\partial G_{\alpha_i}^\theta}{\partial \hat{P}_{k_i}}$$

$$(1.34)$$

for $\alpha = (i_1, \ldots, i_{l-1}, r)$ with $l > 1$ and $(i_1, \ldots, i_{l-1}, r) \in \{0, 1, \ldots, m\}^l$ without duplicated elements. If the multi-index α contains repeated components, then we form a new multi-index α' without any duplicates by associating different subscripts to the repeating numbers, e.g., if $\alpha = (1, 0, 0, 1, 2, 1)$, then $\alpha' = (1_1, 0_1, 0_2, 1_2, 2, 1_3)$. Now we take $G_{(0,0,0)}^1$ as an example to show how to use (1.34). It can be verified that

$$G_{(0,0,0)}^1 = G_{(0_1,0_2,0_3)}^1$$

$$= \sum_{k_1=1}^{n} \frac{\partial H_{0_3}}{\partial q_{k_1}} \frac{\partial G_{(0_1,0_2)}^1}{\partial P_{k_1}} + \sum_{k_1,k_2=1}^{n} \frac{1}{2} \frac{\partial^2 H_{0_3}}{\partial q_{k_1} \partial q_{k_2}} \left(\frac{\partial G_{(0_1)}^1}{\partial P_{k_1}} \frac{\partial G_{(0_2)}^1}{\partial P_{k_2}} + \frac{\partial G_{(0_2)}^1}{\partial P_{k_1}} \frac{\partial G_{(0_1)}^1}{\partial P_{k_2}} \right)$$

$$= \sum_{k_1=1}^{n} \frac{\partial H_0}{\partial q_{k_1}} \frac{\partial G_{(0,0)}^1}{\partial P_{k_1}} + \sum_{k_1,k_2=1}^{n} \frac{\partial^2 H_0}{\partial q_{k_1} \partial q_{k_2}} \frac{\partial G_{(0)}^1}{\partial P_{k_1}} \frac{\partial G_{(0)}^1}{\partial P_{k_2}}.$$

$$(1.35)$$

Substituting

$$G^1_{(0,0)} = G^1_{(0_1,0_2)} = \sum_{k=1}^n \frac{\partial H_{0_2}}{\partial q_k} \frac{\partial H_{0_1}}{\partial P_k} = \sum_{k=1}^n \frac{\partial H_0}{\partial q_k} \frac{\partial H_0}{\partial P_k}$$

into (1.35) yields

$$G^1_{(0,0,0)} = \sum_{k_1,k_2=1}^n \left(\frac{\partial^2 H_0}{\partial q_{k_1} \partial q_{k_2}} \frac{\partial H_0}{\partial P_{k_1}} \frac{\partial H_0}{\partial P_{k_2}} + \frac{\partial H_0}{\partial q_{k_1}} \frac{\partial H_0}{\partial P_{k_2}} \frac{\partial^2 H_0}{\partial q_{k_2} \partial P_{k_1}} + \frac{\partial H_0}{\partial q_{k_1}} \frac{\partial H_0}{\partial q_{k_2}} \frac{\partial^2 H_0}{\partial P_{k_1} \partial P_{k_2}} \right),$$

which is the same as the result in [93]. Similarly, we can deduce

$$G^1_{(1,1,1)} = \sum_{k_1,k_2=1}^n \left(\frac{\partial^2 H_1}{\partial q_{k_1} \partial q_{k_2}} \frac{\partial H_1}{\partial P_{k_1}} \frac{\partial H_1}{\partial P_{k_2}} + \frac{\partial H_1}{\partial q_{k_1}} \frac{\partial H_1}{\partial P_{k_2}} \frac{\partial^2 H_1}{\partial q_{k_2} \partial P_{k_1}} + \frac{\partial H_1}{\partial q_{k_1}} \frac{\partial H_1}{\partial q_{k_2}} \frac{\partial^2 H_1}{\partial P_{k_1} \partial P_{k_2}} \right),$$

$$G^1_{(1,1,0)} = \sum_{k_1,k_2=1}^n \left(\frac{\partial^2 H_0}{\partial q_{k_1} \partial q_{k_2}} \frac{\partial H_1}{\partial P_{k_1}} \frac{\partial H_1}{\partial P_{k_2}} + \frac{\partial H_1}{\partial P_{k_2}} \frac{\partial H_0}{\partial q_{k_1}} \frac{\partial^2 H_1}{\partial q_{k_2} \partial P_{k_1}} + \frac{\partial H_0}{\partial q_{k_1}} \frac{\partial H_1}{\partial q_{k_2}} \frac{\partial^2 H_1}{\partial P_{k_1} \partial P_{k_2}} \right),$$

$$G^1_{(0,1,1)} = \sum_{k_1,k_2=1}^n \left(\frac{\partial^2 H_1}{\partial q_{k_1} \partial q_{k_2}} \frac{\partial H_1}{\partial P_{k_1}} \frac{\partial H_0}{\partial P_{k_2}} + \frac{\partial H_1}{\partial q_{k_1}} \frac{\partial H_0}{\partial P_{k_2}} \frac{\partial^2 H_1}{\partial q_{k_2} \partial P_{k_1}} + \frac{\partial H_1}{\partial q_{k_1}} \frac{\partial H_1}{\partial q_{k_2}} \frac{\partial^2 H_0}{\partial P_{k_1} \partial P_{k_2}} \right),$$

$$G^1_{(1,0,1)} = \sum_{k_1,k_2=1}^n \left(\frac{\partial^2 H_1}{\partial q_{k_1} \partial q_{k_2}} \frac{\partial H_0}{\partial P_{k_1}} \frac{\partial H_1}{\partial P_{k_2}} + \frac{\partial H_1}{\partial q_{k_1}} \frac{\partial H_1}{\partial P_{k_2}} \frac{\partial^2 H_0}{\partial q_{k_2} \partial P_{k_1}} + \frac{\partial H_1}{\partial q_{k_1}} \frac{\partial H_0}{\partial q_{k_2}} \frac{\partial^2 H_1}{\partial P_{k_1} \partial P_{k_2}} \right).$$

Applying the same approach to S_θ with $\theta = \frac{1}{2}$, we get

$$G^{\frac{1}{2}}_{(0)} = H_0, \quad G^{\frac{1}{2}}_{(1)} = H_1, \quad G^{\frac{1}{2}}_{(0,0)} = 0, \quad G^{\frac{1}{2}}_{(1,1)} = 0,$$

$$G^{\frac{1}{2}}_{(1,0)} = \frac{1}{2} (\nabla H_0)^\top J_{2n}^{-1} \nabla H_1, \quad G^{\frac{1}{2}}_{(0,1)} = \frac{1}{2} (\nabla H_1)^\top J_{2n}^{-1} \nabla H_0,$$

$$G^{\frac{1}{2}}_{(0,0,0)} = \frac{1}{4} \left(J_{2n}^{-1} \nabla H_0 \right)^\top \nabla^2 H_0 \left(J_{2n}^{-1} \nabla H_0 \right), \dots$$

In sum, the stochastic generating function S_θ, $\theta \in [0, 1]$, can be expressed as

$$S_\theta(\hat{P}, \hat{Q}, t)$$

$$= H_0(\hat{P}, \hat{Q}) J_{(0)}(t) + \sum_{r=1}^m H_r(\hat{P}, \hat{Q}) J_{(r)}(t) + (2\theta - 1) \sum_{k=1}^n \frac{\partial H_0(\hat{P}, \hat{Q})}{\partial \hat{Q}_k} \frac{\partial H_0(\hat{P}, \hat{Q})}{\partial \hat{P}_k} J_{(0,0)}(t)$$

$$+ \sum_{r=1}^m \sum_{s=1}^m \sum_{k=1}^n \left(\theta \frac{\partial H_r(\hat{P}, \hat{Q})}{\partial \hat{Q}_k} \frac{\partial H_s(\hat{P}, \hat{Q})}{\partial \hat{P}_k} + (\theta - 1) \frac{\partial H_r(\hat{P}, \hat{Q})}{\partial \hat{P}_k} \frac{\partial H_s(\hat{P}, \hat{Q})}{\partial \hat{Q}_k} \right) J_{(s,r)}(t)$$

$$+ \sum_{r=1}^{m} \sum_{k=1}^{n} \left(\theta \frac{\partial H_r(\hat{P}, \hat{Q})}{\partial \hat{Q}_k} \frac{\partial H_0(\hat{P}, \hat{Q})}{\partial \hat{P}_k} + (\theta - 1) \frac{\partial H_r(\hat{P}, \hat{Q})}{\partial \hat{P}_k} \frac{\partial H_0(\hat{P}, \hat{Q})}{\partial \hat{Q}_k} \right) J_{(0,r)}(t)$$

$$+ \sum_{r=1}^{m} \sum_{k=1}^{n} \left((\theta - 1) \frac{\partial H_0(\hat{P}, \hat{Q})}{\partial \hat{P}_k} \frac{\partial H_r(\hat{P}, \hat{Q})}{\partial \hat{Q}_k} + \theta \frac{\partial H_0(\hat{P}, \hat{Q})}{\partial \hat{Q}_k} \frac{\partial H_r(\hat{P}, \hat{Q})}{\partial \hat{P}_k} \right) J_{(r,0)}(t)$$

$$+ \cdots . \tag{1.36}$$

Below there is the invariance property of the coefficient G_α^θ with $\theta \in \{\frac{1}{2}, 1\}$. For any permutation π on $\{1, \ldots, l\}$, $l \geq 1$, and any multi-index $\alpha = (i_1, \ldots, i_l)$ with $l(\alpha) = l$, denote by $\pi(\alpha) := (i_{\pi(1)}, \ldots, i_{\pi(l)})$. By means of the formula (1.34), we have the following theorem which can be found in [12].

Theorem 1.2.5 *For the stochastic Hamiltonian system (1.9) preserving the Hamiltonians, the coefficients G_α^θ with $\theta \in \{\frac{1}{2}, 1\}$ are invariants under permutations, i.e., $G_\alpha^\theta = G_{\pi(\alpha)}^\theta$ for any permutation π on $\{1, \ldots, l\}$, where $l = l(\alpha)$.*

In Chap. 2, we will apply the stochastic θ-generating function theory with $\theta \in [0, 1]$ to constructing both strongly and weakly convergent stochastic symplectic methods for the stochastic Hamiltonian system (1.9).

1.3   Non-canonical Stochastic Hamiltonian Systems

In this section, we discuss non-canonical stochastic Hamiltonian systems, which can be regarded as generalizations of (1.9). They are stochastic forced Hamiltonian systems and stochastic Poisson systems, both of which possess stochastic geometric structure and even some long-time behaviors.

1.3.1   Stochastic Forced Hamiltonian Systems

Stochastic forced Hamiltonian systems can be used to model unresolved physical processes on which the Hamiltonian of the deterministic system might otherwise depend, and has the widespread occurrence in the fields of physics, chemistry, and biology, etc. Particular examples include dissipative particle dynamics (see [224]), molecular dynamics (see [24, 143, 166, 238]), energy localization in thermal equilibrium (see [222]), investigations of the dispersion of passive tracers in turbulent flows (see [230, 252]), lattice dynamics in strongly anharmonic crystals (see [112]), collisional kinetic plasmas (see [149, 240]), and description of noise induced transport in stochastic ratchets (see [168]). Below we shall introduce the

$2n$-dimensional stochastic forced Hamiltonian system

$$
\begin{cases}
dP(t) = \left(-\dfrac{\partial H_0(P(t), Q(t))}{\partial Q} + \tilde{f}_0(P(t), Q(t)) \right) dt \\[2mm]
\qquad\quad + \displaystyle\sum_{r=1}^{m} \left(-\dfrac{\partial H_r(P(t), Q(t))}{\partial Q} + \tilde{f}_r(P(t), Q(t)) \right) \circ dW_r(t), \\[4mm]
dQ(t) = \dfrac{\partial H_0(P(t), Q(t))}{\partial P} dt + \displaystyle\sum_{r=1}^{m} \dfrac{\partial H_r(P(t), Q(t))}{\partial P} \circ dW_r(t),
\end{cases}
\tag{1.37}
$$

where $t \in (0, T]$, $P(0) = p \in \mathbb{R}^n$, $Q(0) = q \in \mathbb{R}^n$, and $W_r(\cdot)$, $r = 1, \ldots, m$, are independent standard Wiener processes. Moreover, $H_r \in \mathbf{C}^\infty(\mathbb{R}^{2n}, \mathbb{R})$ and $\tilde{f}_r \in \mathbf{C}^\infty(\mathbb{R}^{2n}, \mathbb{R})$, $r = 0, 1, \ldots, m$, are Hamiltonians and forcing terms, respectively. Assume that the above coefficients satisfy appropriate conditions such that (1.37) defines a forward stochastic flow of diffeomorphisms $\phi_t \colon \mathbb{R}^{2n} \to \mathbb{R}^{2n}$, where $t \in [0, T]$.

Recall that stochastic Hamiltonian system (1.9) is related to the stochastic variational principle, and the associated stochastic action functional is defined as

$$
\begin{aligned}
\mathbb{S}(P(\cdot), Q(\cdot)) \\
= \int_0^T (P(t))^\top \circ dQ(t) - \int_0^T H_0(P(t), Q(t))dt \\
- \sum_{r=1}^{m} \int_0^T H_r(P(t), Q(t)) \circ dW_r(t)
\end{aligned}
$$

in Sect. 1.2.2. For the stochastic forced Hamiltonian system (1.37), we present its associated stochastic variational principle, which is a variant of the Lagrange–d'Alembert principle in [160].

Theorem 1.3.1 *If* $(P(t), Q(t))$ *satisfies the stochastic forced Hamiltonian system* (1.37) *where* $t \in [0, T]$, *then the pair* $(P(\cdot), Q(\cdot))$ *also satisfies the integral equation*

$$
\begin{aligned}
\delta\mathbb{S}(P(\cdot), Q(\cdot)) + \int_0^T (\tilde{f}_0(P(t), Q(t)))^\top \delta Q(t)dt \\
+ \sum_{r=1}^{m} \int_0^T (\tilde{f}_r(P(t), Q(t)))^\top \delta Q(t) \circ dW_r(t) = 0
\end{aligned}
$$

for all variations $(\delta P(\cdot), \delta Q(\cdot))$ *with* $\delta Q(0) = \delta Q(T) = 0$, *a.s.*

Proof Following the proof of Theorem 1.2.2, we have

$$\delta \mathbb{S}(P(\cdot), Q(\cdot))$$

$$= \int_0^T (\delta P(t))^\top \Big(\circ\, dQ(t) - \frac{\partial H_0}{\partial P}(P(t), Q(t)) dt - \sum_{r=1}^m \frac{\partial H_r}{\partial P}(P(t), Q(t)) \circ dW_r(t) \Big)$$

$$- \int_0^T (\delta Q(t))^\top \Big(\circ\, dP(t) + \frac{\partial H_0}{\partial Q}(P(t), Q(t)) dt + \sum_{r=1}^m \frac{\partial H_r}{\partial Q}(P(t), Q(t)) \circ dW_r(t) \Big).$$

Then a straight computation yields

$$\delta \mathbb{S}(P(\cdot), Q(\cdot)) + \int_0^T (\tilde{f}_0(P(t), Q(t)))^\top \delta Q(t) dt$$

$$+ \sum_{r=1}^m \int_0^T (\tilde{f}_r(P(t), Q(t)))^\top \delta Q(t) \circ dW_r(t)$$

$$= \int_0^T (\delta P(t))^\top \Big(\frac{\partial H_0}{\partial P}(P(t), Q(t)) dt + \sum_{r=1}^m \frac{\partial H_r}{\partial P}(P(t), Q(t)) \circ dW_r(t)$$

$$- \frac{\partial H_0}{\partial P}(P(t), Q(t)) dt - \sum_{r=1}^m \frac{\partial H_r}{\partial P}(P(t), Q(t)) \circ dW_r(t) \Big)$$

$$+ \int_0^T (\delta Q(t))^\top \Big(\frac{\partial H_0}{\partial Q}(P(t), Q(t)) dt + \tilde{f}_0(P(t), Q(t)) dt$$

$$+ \sum_{r=1}^m \Big(\frac{\partial H_r}{\partial Q}(P(t), Q(t)) + \tilde{f}_r(P(t), Q(t)) \Big) \circ dW_r(t)$$

$$- \tilde{f}_0(P(t), Q(t)) dt - \frac{\partial H_0}{\partial Q}(P(t), Q(t)) dt$$

$$- \sum_{r=1}^m \Big(\frac{\partial H_r}{\partial Q}(P(t), Q(t)) + \tilde{f}_r(P(t), Q(t)) \Big) \circ dW_r(t) \Big) = 0,$$

which finishes the proof.   □

As shown in Sect. 1.2, the phase flow of stochastic Hamiltonian system preserves canonical stochastic symplectic structure. However, this geometric property does not hold for the stochastic forced Hamiltonian system (1.37). In certain situations, the phase flow of (1.37) may preserve the *stochastic conformal symplectic structure* in the sense that for all $t \in [0, T]$, there exists $c(t)$ such that

$$\phi_t^* \omega_2 = c(t)\omega_2, \quad a.s.$$

Specifically, the conformal symplecticity is considered in [34, 128, 160] for the stochastic Langevin equation, which is widely used to describe the dissipative system which interacts with the environment, especially in the fields of molecular simulations, quantum systems and chemical interactions (see e.g., [62, 110, 232] and references therein). More precisely, let $\tilde{f}_0(P, Q) = -\upsilon P$, $\upsilon > 0$ and $\tilde{f}_r(P, Q) = 0$ for $r = 1, \ldots, m$, where $m \geq n$. Given $H_0(P, Q) = F(Q) + \frac{1}{2}P^{\top}MP$ with $F \in \mathbf{C}^{\infty}(\mathbb{R}^n, \mathbb{R})$ and $M \in \mathbb{R}^{n \times n}$ being a positive definite symmetric matrix, and $H_r(P, Q) = \sigma_r^{\top}Q$ with $\sigma_r \in \mathbb{R}^n$, $r \in \{1, \ldots, m\}$, then (1.37) becomes the stochastic Langevin equation

$$\begin{cases} dP(t) = -\dfrac{\partial F(Q(t))}{\partial Q}dt - \upsilon P(t)dt - \displaystyle\sum_{r=1}^{m} \sigma_r \circ dW_r(t), \quad P(0) = p \in \mathbb{R}^n, \\ dQ(t) = MP(t)dt, \quad Q(0) = q \in \mathbb{R}^n. \end{cases}$$

$$(1.38)$$

Suppose that $\mathrm{rank}\{\sigma_1, \ldots, \sigma_m\} = n$ and $f := \frac{\partial F}{\partial Q}$ satisfies the global Lipschitz condition. Then there exists a unique solution $X(t) = (P(t)^{\top}, Q(t)^{\top})^{\top}$ such that for any $\gamma \geq 1$,

$$\mathbb{E}\left[\sup_{0 \leq t \leq T} \|X(t)\|^{\gamma}\right] \leq C.$$

where $C := C(p, q, T, \gamma)$ is a positive constant. Below we demonstrate the property of conformal symplecticity for (1.38).

Theorem 1.3.2 *The phase flow of the stochastic Langevin equation* (1.38) *preserves the stochastic conformal symplectic structure*

$$dP(t) \wedge dQ(t) = \exp(-\upsilon t)dp \wedge dq, \quad a.s., \tag{1.39}$$

where $t \in [0, T]$.

Proof By calculating the differential 2-form, we deduce

$$
\begin{aligned}
&d(dP(t) \wedge dQ(t)) \\
&\quad = d(dP(t)) \wedge dQ(t) + dP(t) \wedge d(dQ(t)) \\
&\quad = d\left(\left(-\frac{\partial F(Q(t))}{\partial Q} - vP(t)\right) dt - \sum_{r=1}^{m} \sigma_r \circ dW(t)\right) \wedge dQ(t) \\
&\qquad + dP(t) \wedge d(MP(t)dt) \\
&\quad = \left(\left(-\frac{\partial^2 F(Q(t))}{\partial Q^2} dQ(t) - vdP(t)\right) \wedge dQ(t) + dP(t) \wedge d(MP(t))\right) dt \\
&\quad = -v(dP(t) \wedge dQ(t))dt,
\end{aligned}
$$

based on the fact that $AdQ \wedge dQ = 0$ for any symmetric matrix $A \in \mathbb{R}^{n \times n}$. This completes the proof. $\qquad\square$

From (1.39) it can be seen that the symplectic form of (1.38) dissipates exponentially. As a direct consequence of the conformal symplecticity, the phase volume $\mathrm{Vol}(t)$, $t \in [0, T]$, of (1.38) also dissipates exponentially. In detail, let $D_0 \in \mathbb{R}^{2n}$ be a random domain with finite volume and independent of $W_r(t)$, $t \in (0, T]$, $r = 1, \ldots, m$. The transformation $(p^\top, q^\top)^\top \mapsto ((P(t))^\top, (Q(t))^\top)^\top$ maps D_0 into the domain $D_t \in \mathbb{R}^{2n}$ for $t \in [0, T]$. One can check that the volume $\mathrm{Vol}(t)$, $t \in [0, T]$, of the domain D_t is as follows

$$
\begin{aligned}
\mathrm{Vol}(t) &= \int_{D_t} dP_1 \cdots dP_n dQ_1 \cdots dQ_n \\
&= \int_{D_0} \det\left(\frac{\partial(P_1, \ldots, P_n, Q_1, \ldots, Q_n)}{\partial(p_1, \ldots, p_n, q_1, \ldots, q_n)}\right) dp_1 \cdots dp_n dq_1 \cdots dq_n,
\end{aligned}
$$

where the determinant of Jacobian matrix satisfies $\det\left(\frac{\partial(P_1,\ldots,P_n,Q_1,\ldots,Q_n)}{\partial(p_1,\ldots,p_n,q_1,\ldots,q_n)}\right) = \exp(-vtn)$ with n being the dimension of P (see [194]). It means that $\mathrm{Vol}(t) = \exp(-vtn)\mathrm{Vol}(0)$, $t \in [0, T]$.

As for another long-time behavior of the considered system, the ergodicity of (1.38) is given in [34, 131, 188] and references therein, by proving that (1.38) possesses a unique invariant measure μ, when $M = I_n$. A probability measure $\mu \in \mathscr{P}(\mathbb{R}^{2n})$ is said to be *invariant* for a Markov semigroup P_t, $t \geq 0$, if

$$
\int_{\mathbb{R}^{2n}} P_t \varphi(x) \mu(dx) = \int_{\mathbb{R}^{2n}} \varphi d\mu
$$

for all $\varphi \in \mathbf{B}_b(\mathbb{R}^{2n})$, where $\mathscr{P}(\mathbb{R}^{2n})$ and $\mathbf{B}_b(\mathbb{R}^{2n})$ denote the space of all the probability measures on $\mathbb{R}^{2n}$ and the space of all measurable and bounded functions

defined on $\mathbb{R}^{2n}$, respectively. To obtain the existence of the invariant measure of
(1.38), [188] gives the following assumption.

Assumption 1.3.1 *Let* $F \in \mathbf{C}^{\infty}(\mathbb{R}^n, \mathbb{R})$ *satisfy that*

(1) $F(u) \geq 0$, *for all* $u \in \mathbb{R}^n$;
(2) *there exist* $\alpha > 0$ *and* $\beta \in (0, 1)$ *such that for all* $u \in \mathbb{R}^n$, *it holds*

$$\frac{1}{2} u^{\top} f(u) \geq \beta F(u) + v^2 \frac{\beta(2 - \beta)}{8(1 - \beta)} \|u\|^2 - \alpha.$$

It can be verified that (1.38) satisfies the hypoelliptic setting

$$\text{span}\{U_i, [U_0, U_i], i = 1, \ldots, m\} = \mathbb{R}^{2n} \tag{1.40}$$

with vector fields $U_0 = ((-f(Q) - vP)^{\top}, P^{\top})^{\top}$ and $U_i = (\sigma_i^{\top}, 0)^{\top}, i = 1, \ldots, m$. Thus, the solution of (1.38) is a strong Feller process via Hömander's theorem. Combining (1.40) with Assumption 1.3.1 yields the ergodicity of (1.38) (see [188] for more details). Intuitively speaking, the ergodicity of (1.38) reads that the temporal averages associated with $P(t)$ and $Q(t)$ will converge to the spatial average with respect to the invariant measure μ. More precisely, for any $\phi \in \mathbf{C}_b(\mathbb{R}^{2n}, \mathbb{R})$,

$$\lim_{T \to \infty} \frac{1}{T} \int_0^T \mathbb{E}^{(p,q)}[\phi(P(t), Q(t))]\, dt = \int_{\mathbb{R}^{2n}} \phi\, d\mu \quad \text{in } \mathbf{L}^2(\mathbb{R}^{2n}, \mu), \tag{1.41}$$

where $\mathbb{E}^{(p,q)}[\cdot]$ denotes the expectation with respect to $P(0) = p$ and $Q(0) = q$. One can find that the Markov semigroup here is given by $P_t \phi(\cdot) = \mathbb{E}^{(p,q)}[\phi(\cdot)]$ with $t \geq 0$ and $\phi \in \mathbf{C}_b(\mathbb{R}^{2n}, \mathbb{R})$.

Remark 1.3.1 If $n = 1$, $M = 1$, and $F(Q) = Q^4$, then the stochastic Langevin equation (1.38), which includes non-globally Lipschitz continuous coefficients, defines a forward stochastic flow of $\mathbf{C}^{\infty}$-diffeomorphisms (see [269]). Denote $U(P, Q) := K(\frac{1}{2}P^2 + Q^4 + 1)$ with $K \geq 1$. For any $\beta \geq K\left(\sum_{r=1}^{m} \sigma_r^2 - 2v\right)$, it holds that

$$\sup_{t \in [0,T]} \mathbb{E}\left[\exp\left(\frac{U(P(t), Q(t))}{\exp(\beta t)}\right)\right] \leq C(\beta, T) \exp(U(p, q)). \tag{1.42}$$

The above property is also called the exponential integrability of the exact solution. Furthermore, [82] shows that the exact solution $(P(t), Q(t))$ admits an infinitely differentiable density function.

1.3.2 Stochastic Poisson Systems

There are also many circumstances that the behavior of the system is applicable to be characterized by neither stochastic Hamiltonian systems nor stochastic forced Hamiltonian systems. Instead, the stochastic Poisson system, as an extension of the stochastic Hamiltonian system to arbitrary dimension and variable structure matrices, is taken into consideration and widely used in the fields of plasma physics, mathematical biology and so on (see e.g., [17, 45, 64, 112, 114, 228] and references therein). It includes the harmonic oscillator in presence of nonequilibrium environment, stochastic Lotka–Volterra (LV) equations describing the dynamics of biological systems with several interacting species, and the stochastic Ablowitz–Ladik model which is used to depict the self trapping on a dimer, dynamics of anharmonic lattices and pulse dynamics in nonlinear optics with random perturbations.

A d-dimensional stochastic Poisson system reads

$$\begin{cases} dX(t) = B(X(t))\Big(\nabla K_0(X(t))dt + \sum_{r=1}^{m} \nabla K_r(X(t)) \circ dW_r(t)\Big), \\ X(0) = x, \end{cases} \tag{1.43}$$

where $t \in (0, T]$, $x \in \mathbb{R}^d$, $B \in \mathbf{C}^\infty(\mathbb{R}^d, \mathbb{R}^{d\times d})$, $K_r \in \mathbf{C}^\infty(\mathbb{R}^d, \mathbb{R})$, $r = 0, 1, \ldots, m$, and $W(\cdot) = (W_1(\cdot), \ldots, W_m(\cdot))$ is an m-dimensional standard Wiener process. Here, the structure matrix $B(X) = (b_{ij}(X)) \in \mathbb{R}^{d\times d}$ satisfies

$$b_{ij}(X) = -b_{ji}(X), \quad X \in \mathbb{R}^d, \tag{1.44}$$

and

$$\sum_{s=1}^{d} \left(\frac{\partial b_{ij}(X)}{\partial X_s} b_{sk}(X) + \frac{\partial b_{jk}(X)}{\partial X_s} b_{si}(X) + \frac{\partial b_{ki}(X)}{\partial X_s} b_{sj}(X) \right) = 0, \quad X \in \mathbb{R}^d \tag{1.45}$$

for all $i, j, k \in \{1, \ldots, d\}$, and is of constant rank $2n = d - l$ with $0 \le l < d$ and $n \in \mathbb{N}_+$. Further, suppose that the coefficients given by

$$a(X) := B(X)\nabla K_0(X), \quad b_r(X) := B(X)\nabla K_r(X), \quad a_r(X) := \frac{1}{2}\frac{\partial b_r(X)}{\partial X} b_r(X)$$

with $r = 1, \ldots, m$, satisfy the global Lipschitz condition, i.e., there exists $C > 0$ such that for any $X_1, X_2 \in \mathbb{R}^d$,

$$\|a(X_1)-a(X_2)\| + \sum_{r=1}^{m} (\|a_r(X_1)-a_r(X_2)\| + \|b_r(X_1)-b_r(X_2)\|) \le C\|X_1-X_2\|,$$

which guarantees the existence and uniqueness of the solution of (1.43), and the existence of the Jacobian matrix $\frac{\partial X}{\partial x}$ of X. It can be verified that the stochastic Poisson system (1.43) degenerates to a stochastic Hamiltonian system, when $B(X) = J_{2n}^{-1}$ with $d = 2n$ and $X \in \mathbb{R}^d$.

The structure matrix $B(X)$ characterized by both (1.44) and (1.45) defines the Poisson bracket as

$$\{F, G\}(X) := (\nabla F(X))^\top B(X) \nabla G(X) = \sum_{i,j=1}^d \frac{\partial F(X)}{\partial X_i} b_{ij}(X) \frac{\partial G(X)}{\partial X_j}, \qquad (1.46)$$

where $F \in \mathbf{C}^\infty(\mathbb{R}^d, \mathbb{R})$ and $G \in \mathbf{C}^\infty(\mathbb{R}^d, \mathbb{R})$. The Poisson bracket $\{\cdot, \cdot\}$ is bilinear, skew-symmetric, and satisfies the Jacobi identity

$$\{\{F, G\}, H\} + \{\{G, H\}, F\} + \{\{H, F\}, G\} = 0,$$

and the Leibniz rule

$$\{F \cdot G, H\} = F \cdot \{G, H\} + G \cdot \{H, F\}$$

with $H, F, G \in \mathbf{C}^\infty(\mathbb{R}^d, \mathbb{R})$. A differentiable mapping $\varphi \colon U \to \mathbb{R}^d$ with U being an open set in $\mathbb{R}^d$ is called a *Poisson mapping* if it commutes with the Poisson bracket, namely,

$$\{F \circ \varphi, G \circ \varphi\}(X) = \{F, G\}(\varphi(X)), \qquad (1.47)$$

or equivalently,

$$\frac{\partial \varphi(X)}{\partial X} B(X) \left(\frac{\partial \varphi(X)}{\partial X}\right)^\top = B(\varphi(X)) \qquad (1.48)$$

for all smooth functions F, G defined on $\varphi(U)$.

Stochastic Poisson Structure The phase flow of the stochastic Poisson system preserves the stochastic Poisson structure, i.e.,

$$\frac{\partial \phi_t(x)}{\partial x} B(x) \left(\frac{\partial \phi_t(x)}{\partial x}\right)^\top = B(\phi_t(x)), \quad a.s. \qquad (1.49)$$

for $t \in [0, T]$. Below we first introduce the Darboux–Lie theorem, which is applied to proving the result concerning the stochastic Poisson structure of (1.43).

Theorem 1.3.3 (See [88, 116, 175]) *Suppose that the matrix $B(X)$ defines a Poisson bracket and is of constant rank $d - l = 2n$ in a neighborhood of $x \in \mathbb{R}^d$. Then there exist functions $P_1(X), \ldots, P_n(X), Q_1(X), \ldots, Q_n(X)$, and $C_1(X), \ldots, C_l(X)$ satisfying*

$$\begin{aligned}
\{P_i, P_j\} &= 0, & \{P_i, Q_j\} &= -\delta_{ij}, & \{P_i, C_s\} &= 0, \\
\{Q_i, P_j\} &= \delta_{ij}, & \{Q_i, Q_j\} &= 0, & \{Q_i, C_s\} &= 0, \\
\{C_k, P_j\} &= 0, & \{C_k, Q_j\} &= 0, & \{C_k, C_s\} &= 0
\end{aligned} \tag{1.50}$$

for $i = 1, \ldots, n, j = 1, \ldots, n, k = 1, \ldots, l, s = 1, \ldots, l$, on a neighborhood of x. The gradients of P_i, Q_i, C_k, are linearly independent such that $X \mapsto (P_i(X), Q_i(X), C_k(X))$ constitutes a local change of coordinates to canonical form.

Theorem 1.3.4 (See [132]) *The phase flow $\phi_t : x \mapsto X(t)$ of the stochastic Poisson system* (1.43) *preserves the stochastic Poisson structure for $t \in [0, T]$.*

Proof It is known that $B(X)$ of (1.43), under the conditions (1.44) and (1.45), defines a Poisson bracket and is of constant rank $d - l = 2n$. It follows from Theorem 1.3.3 that there exists a coordinate transformation

$$Y = \Theta(X) = (P_1(X), \ldots, P_n(X), Q_1(X), \ldots, Q_n(X), C_1(X), \ldots, C_l(X))^\top$$

such that

$$A(X)B(X)A(X)^\top = \begin{bmatrix} J_{2n}^{-1} & 0 \\ 0 & 0 \end{bmatrix} = \begin{bmatrix} 0 & -I_n & 0 \\ I_n & 0 & 0 \\ 0 & 0 & 0 \end{bmatrix} \tag{1.51}$$

with the Jacobian matrix $A(X) = \frac{\partial Y}{\partial X}$. Based on (1.51) and (1.43) can be transformed to

$$\begin{aligned}
dY(t) &= \frac{\partial Y(t)}{\partial X}\left(B(X(t))\left(\nabla K_0(X(t))dt + \sum_{r=1}^{m} \nabla K_r(X(t)) \circ dW_r(t)\right)\right) \\
&= \frac{\partial Y(t)}{\partial X}\left(B(X(t))\left(\left(\frac{\partial Y(t)}{\partial X}\right)^\top \nabla H_0(Y(t))dt + \left(\frac{\partial Y(t)}{\partial X}\right)^\top \sum_{r=1}^{m} \nabla H_r(Y(t)) \circ dW_r(t)\right)\right) \\
&= \begin{bmatrix} J_{2n}^{-1} & 0 \\ 0 & 0 \end{bmatrix}\left(\nabla H_0(Y(t))dt + \sum_{r=1}^{m} \nabla H_r(Y(t)) \circ dW_r(t)\right), \tag{1.52}
\end{aligned}$$

where $H_i(Y) = K_i(X)$, $i = 0, 1, \ldots, m$, and $Y(0) = y = \Theta(x)$. Denote by $Y = (Z^\top, \hat{C}^\top)^\top$ and $Z = (P^\top, Q^\top)^\top$ with

$$P = (P_1, \ldots, P_n)^\top, \quad Q = (Q_1, \ldots, Q_n)^\top, \quad \hat{C} = (C_1, \ldots, C_l)^\top.$$

Then (1.52) can be rewritten as

$$
\begin{cases}
dZ(t) = J_{2n}^{-1}\left(\nabla_Z H_0(Z(t), \hat{C}(t))dt + \sum_{r=1}^{m} \nabla_Z H_r(Z(t), \hat{C}(t)) \circ dW_r(t)\right), \\
d\hat{C}(t) = 0.
\end{cases}
$$

(1.53)

The first equation of (1.53) is a $2n$-dimensional stochastic Hamiltonian system, whose phase flow preserves the symplecticity. It can be derived that the stochastic flow $\psi_t : y \mapsto Y(t)$ of (1.52) satisfies $\psi_t(y) = \Theta(\phi_t(x))$ and

$$
\frac{\partial \psi_t(y)}{\partial y} \begin{bmatrix} J_{2n}^{-1} & 0 \\ 0 & 0 \end{bmatrix} \left(\frac{\partial \psi_t(y)}{\partial y}\right)^\top = \begin{bmatrix} J_{2n}^{-1} & 0 \\ 0 & 0 \end{bmatrix}.
$$

Moreover, according to

$$
\frac{\partial \psi_t(y)}{\partial y} = \Theta'(\phi_t(x)) \frac{\partial \phi_t(x)}{\partial x} A(x)^{-1}
$$

and (1.51), we obtain

$$
\Theta'(\phi_t(x)) \frac{\partial \phi_t(x)}{\partial x} B(x) \left(\frac{\partial \phi_t(x)}{\partial x}\right)^\top \left(\Theta'(\phi_t(x))\right)^\top = \begin{bmatrix} J_{2n}^{-1} & 0 \\ 0 & 0 \end{bmatrix}.
$$

Since $\Theta'(\phi_t(x)) = A(\phi_t(x))$, replacing X in (1.51) by $\phi_t(x)$ yields (1.49). This ends the proof. □

Remark 1.3.2 The stochastic Poisson structure can also be achieved by mimicking the analysis of the deterministic case and using the stochastic variational equation.

In the sequel, we call (1.52) the canonical form of (1.43), and Θ the canonical transformation.

Casimir Function A function $\mathscr{C} : \mathbb{R}^d \to \mathbb{R}$ is called a *Casimir function* of the system (1.43) if

$$
\nabla \mathscr{C}(X)^\top B(X) = 0 \quad \forall \, X \in \mathbb{R}^d.
$$

(1.54)

By means of (1.43) and the Stratonovich chain rule, we have

$$
d\mathscr{C}(X(t)) = (\nabla \mathscr{C}(X(t)))^\top B(X(t)) \left(\nabla K_0(X(t))dt + \sum_{r=1}^{m} \nabla K_r(X(t)) \circ dW_r(t)\right) = 0,
$$

which means that $\mathscr{C}$ is an invariant of its corresponding stochastic Poisson system. The existence and concrete forms of Casimir functions depend on the structure matrix $B(X)$ of (1.43), where $X \in \mathbb{R}^d$.

Below we present some concrete models of stochastic Poisson systems.

Example 1.3.1 (See [132]) Consider the system

$$
\begin{cases}
dX(t) = B(X(t))\nabla K(X(t))\Big(dt + \sum_{r=1}^{m} \sigma_r \circ dW_r(t)\Big), \\
X(0) = x,
\end{cases}
\tag{1.55}
$$

where $X = (X_1, X_2, X_3)^\top$, $x = (x_1, x_2, x_3)^\top$ is a constant vector, $B = \begin{bmatrix} 0 & -1 & 0 \\ 1 & 0 & -1 \\ 0 & 1 & 0 \end{bmatrix}$, $K(X) = \sin(X_1) - \cos(X_2) + \sin(X_3)$, and σ_r, $r = 1, \ldots, m$, are constants. The equivalent Itô form of the system (1.55) is

$$
\begin{cases}
dX(t) = \Big(\bar{a}(X(t)) + \sum_{r=1}^{m} \bar{a}_r(X(t))\Big)dt + \sum_{r=1}^{m} \bar{b}_r(X(t))dW_r(t), \\
X(0) = x,
\end{cases}
$$

whose the coefficients

$$\bar{a}(X) = (-\sin(X_2), \cos(X_1) - \cos(X_3), \sin(X_2))^\top,$$

$$\bar{a}_r(X) = \frac{1}{2}((\cos(X_3) - \cos(X_1))\cos(X_2), (\sin(X_1) + \sin(X_3))\sin(X_2),$$

$$(\cos(X_1) - \cos(X_3))\cos(X_2))^\top,$$

$$\bar{b}_r(X) = \sigma_r \bar{a}(X)$$

with $r = 1, \ldots, m$, satisfy the global Lipschitz condition. Meanwhile, the matrix B satisfies (1.44) and (1.45). Therefore, (1.55) is a stochastic Poisson system, whose phase flow preserves the stochastic Poisson structure. Moreover, it can be verified that the function $\mathscr{C}: \mathbb{R}^3 \to \mathbb{R}$ defined by $\mathscr{C}(X) = X_1 + X_3$ is a Casimir function of (1.55).

Example 1.3.2 (Stochastic Rigid Body System) Now we consider the stochastic rigid body system (see e.g., [17, 64, 132] and references therein)

$$
\begin{cases}
dX(t) = B(X(t))\nabla K(X(t))(dt + c \circ dW(t)), \\
X(0) = x,
\end{cases}
\tag{1.56}
$$

where $c \in \mathbb{R}$, $t \in (0, T]$, $X = (X_1, X_2, X_3)^\top$, $K(X) := \frac{1}{2}\left(\frac{x_1^2}{c_1} + \frac{x_2^2}{c_2} + \frac{x_3^2}{c_3}\right)$ with c_1, c_2, c_3 being non-zero constants and

$$B(X) = \begin{bmatrix} 0 & -X_3 & X_2 \\ X_3 & 0 & -X_1 \\ -X_2 & X_1 & 0 \end{bmatrix}.$$

It can be verified that the equivalent Itô form of (1.56) is

$$\begin{cases} dX(t) = f_1(X(t))dt + f_2(X(t))dW(t), \\ X(0) = x, \end{cases}$$

where

$$f_1(X) = \frac{c^2}{2} \begin{bmatrix} X_1 X_3^2 \left(\frac{1}{c_3} - \frac{1}{c_2}\right)\left(\frac{1}{c_1} - \frac{1}{c_3}\right) + X_1 X_2^2 \left(\frac{1}{c_3} - \frac{1}{c_2}\right)\left(\frac{1}{c_2} - \frac{1}{c_1}\right) \\ X_2 X_3^2 \left(\frac{1}{c_1} - \frac{1}{c_3}\right)\left(\frac{1}{c_3} - \frac{1}{c_2}\right) + X_1^2 X_2 \left(\frac{1}{c_1} - \frac{1}{c_3}\right)\left(\frac{1}{c_1} - \frac{1}{c_1}\right) \\ X_2^2 X_3 \left(\frac{1}{c_2} - \frac{1}{c_1}\right)\left(\frac{1}{c_3} - \frac{1}{c_2}\right) + X_1^2 X_3 \left(\frac{1}{c_2} - \frac{1}{c_1}\right)\left(\frac{1}{c_1} - \frac{1}{c_3}\right) \end{bmatrix}$$
$$+ \left[X_2 X_3 \left(\frac{1}{c_3} - \frac{1}{c_2}\right), X_1 X_3 \left(\frac{1}{c_1} - \frac{1}{c_3}\right), X_1 X_2 \left(\frac{1}{c_2} - \frac{1}{c_1}\right)\right]^\top,$$

and $f_2(X) = c\left[X_2 X_3 \left(\frac{1}{c_3} - \frac{1}{c_2}\right), X_1 X_3 \left(\frac{1}{c_1} - \frac{1}{c_3}\right), X_1 X_2 \left(\frac{1}{c_2} - \frac{1}{c_1}\right)\right]^\top$. One can check that f_1 and f_2 satisfy the local Lipschitz condition, i.e., for each $N \in \mathbb{N}_+$, there exists $C_N > 0$ for which

$$\|f_1(X_1) - f_1(X_2)\| + \|f_2(X_1) - f_2(X_2)\| \le C_N \|X_1 - X_2\|$$

with $\|X_1\| \le N$ and $\|X_2\| \le N$. In addition,

$$X^\top f_1(X) + \frac{1}{2}\|f_2(X)\|^2$$
$$= X_1 X_2 X_3 \left(\frac{1}{c_3} - \frac{1}{c_2} + \frac{1}{c_1} - \frac{1}{c_3} + \frac{1}{c_2} - \frac{1}{c_1}\right)$$
$$+ \frac{c^2}{2}\left(-X_1^2 X_3^2 \left(\frac{1}{c_1} - \frac{1}{c_3}\right)^2 - X_1^2 X_2^2 \left(\frac{1}{c_1} - \frac{1}{c_2}\right)^2 - X_2^2 X_3^2 \left(\frac{1}{c_2} - \frac{1}{c_3}\right)^2\right)$$
$$+ \frac{c^2}{2}\left(X_1^2 X_3^2 \left(\frac{1}{c_1} - \frac{1}{c_3}\right)^2 + X_1^2 X_2^2 \left(\frac{1}{c_1} - \frac{1}{c_2}\right)^2 + X_2^2 X_3^2 \left(\frac{1}{c_2} - \frac{1}{c_3}\right)^2\right) = 0,$$

which implies that the Khasminskii-type condition is satisfied. According to [183, Theorem 3.5 of Chapter 2], one can obtain the well-posedness of (1.56).

Given the initial value $x = (x_1, x_2, x_3)^\top$, the Casimir function is a quadratic function

$$\mathscr{C}(X) := \frac{1}{2}\left(X_1^2 + X_2^2 + X_3^2\right).$$

Based on the Casimir function and [16, Theorem 7.3.6], it can be verified that the exact solution X is differential with respect to x under the non-global Lipschitz condition. Moreover, it is obvious that the matrix $B(X)$ satisfies (1.44) and (1.45), and is of constant rank 2, since its kernel can be easily found to be of dimension 1. Therefore, the stochastic rigid body system (1.56) is a stochastic Poisson system, whose phase flow preserves the stochastic Poisson structure.

1.4   Rough Hamiltonian Systems

Standard Wiener processes have been of vital significance in the solution of many stochastic problems, including estimation, filtering and mutual information. For the modeling of physical and social phenomena by stochastic systems, it has been demonstrated empirically in some cases that the standard Wiener process is inadequate to capture some important properties of the phenomenon. Nowadays noises rougher than standard Wiener processes have attracted wide attention. A well-known example is the fractional Brownian motion (except the case of Hurst parameter $\mathbf{H} = \frac{1}{2}$ when it is a standard Brownian motion), which is suitable in simulating the prices of electricity in a liberated electricity market, and the empirical volatility of a stock, and other turbulence phenomena (see e.g., [182, 257] and references therein). From Fig. 1.3, it can be observed that the regularity of the fractional Brownian motion with $\mathbf{H} = \frac{1}{5}$ is rougher than that of the standard Wiener process. At this time, stochastic integrals with respect to the standard Wiener process are no longer applicable. The rough path theory and stochastic differential equation driven by rough path are introduced (see e.g., [105, 179] and references therein), and complement the areas where the Itô's theory runs into difficulty. In this part, we consider the $2n$-dimensional Hamiltonian systems

$$\begin{cases} dP_t = -\dfrac{\partial H_0(P_t, Q_t)}{\partial Q}dt - \displaystyle\sum_{r=1}^{m}\dfrac{\partial H_r(P_t, Q_t)}{\partial Q}d\mathbb{X}_t^r, & P_0 = p \in \mathbb{R}^n, \\[4mm] dQ_t = \dfrac{\partial H_0(P_t, Q_t)}{\partial P}dt + \displaystyle\sum_{r=1}^{m}\dfrac{\partial H_r(P_t, Q_t)}{\partial P}d\mathbb{X}_t^r, & Q_0 = q \in \mathbb{R}^n, \end{cases} \tag{1.57}$$

where $t \in (0, T]$, and $\mathbb{X}_t^r$, $r = 1, \ldots, m$, are Gaussian noises under the following assumption.

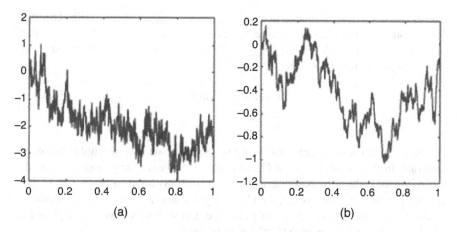

Fig. 1.3 Trajectories of fractional Brownian motion with Hurst parameter **H**. (a) **H** $= \frac{1}{5}$. (b) **H** $= \frac{1}{2}$

Assumption 1.4.1 (See [130]) *Let* $\mathbb{X}_t^r : [0, T] \to \mathbb{R}, r = 1, \ldots, m$, *be independent centered Gaussian processes with continuous sample paths. There exist some* $\rho \in [1, 2)$ *and* $C > 0$ *such that for any* $0 \le s < t \le T$, *the covariance of* $\mathbb{X} := (\mathbb{X}^0, \mathbb{X}^1, \ldots, \mathbb{X}^m)^\top$ *satisfies*

$$\sup_{(t_k),(t_i) \in \mathscr{D}([s,t])} \left(\sum_{k,i} \left| \mathbb{E}\big[(\mathbb{X}_{t_k,t_{k+1}})^\top \mathbb{X}_{t_i,t_{i+1}} \big] \right|^\rho \right)^{1/\rho} \le C|t-s|^{1/\rho},$$

where $\mathscr{D}([s,t])$ *denotes the set of all dissections of* $[s,t]$ *and* $\mathbb{X}_{t_k,t_{k+1}} := \mathbb{X}_{t_{k+1}} - \mathbb{X}_{t_k}$.

Let $Y = (P^\top, Q^\top)^\top$ with $P = (P^1, \ldots, P^n)^\top$ and $Q = (Q^1, \ldots, Q^n)^\top$, $y = (p^\top, q^\top)^\top$ with $p = (p^1, \cdots, p^n)^\top$ and $q = (q^1, \cdots, q^n)^\top$, $\mathbb{X}_t^0 = t$ and

$$V_r = (V_r^1, \ldots, V_r^{2n})^\top = \left(-\frac{\partial H_r}{\partial Q^1}, \ldots, -\frac{\partial H_r}{\partial Q^n}, \frac{\partial H_r}{\partial P^1}, \ldots, \frac{\partial H_r}{\partial P^n} \right)^\top$$

for $r = 0, 1, \ldots, m$. Denoting $\mathbb{X} = (\mathbb{X}^0, \mathbb{X}^1, \ldots, \mathbb{X}^m)^\top$ and $V = (V_0, V_1, \ldots, V_m)$, we have the compact form for (1.57) as follows

$$\begin{cases} dY_t = \displaystyle\sum_{r=0}^m V_r(Y_t) d\mathbb{X}_t^r = V(Y_t) d\mathbb{X}_t, \\ Y_0 = y. \end{cases} \tag{1.58}$$

For any $\gamma > 2\rho$, the Gaussian process $\mathbb{X}$ under Assumption 1.4.1 can be naturally lifted to a forthcoming Hölder-type weak geometric γ-rough path

$\mathbf{X}\colon [0, T] \to G^{[\gamma]}(\mathbb{R}^{m+1})$ such that $\pi_1(\mathbf{X}_{s,t}) = \mathbb{X}_t - \mathbb{X}_s$ (see e.g., [105, Theorem 15.33] and references therein), where $\mathbf{X}_{s,t} := \mathbf{X}_s^{-1} \otimes \mathbf{X}_t$, and $[\gamma]$ is the integer part of γ, i.e., $[\gamma] \in \mathbb{N}_+$ with $\gamma - 1 < [\gamma] \le \gamma$. Thus, (1.57) or (1.58) can be transformed into a rough differential equation

$$\begin{cases} dY_t = V(Y_t)d\mathbf{X}_t, \\ Y_0 = y, \end{cases} \tag{1.59}$$

and interpreted via the rough path theory where the chain rule also holds for rough integrals. In this sense, (1.57) or (1.58) is called a *rough Hamiltonian system*. As a consequence, the well-posedness of (1.57) or (1.58) is given by that of (1.59). When $\mathbb{X}_t^r$, $r = 1, \dots, m$, are standard Wiener processes, Y solves the corresponding stochastic Hamiltonian system (1.9). Below we start with some basic concepts in the rough path theory, and refer to [105] for more details.

Definition 1.4.1 Let $(E, \mathbf{d})$ be a metric space. A path $x\colon [0, T] \to E$ is said to be

(1) Hölder continuous with exponent $\alpha \ge 0$, or simply α-Hölder, if

$$\|x\|_{\alpha\text{-Höl};[0,T]} := \sup_{0 \le s < t \le T} \frac{\mathbf{d}(x_s, x_t)}{|t - s|^\alpha} < \infty;$$

(2) of finite p-variation for some $p > 0$, if

$$\|x\|_{p\text{-var};[0,T]} := \sup_{(t_i) \in \mathscr{D}([0,T])} \left(\sum_i \mathbf{d}\left(x_{t_i}, x_{t_{i+1}}\right)^p \right)^{1/p} < \infty,$$

where $\mathscr{D}([0, T])$ is the set of all dissections of $[0, T]$.

The notations $\mathbf{C}^{\alpha\text{-Höl}}([0, T], E)$ and $\mathbf{C}^{p\text{-var}}([0, T], E)$ are used for the set of α-Hölder paths and the set of continuous paths of finite p-variation, respectively. Here we only focus on the Euclidean space case with $E = \mathbb{R}^{m+1}$. Given two vectors $a, b \in \mathbb{R}^{m+1}$ with coordinates $(a_i)_{i=1,\dots,m+1}$ and $(b_i)_{i=1,\dots,m+1}$, one can construct the matrix $(a_i b_j)_{i,j=1,\dots,m+1}$ and the 2-tensor

$$a \otimes b := \sum_{i,j=1}^{m+1} (a_i b_j) e_i \otimes e_j \in \mathbb{R}^{m+1} \otimes \mathbb{R}^{m+1}.$$

More generally, for $a = \sum_{i_1,\dots,i_k} a_{i_1,\dots,i_k} e_{i_1} \otimes \cdots \otimes e_{i_k} \in \left(\mathbb{R}^{m+1}\right)^{\otimes k}$ and similar notation $b \in \left(\mathbb{R}^{m+1}\right)^{\otimes l}$, where $k, l \in \mathbb{N}_+$, we achieve

$$a \otimes b := \sum a_{i_1,\dots,i_k} b_{j_1,\dots,j_l} e_{i_1} \otimes \cdots \otimes e_{i_k} \otimes e_{j_1} \otimes \cdots \otimes e_{j_l}$$

$$\in \left(\mathbb{R}^{m+1}\right)^{\otimes k} \otimes \left(\mathbb{R}^{m+1}\right)^{\otimes l} \cong \left(\mathbb{R}^{m+1}\right)^{\otimes (k+l)}.$$

Notice that $\left(\mathbb{R}^{m+1}\right)^{\otimes k} \cong \mathbb{R}^{(m+1)^k}$ for $k \in \mathbb{N}_+$. It is a convenient convention to set

$$\left(\mathbb{R}^{m+1}\right)^{\otimes 0} := \mathbb{R}.$$

Definition 1.4.2 The step-N signature of $x \in \mathbf{C}^{1\text{-var}}\left([s,t],\mathbb{R}^{m+1}\right)$ with $N \in \mathbb{N}$ is given by

$$S_N(x)_{s,t} = \left(1, \int_{s \leq u_1 \leq t} dx_{u_1}, \dots, \int_{s \leq u_1 < \cdots < u_N \leq t} dx_{u_1} \otimes \cdots \otimes dx_{u_N}\right)$$

$$\in \oplus_{k=0}^{N} \left(\mathbb{R}^{m+1}\right)^{\otimes k}.$$

The path $t \mapsto S_N(x)_{s,t}$ is called the step-N lift of x.

Now we define

$$T^N\left(\mathbb{R}^{m+1}\right) := \oplus_{k=0}^{N}\left(\mathbb{R}^{m+1}\right)^{\otimes k}$$

and $\pi_k \colon T^N\left(\mathbb{R}^{m+1}\right) \to \left(\mathbb{R}^{m+1}\right)^{\otimes k}$ as the projection to the kth tensor level. One can observe that $S_N(x)_{s,t}$ contains the information about iterated integrals of x up to N-level. If $a \in \mathbb{R}^{m+1}$, then the step-N signature of $x(\cdot)$ given by $t \mapsto ta$, $t \in [0,1]$, equals

$$S_N(x)_{0,1} = 1 + \sum_{k=1}^{N} \int_{0 \leq r_1 < \cdots < r_k \leq 1} dx_{r_1} \otimes \cdots \otimes dx_{r_k}$$

$$= 1 + \sum_{k=1}^{N} a^{\otimes k} \int_{0 \leq r_1 < \cdots < r_k \leq 1} dr_1 \dots dr_k$$

$$= 1 + \sum_{k=1}^{N} \frac{a^{\otimes k}}{k!}.$$

Definition 1.4.3 The set of all step-N signatures of continuous paths of finite length,

$$G^N\left(\mathbb{R}^{m+1}\right) := \left\{S_N(x)_{0,1} : x \in \mathbf{C}^{1\text{-var}}\left([0,1],\mathbb{R}^{m+1}\right)\right\}$$

is called the free nilpotent group of step N over $\mathbb{R}^{m+1}$.

Definition 1.4.4 Let $x: [0, T] \to E$ be a continuous path with finite p-variation for some $p < 2$. The signature of x is defined by

$$S(x)_{0,T}$$

$$:= \left(1, \int_{0 \le u_1 \le T} dx_{u_1}, \int_{0 \le u_1 < u_2 \le T} dx_{u_1} \otimes dx_{u_2}, \int_{0 \le u_1 < \cdots < u_3 \le T} dx_{u_1} \otimes \cdots \otimes dx_{u_3}, \ldots \right).$$

Definition 1.4.5 Let $x: [0, s] \to E$ and $y: [s, T] \to E$ be two continuous paths. Their concatenation is the path $x * y: [0, T] \to E$ defined by

$$(x * y)_u = \begin{cases} x_u, & \text{if} \quad u \in [0, s]; \\ x_s + y_u - y_s, & \text{if} \quad u \in [s, T]. \end{cases}$$

The following fundamental theorem (Chen's theorem) asserts that the signature is a homomorphism.

Theorem 1.4.1 (See [179]) Let $x: [0, s] \to E$ and $y: [s, T] \to E$ be two continuous paths with finite one-variation. Then

$$S(x * y)_{0,T} = S(x)_{0,s} \otimes S(y)_{s,T}.$$

Remark 1.4.1 (See [105]) The notation $S_N(x)_{s,t}$ can be justified by Chen's theorem below: the step-N lift $t \mapsto S_N(x)_{0,t}$ of some $x \in \mathbf{C}^{1\text{-var}} \left([0, T], \mathbb{R}^{m+1} \right)$ takes values in $G^N \left(\mathbb{R}^{m+1} \right)$ so that $S_N(x)_{s,t}$ is the natural increment of this path, i.e., the product of $\left(S_N(x)_{0,s} \right)^{-1}$ with $S_N(x)_{0,t}$.

For $\mathbf{a} = \left(a^0, a^1, \ldots, a^{[\gamma]} \right)$, $\mathbf{b} = \left(b^0, b^1, \ldots, b^{[\gamma]} \right) \in G^{[\gamma]} \left(\mathbb{R}^d \right)$, we consider the operation

$$\mathbf{a} \otimes \mathbf{b} := \left(c^0, c^1, \ldots, c^{[\gamma]} \right), \quad c^i := \sum_{k=0}^{i} a^k \otimes b^{i-k} \quad \forall i = 0, 1, \ldots, [\gamma].$$

If $\mathbf{a} \otimes \mathbf{b} = (1, 0, \ldots, 0)$, then $\mathbf{b} := \mathbf{a}^{-1}$. Denote by $\left(G^{[\gamma]}(\mathbb{R}^{m+1}), \text{dist} \right)$ the free step-$[\gamma]$ nilpotent Lie group of $\mathbb{R}^{m+1}$ equipped with the Carnot–Carathéodory metric

$$\text{dist}(\mathbf{a}, \mathbf{b}) := \max_{i=1,\ldots,[\gamma]} \left\| \left(\mathbf{a}^{-1} \otimes \mathbf{b} \right)^i \right\|^{1/i}, \quad \mathbf{a}, \mathbf{b} \in G^{[\gamma]} \left(\mathbb{R}^{m+1} \right).$$

Definition 1.4.6 A continuous map $\mathbf{X} \colon [0, T] \to G^{[\gamma]}(\mathbb{R}^{m+1}) \subset T^{[\gamma]}(\mathbb{R}^{m+1})$ is called γ-rough path if

$$\|\mathbf{X}\|_{\gamma\text{-var};[0,T]} := \sup_{(t_k) \in \mathscr{D}([0,T])} \left(\sum_k \text{dist}(\mathbf{X}_{t_k}, \mathbf{X}_{t_{k+1}})^\gamma \right)^{1/\gamma} < \infty.$$

Moreover, $\mathbf{X}$ is of Hölder-type if

$$\|\mathbf{X}\|_{\frac{1}{\gamma}\text{-Höl};[0,T]} := \sup_{0 \le s < t \le T} \frac{\text{dist}(\mathbf{X}_s, \mathbf{X}_t)}{|t - s|^{1/\gamma}} < \infty.$$

Since $\mathbf{X}$ in (1.59) takes values in $G^{[\gamma]}(\mathbb{R}^{m+1})$ with $\gamma > 2\rho$, Definition 1.4.6 implies that $\|\pi_1(\mathbf{X}_{s,t})\| \le \|\mathbf{X}\|_{\frac{1}{\gamma}\text{-Höl};[0,T]}|t - s|^{\frac{1}{\gamma}}$ for any $0 \le s < t \le T$. Similar to the $\frac{1}{\gamma}$-Hölder continuity in classical case, a larger γ implies lower regularity of $\mathbf{X}$. In the following, we introduce the definition of the solution of (1.59).

Definition 1.4.7 Let $\eta \in [1, \infty)$ and $\mathbf{X}$ be an η-rough path. Suppose that there exists a sequence of bounded variation functions $\{x^n\}_{n=1}^\infty$ on $\mathbb{R}^{m+1}$ such that

$$\sup_{n \in \mathbb{N}} \|S_{[\eta]}(x^n)\|_{\eta\text{-var};[0,T]} < \infty \quad \text{and} \quad \lim_{n \to \infty} \sup_{0 \le s < t \le T} \text{dist}\big(S_{[\eta]}(x^n)_{s,t}, \mathbf{X}_{s,t}\big) = 0.$$

Suppose in addition that y^n, $n \ge 1$, are solutions of equations $dy_t^n = V(y_t^n)dx_t^n$ in the Riemann–Stieltjes integral sense, with the same initial value y as in (1.59). If y_t^n converges to Y_t in the $\mathbf{L}^\infty([0, T])$-norm, i.e.,

$$\lim_{n \to \infty} \sup_{0 \le t \le T} \|y_t^n - Y_t\| = 0,$$

then we call Y_t a solution of (1.59).

To ensure the well-posedness of a rough differential equation, proper assumptions are needed for V, which are described by the notation Lip^η.

Definition 1.4.8 Let $\eta > 0$, and $\lfloor \eta \rfloor$ be the largest integer strictly smaller than η, i.e., $\eta - 1 \le \lfloor \eta \rfloor < \eta$. We say that $V \in Lip^\eta$, if $V \colon \mathbb{R}^{2n} \to \mathbb{R}^{2n \times (m+1)}$ is $\lfloor \eta \rfloor$ times continuously differentiable and there exists some constant $C > 0$ such that

$$\|D^k V(y)\| \le C \quad \forall k = 0, 1, \ldots, \lfloor \eta \rfloor \text{ and } y \in \mathbb{R}^{2n},$$

$$\|D^{\lfloor \eta \rfloor} V(y_1) - D^{\lfloor \eta \rfloor} V(y_2)\| \le C\|y_1 - y_2\|^{\eta - \lfloor \eta \rfloor} \quad \forall y_1, y_2 \in \mathbb{R}^{2n},$$

where $D^k V$ denotes the kth derivative of V. The smallest constant C satisfying the above inequalities is denoted by $\|V\|_{Lip^\eta}$.

In the sequel, we give a theorem of the well-posedness of a random differential equation.

Theorem 1.4.2 (See [105, 130]) *Let* $\gamma \in [1, \infty)$ *and* X *be a* γ-*rough path. If* $V \in$ *Lip*$^\eta$ *with* $\eta > \gamma$, *or* V *is linear, then* (1.59) *has a unique solution. In addition, the Jacobian matrix* $\frac{\partial Y_t}{\partial y}$ *exists and satisfies the linear rough differential equation*

$$
\begin{cases}
d\dfrac{\partial Y_t}{\partial y} = \displaystyle\sum_{r=0}^{m} DV_r(Y_t)\dfrac{\partial Y_t}{\partial y} d\mathbf{X}_t^r, \quad t \in (0, T], \\[2ex]
\dfrac{\partial Y_0}{\partial y} = I_{2n} \in \mathbb{R}^{2n \times 2n}.
\end{cases}
$$

Assume that the coefficient $V \in Lip^\eta$ with $\eta > \gamma > 2\rho$, or is linear, it follows from Theorem 1.4.2 that (1.57) or (1.58) shares a characteristic property of Hamiltonian systems.

Theorem 1.4.3 (See [130]) *The phase flow of rough Hamiltonian system* (1.57) *preserves the symplectic structure almost surely, i.e., for all* $t \in [0, T]$,

$$
dP_t \wedge dQ_t = dp \wedge dq, \quad a.s.
$$

Proof As in the proof of Theorem 1.2.1, we denote $P_p^{jk} := \frac{\partial P^j}{\partial p^k}$, $Q_p^{jk} := \frac{\partial Q^j}{\partial p^k}$, $P_q^{jk} := \frac{\partial P^j}{\partial q^k}$ and $Q_q^{jk} := \frac{\partial Q^j}{\partial q^k}$, $j, k \in \{1, \dots, n\}$, where p^j, q^j, P^j, Q^j, are the jth component of p, q, P, Q, respectively. By Theorem 1.4.2, one obtains

$$
d\left(\sum_{j=1}^{n} \left(P_p^{jk} Q_q^{jl} - P_q^{jl} Q_p^{jk} \right) \right) = 0 \quad \forall\, k, l \in \{1, \dots, n\},
$$

$$
d\left(\sum_{j=1}^{n} \left(P_p^{jk} Q_p^{jl} - P_p^{jl} Q_p^{jk} \right) \right) = 0 \quad \forall\, k \neq l,
$$

$$
d\left(\sum_{j=1}^{n} \left(P_q^{jk} Q_q^{jl} - P_q^{jl} Q_q^{jk} \right) \right) = 0 \quad \forall\, k \neq l,
$$

which completes the proof.   □

1.5 End Notes

This chapter provides an introduction to the geometric structure and long-time behaviors of stochastic flows associated with several Stratonovich stochastic differential equations. The stochastic flow of stochastic differential equations with globally Lipschitz continuous coefficients has been taken into consideration in [97, 142, 163, 243] and references therein. For the stochastic flow of stochastic differential equations with non-globally Lipschitz continuous coefficients, we refer to [99, 223, 268] and references therein. Especially, [269] studies the stochastic flow and Bismut formula for a stochastic Hamiltonian system which is hypoelliptic and has degenerate noises and superlinearly growing coefficients.

For the general stochastic Hamiltonian system (1.9), we have shown that it extremizes the stochastic action functional $\mathbb{S}(P(\cdot), Q(\cdot))$. In the converse direction, it is natural to expect that if $(P(\cdot), Q(\cdot))$ is a critical point of the stochastic action functional, then $(P(t), Q(t))$ would be a solution to the stochastic Hamiltonian system. Some variants of such result have been given in [35, 169, 241, 256] and references therein. For instance, [256] proves the special case when diffusion coefficients are independent of P. Street and Crisan in [241] investigates this direction by restricting the domain of the action functionals. For more general situations, [122, 160] leave such a conjecture as an open problem.

The underlying qualitative features including stochastic symplectic and conformal symplectic geometric structure of stochastic systems can be characterized via the stochastic action functional. As a consequence, one may appeal to the stochastic generating function theory as a guideline for exploring those geometric properties. We would like to mention that the generating function theory has been applied to many active research fields, such as optimal transport (see [70, 80]), optimal control (see [152, 213, 214]) and machine learning (see [48, 237]), etc. Nevertheless, the generating function theory associated with stochastic Poisson systems and rough Hamiltonian systems is still far from being well understood.

By means of the Kolmogorov–Arnold–Moser (KAM) theory, it is known in [4] that periodic orbits and stable fixed points exist for a dense set of initial conditions, and that the solution of deterministic Hamiltonian systems is quasiperiodic on a Cantor set of invariant tori. Furthermore, there are also orbits possessing different nature, and a phenomenon of diffusive behavior, that is, 'Arnold diffusion' occurs in high dimensional case (see [4]). However, as the stochastic perturbation strength of Hamiltonian systems increases, the extraordinarily complicated behavior gets reinforced in general. Compared with the deterministic case, little has been done up to now for the study of dynamical behaviors of stochastic flows for Hamiltonian systems. These behaviors deserve to be deeply explored in both theoretical and numeric aspects.

Chapter 2
Stochastic Structure-Preserving Numerical Methods

Stochastic structure-preserving numerical methods, which preserve intrinsic features of stochastic differential equations, admit stabilities in long-time computations, and enhance abilities on tracking and prediction of important probabilistic information and dynamical behaviours in the stochastic phenomenon. This chapter is devoted to stochastic structure-preserving numerical methods, especially stochastic numerical methods preserving symplecticity. In this chapter, given an interval $[0, T]$, we take a partition $0 = t_0 \leq t_1 \leq \cdots \leq t_N = T$ with a uniform time-step size $h = t_{k+1} - t_k$, $k = 0, 1, \ldots, N - 1$.

2.1 Stochastic Numerical Methods

In this section, we consider stochastic numerical methods for the d-dimensional Stratonovich stochastic differential equation

$$\begin{cases} dX(t) = \sigma_0(X(t))dt + \sum_{r=1}^{m} \sigma_r(X(t)) \circ dW_r(t), \quad t \in (0, T], \\ X(0) = x, \end{cases} \quad (2.1)$$

where $W_r(\cdot)$, $r = 1, \ldots, m$, are independent standard Wiener processes. There exists a unique solution $X: \Omega \times [0, T] \to \mathbb{R}^d$, which is a Markov process, when

© The Author(s), under exclusive license to Springer Nature Singapore Pte Ltd. 2022
J. Hong, L. Sun, *Symplectic Integration of Stochastic Hamiltonian Systems*,
Lecture Notes in Mathematics 2314, https://doi.org/10.1007/978-981-19-7670-4_2

$\mathbb{E}[\|x\|^2] < \infty$ and coefficients $\sigma_r : \mathbb{R}^d \to \mathbb{R}^d$, $r = 0, 1, \ldots, m$, satisfy (1.2) or the following conditions (see e.g., [16, 148, 183] and references therein):

(1) (linear growth) there exists $C > 0$ such that for any $y \in \mathbb{R}^d$,

$$\|a(y)\| + \sum_{r=1}^{m} \|\sigma_r(y)\| \le C(1 + \|y\|),$$

where $a(X) = \sigma_0(X) + \frac{1}{2} \sum_{r=1}^{m} \left(\frac{\partial \sigma_r}{\partial x} \sigma_r \right)(X)$;

(2) (local Lipschitz condition) for each $N \in \mathbb{N}_+$, there exists $C_N > 0$ such that

$$\|a(y_1) - a(y_2)\| + \sum_{r=1}^{m} \|\sigma_r(y_1) - \sigma_r(y_2)\| \le C_N \|y_1 - y_2\|$$

with $\|y_1\| \le N$, $\|y_2\| \le N$ and $y_1 \in \mathbb{R}^d$, $y_2 \in \mathbb{R}^d$.

Consider (2.1) over $[t_k, t_{k+1}]$:

$$X(t_{k+1}) = X(t_k) + \int_{t_k}^{t_{k+1}} \sigma_0(X(s))ds + \sum_{r=1}^{m} \int_{t_k}^{t_{k+1}} \sigma_r(X(s)) \circ dW_r(s)$$

for $k \in \{0, 1, \ldots, N - 1\}$. According to the definition of the Stratonovich integral, a 'natural' implicit method for (2.1) is

$$X_{k+1} = X_k + \sigma_0 \left(\frac{X_k + X_{k+1}}{2} \right) h + \sum_{r=1}^{m} \sigma_r \left(\frac{X_k + X_{k+1}}{2} \right) \Delta_k W_r \qquad (2.2)$$

with the increments of Wiener processes

$$\Delta_k W_r = W_r(t_{k+1}) - W_r(t_k) = \int_{t_k}^{t_{k+1}} dW_r(s), \quad r = 1, \ldots, m, \ k = 0, 1, \ldots, N - 1.$$

The above method applied to the 1-dimensional linear stochastic differential equation

$$dX(t) = \alpha X(t)dt + \sigma X(t) \circ dW(t), \quad X(0) = x$$

with $\alpha, \sigma \in \mathbb{R}$ takes the form of

$$X_{k+1} = X_k + \alpha \frac{X_k + X_{k+1}}{2} h + \sigma \frac{X_k + X_{k+1}}{2} \Delta_k W, \qquad (2.3)$$

where $k \in \{0, 1, \ldots, N - 1\}$. For the formal value of X_{k+1}, $k = 0, 1, \ldots, N - 1$, from (2.3)

$$X_{k+1} = \frac{2 + \alpha h + \sigma \Delta_k W}{2 - \alpha h - \sigma \Delta_k W} X_k,$$

we have $\mathbb{E}[\|X_{k+1}\|] = \infty$, based on the unboundedness of the random variable $\Delta_k W$ for arbitrarily small h. Further, this method cannot be realized since $2 - \alpha h - \sigma \Delta_k W$ can vanish for any small h and any $k \in \{0, 1, \ldots, N - 1\}$. Thus the approximation (2.2) is not suitable for (2.1) (see [193]). Motivated by the fact, it is natural to ask the following question.

Question 2.1.1 *Given a stochastic numerical method, how can one characterize its convergence, and how fast does it converge?*

Two popular ways of measuring the accuracy of stochastic algorithms are strong convergence and weak convergence. Strong convergence measures the error in $L^p(\Omega)$-norm, while the weak convergence is considered if one is interested in the approximation of distributional properties of $X(\cdot)$. In the following subsection, we shall review some basic concepts, and introduce the fundamental theorem on strong and weak convergence.

2.1.1 Strong and Weak Convergence

As usual, let $X_{s,y}(t)$ be the solution of the stochastic differential Eq. (2.1) satisfying the initial condition $X(s) = y$ for $0 \le s \le t \le T$. In the following, we consider the one-step approximation $\bar{X}_{t,y}(t + h)$ depending on y, t, h and $\{W_r(\theta) - W_r(t), r = 1, \ldots, m, 0 \le t \le \theta \le t + h \le T\}$ of the type

$$\bar{X}_{t,y}(t + h) = y + A(t, y, h, W_r(\theta) - W_r(t), r = 1, \ldots, m, t \le \theta \le t + h),$$

where A is a vector-valued function of dimension d. Define recursively the sequence

$$X_0 = x,$$
$$X_{k+1} = \bar{X}_{t_k, X_k}(t_{k+1})$$
$$= X_k + A(t_k, X_k, h, W_r(\theta) - W_r(t_k), r = 1, \ldots, m, t_k \le \theta \le t_{k+1})$$

for $k = 0, 1, \ldots, N - 1$.

Definition 2.1.1 (see [151, 270]) Let X_k be the numerical approximation to $X(t_k)$ with constant time-step size h for $k \in \{1, \ldots, N\}$. Then $X_k, k = 1, \ldots, N$, is said to have a strong convergence order γ in $L^p(\Omega, \mathbb{R})$ with $p \ge 1$, if there exists a constant $C > 0$ independent of h such that

$$\sup_{k \in \{1,\dots,N\}} \mathbb{E}\left[\|X_k - X(t_k)\|^p\right] \le C h^{p\gamma}$$

for sufficiently small h.

In many situations, the strong convergence refers to convergence in the mean-square sense, i.e., $p = 2$. A strongly convergent numerical method ensures that important features of single sample paths of the exact solution are reproduced by its approximation (see [161]). The following two theorems give the relationship between local errors and global errors of numerical methods for (2.1) whose coefficients satisfy the condition (1.2) and initial value satisfies $\mathbb{E}[\|x\|^2] < \infty$. The first one is called the fundamental theorem on the mean-square order of convergence.

Theorem 2.1.1 (See [193, 197]) *Suppose that the one-step approximation* $\bar{X}_{t,y}(t + h)$ *has order of accuracy* p_1 *for the expectation of the deviation and order of accuracy* p_2 *for the mean-square deviation. More precisely, for arbitrary* $0 \le t \le T - h$ *and* $y \in \mathbb{R}^d$, *the following inequalities hold*

$$\left\| \mathbb{E}\left[X_{t,y}(t + h) - \bar{X}_{t,y}(t + h)\right] \right\| \le K \left(1 + \|y\|^2\right)^{1/2} h^{p_1},$$

$$\left(\mathbb{E}\left[\|X_{t,y}(t + h) - \bar{X}_{t,y}(t + h)\|^2\right]\right)^{1/2} \le K \left(1 + \|y\|^2\right)^{1/2} h^{p_2}.$$

Also, let

$$p_2 \ge \frac{1}{2}, \ p_1 \ge p_2 + \frac{1}{2}.$$

Then for any $N \in \mathbb{N}$ *and* $k = 0, 1, \dots, N$, *the following inequality holds*

$$\left(\mathbb{E}\left[\|X_{0,x}(t_k) - X_k\|^2\right]\right)^{1/2} \le K \left(1 + \mathbb{E}[\|x\|^2]\right)^{1/2} h^{p_2 - 1/2}, \tag{2.4}$$

i.e., the order of accuracy of the method constructed using the one-step approximation $\bar{X}_{t,y}(t + h)$ *is* $\gamma = p_2 - 1/2$.

Notice that all the constants K mentioned above and those appearing in the following, do not rely on the initial value and time-step size h, but may depend on the system and the numerical approximation. The above theorem shows the mean-square convergence order of the numerical method by comparing the numerical solution with the exact solution, and implies that a numerical method of mean-square order $\gamma \in \mathbb{R}_+$ would have local mean-square error behaving like $O\left(h^{\gamma + \frac{1}{2}}\right)$. Here, γ can take fractional values since the solution to (2.1) is usually Hölder continuous with exponent $1/2 - \epsilon$ with $\epsilon \in (0, 1/2)$ (see [270]). Despite the exact solution is usually unknown in practice, we can also obtain the mean-square order of a numerical method by comparing it with another numerical method whose mean-square convergence order is known, by making use of the following theorem as a corollary of Theorem 2.1.1.

Theorem 2.1.2 (See [198]) *Let the one-step approximation $\bar{X}_{t,y}(t+h)$ satisfy the conditions of Theorem 2.1.1. Suppose that $\tilde{X}_{t,y}(t+h)$ meets the following relations*

$$\left\| \mathbb{E}\left[\tilde{X}_{t,y}(t+h) - \bar{X}_{t,y}(t+h) \right] \right\| = O(h^{p_1}),$$

$$\left(\mathbb{E}\left[\|\tilde{X}_{t,y}(t+h) - \bar{X}_{t,y}(t+h)\|^2 \right] \right)^{1/2} = O(h^{p_2})$$

with the same p_1 and p_2 as in Theorem 2.1.1. Then the numerical method based on the one-step approximation $\tilde{X}_{t,y}(t+h)$ has the same mean-square order of accuracy as the numerical method based on $\bar{X}_{t,y}(t+h)$, i.e., its mean-square order is $\gamma = p_2 - 1/2$.

The condition (1.2) associated with the Stratonovich stochastic differential Eq. (2.1) is a significant limitation, while we are taking into account that most of the models of applicable interest have coefficients growing faster at infinity than a linear function. For instance, in the following stochastic differential equation

$$dX(t) = -X^3(t)dt + dW(t), \quad X(0) = x,$$

the drift coefficient $-X^3$ is not globally Lipschitz continuous as it grows cubically. [255] presents a generalization of Theorem 2.1.1 from the globally to non-globally Lipschitz continuous case for the Itô stochastic differential equation

$$dX(t) = \left(\left(\sigma_0 + \frac{1}{2} \sum_{r=1}^{m} \left(\frac{\partial \sigma_r}{\partial x} \sigma_r \right) \right) (X(t)) \right) dt + \sum_{r=1}^{m} \sigma_r(X(t))dW_r(t)$$

$$=: a(X(t))dt + \sum_{r=1}^{m} \sigma_r(X(t))dW_r(t), \quad X(0) = x, \qquad (2.5)$$

which is equivalent to (2.1).

Assumption 2.1.1 (See [255]) *Let $a, \sigma_r: \mathbb{R}^d \to \mathbb{R}^d, r = 1, \ldots, m$, satisfy that*

(1) *for a sufficiently large $\tilde{p}_0 \geq 1$, there exists $C \geq 0$ such that for any $y_1, y_2 \in \mathbb{R}^d$,*

$$\langle y_1 - y_2, a(y_1) - a(y_2) \rangle + \frac{2\tilde{p}_0 - 1}{2} \sum_{r=1}^{m} \|\sigma_r(y_1) - \sigma_r(y_2)\|^2 \leq C\|y_1 - y_2\|^2;$$

(2) *there exist $C \geq 0$ and $\tilde{p}_1 \geq 1$ such that for any $y_1, y_2 \in \mathbb{R}^d$,*

$$\|a(y_1) - a(y_2)\|^2 \leq C \left(1 + \|y_1\|^{2\tilde{p}_1 - 2} + \|y_2\|^{2\tilde{p}_1 - 2}\right) \|y_1 - y_2\|^2.$$

Moreover, the initial value x satisfies $\mathbb{E}\left[\|x\|^{2\tilde{p}}\right] \leq C < \infty$ for all $\tilde{p} \geq 1$.

Notice that the first inequality in Assumption 2.1.1 implies that for any $y \in \mathbb{R}^d$,

$$\langle y, a(y) \rangle + \frac{2\tilde{p}_0 - 1 - \epsilon}{2} \sum_{r=1}^{m} \|\sigma_r(y)\|^2 \leq C_1 + C_2 \|y\|^2$$

with $\epsilon > 0$, $C_1 = \frac{\|a(0)\|^2}{2} + \frac{(2\tilde{p}_0 - 1 - \epsilon)(2\tilde{p}_0 - 1)}{2\epsilon} \sum_{r=1}^{m} \|\sigma_r(0)\|^2$ and $C_2 = C + \frac{1}{2}$.
Together with the initial condition, [148] proves the finiteness of moments of the exact solution, i.e., there exists $K > 0$ such that for any $\tilde{p} \in [1, \tilde{p}_0 - 1]$ and $t \in [0, T]$,

$$\mathbb{E}\left[\|X_{0,x}(t)\|^{2\tilde{p}}\right] < K \left(1 + \mathbb{E}[\|x\|^{2\tilde{p}}]\right).$$

Theorem 2.1.3 (See [255]) *Suppose that*

(1) *Assumption 2.1.1 holds;*
(2) *the one-step approximation $\bar{X}_{t,y}(t + h)$ has orders of accuracy: for some $p_3 \geq 1$, there exist $\alpha \geq 1$, $\delta > 0$ and $K > 0$ such that for arbitrary $t \in [0, T - h]$, $y \in \mathbb{R}^d$ and $h \in (0, \delta]$,*

$$\left\|\mathbb{E}[X_{t,y}(t + h) - \bar{X}_{t,y}(t + h)]\right\| \leq K \left(1 + \|y\|^{2\alpha}\right)^{1/2} h^{q_1},$$

$$\left(\mathbb{E}\left[\|X_{t,y}(t + h) - \bar{X}_{t,y}(t + h)\|^{2p_3}\right]\right)^{1/(2p_3)} \leq K \left(1 + \|y\|^{2\alpha p_3}\right)^{1/(2p_3)} h^{q_2}$$

with $q_2 \geq \frac{1}{2}$ and $q_1 \geq q_2 + 1/2$;
(3) *the numerical approximation X_k has finite moments, i.e., for some $p_3 \geq 1$, there exist $\beta \geq 1$, $\delta > 0$, and $K > 0$ such that for any $h \in (0, \delta]$ and $k = 0, 1, \ldots, N$,*

$$\mathbb{E}\left[\|X_k\|^{2p_3}\right] < K \left(1 + \mathbb{E}\left[\|x\|^{2p_3\beta}\right]\right).$$

Then for any $N \in \mathbb{N}$ and $k = 0, 1, \ldots, N$, the following inequality holds

$$\left(\mathbb{E}\left[\|X_{0,x}(t_k) - X_k\|^{2p_3}\right]\right)^{1/(2p_3)} \leq K \left(1 + \mathbb{E}\left[\|x\|^{2\eta p_3}\right]\right)^{1/(2p_3)} h^{q_2 - 1/2}, \tag{2.6}$$

where $K > 0$ and $\eta \geq 1$ do not depend on h and k, i.e., the order of accuracy of the numerical method based on the one-step approximation $\bar{X}_{t,y}(t+h)$ is $\gamma = q_2 - \frac{1}{2}$.

Corollary 2.1.1 (See [115]) *In the setting of Theorem 2.1.3 for $p_3 \geq 1/(2\gamma)$ in (2.6), there exist $\varepsilon \in (0, \gamma)$ and an almost surely finite random variable $C := C(\omega) > 0$ such that*

$$\sup_{k \in \{1,\dots,N\}} \left\| X_{0,x}(t_k) - X_k \right\| \leq C h^{\gamma - \varepsilon}$$

i.e., the method converges with order $\gamma - \varepsilon$ almost surely.

The above corollary can be proved by taking advantage of the Borel–Cantelli type of arguments (see e.g., [115, 196] and references therein).

Remark 2.1.1 In the case of non-global Lipschitz Itô stochastic differential equations, checking the condition on moments of the numerical approximation is often rather difficult, and each numerical method requires a special consideration. Moreover, many standard numerical methods (explicit numerical methods) will fail to converge (see e.g., [117, 136, 195] and references therein). For some numerical methods shown in [117, 136, 140, 248] and references therein, the boundedness of moments in non-global Lipschitz case is proved.

In many applications of the stochastic differential equations, such as molecular dynamics, financial engineering, mathematical physics and so on, one is interested in simulating averages $\mathbb{E}[\phi(X)]$ of the solution to stochastic differential equations for some function ϕ, which prompts the development of the weakly convergent numerical methods.

Definition 2.1.2 (See [270]) Let X_k be the numerical approximation to $X(t_k)$ with constant time-step size h for $k \in \{1, \dots, N\}$, where $Nh = T$. Then X_k, $k = 1, \dots, N$, is said to converge weakly with order γ to $X(T)$ if for each $\phi \in \mathbf{C}_P^{2(\gamma+1)}(\mathbb{R}^d, \mathbb{R})$, there exists $C > 0$ independent of h such that

$$\sup_{k \in \{1,\dots,N\}} |\mathbb{E}[\phi(X_k)] - \mathbb{E}[\phi(X(t_k))]| \leq C h^\gamma$$

for sufficiently small h.

For calculating a numerical approximation converging to the Stratonovich stochastic differential Eq. (2.1) under condition (1.2) in the weak convergent sense with some desired order γ, we make use of the main theorem on convergence of weak approximations in [197]. For the sake of simplicity, we rewrite $\Delta = X_{t,y}(t+h) - y$ and $\bar{\Delta} = \bar{X}_{t,y}(t+h) - y$, and denote the ith component of Δ and $\bar{\Delta}$ by Δ_i and $\bar{\Delta}_i$, for $i \in \{1, \dots, d\}$, respectively.

Definition 2.1.3 (See [192]) We say that a function $\phi \colon \mathbb{R}^d \to \mathbb{R}$ belongs to the class **F**, if there exist constants $C, \kappa > 0$ such that for $y \in \mathbb{R}^d$, the following inequality holds

$$|\phi(y)| \le C \left(1 + \|y\|^{\kappa}\right).$$

Theorem 2.1.4 (See [192]) *Suppose that*

(1) *coefficients in (2.5) are Lipschitz continuous and together with their partial derivatives of order up to $2\gamma + 2$, inclusively, belong to* **F**;
(2) *the numerical approximation $\bar{X}_{t,y}$ satisfies*

$$\left\| \mathbb{E}\left[\prod_{j=1}^{s} \Delta_{i_j} - \prod_{j=1}^{s} \bar{\Delta}_{i_j} \right] \right\| \le \tilde{C}(y)h^{\gamma+1}, \quad s = 1, \ldots, 2\gamma + 1,$$

$$\mathbb{E}\left[\prod_{j=1}^{2\gamma+2} \|\bar{\Delta}_{i_j}\| \right] \le \tilde{C}(y)h^{\gamma+1}, \quad i_j = 1, \ldots, d$$

*with $\tilde{C} \in$ **F**;*
(3) *the function ϕ together with its partial derivatives of order up to $2\gamma + 2$, inclusively, belongs to* **F**;
(4) *for a sufficiently large $m > 0$, the expectation $\mathbb{E}\left[\|X_k\|^{2m}\right]$ exists and is uniformly bounded with respect to N for $k \in \{1, \ldots, N\}$.*

Then, for all N and $k = 0, 1, \ldots, N$, the following inequality holds

$$\left| \mathbb{E}[\phi(X_{0,x}(t_k))] - \mathbb{E}[\phi(X_k)] \right| \le Ch^{\gamma}, \tag{2.7}$$

i.e., the numerical method has order of accuracy γ in the sense of weak approximations.

Remark 2.1.2 The proof of Theorem 2.1.4 in [192] allows to infer the weak order from the local error. Assuming that coefficients $a, \sigma_r \in \mathbf{C}_P^{2(\gamma+1)}\left(\mathbb{R}^d, \mathbb{R}^d\right)$, $r = 1, \ldots, m$, in (2.5) are globally Lipschitz continuous, the moments of the exact solution of (2.5) exist and are bounded up to a sufficiently high order, and $\phi \in \mathbf{C}_P^{2(\gamma+1)}\left(\mathbb{R}^d, \mathbb{R}\right)$. Then the local error bound

$$|\mathbb{E}[\phi(X_1)] - \mathbb{E}[\phi(X(t_1))]| \le Ch^{\gamma+1}$$

implies the global error bound (2.7). Here, the constant C is independent of h.

2.1.2 Numerical Methods via Taylor Expansions

Motivated by the fact that both strong and weak convergence for the stochastic numerical method for (2.1) can be obtained with the help of the equivalent Itô stochastic differential Eq. (2.5), now we investigate the construction of numerical methods for (2.1) via (2.5):

$$
\begin{cases}
dX(t) = a(X(t))dt + \displaystyle\sum_{r=1}^{m} \sigma_r(X(t))dW_r(t), \\
X(0) = x,
\end{cases}
$$

whose integral form is

$$
X(t) = x + \int_0^t a(X(s))ds + \sum_{r=1}^{m} \int_0^t \sigma_r(X(s))dW_r(s). \tag{2.8}
$$

Define the operators Λ_r, $r = 1, \ldots, m$, and $\mathbb{L}$ as

$$
\Lambda_r = \sigma_r^{\mathrm{T}} \frac{\partial}{\partial x} = \sum_{i=1}^{d} \sigma_r^i \frac{\partial}{\partial x_i},
$$

$$
\mathbb{L} = a^{\mathrm{T}} \frac{\partial}{\partial x} + \frac{1}{2} \sum_{r=1}^{m} \sum_{i=1}^{d} \sum_{j=1}^{d} \sigma_r^i \sigma_r^j \frac{\partial^2}{\partial x_i \partial x_j},
$$

where σ_r^i is the ith component of σ_r for $r \in \{1, \ldots, m\}$ and $i \in \{1, \ldots, d\}$. Moreover, denote by

$$
I_{(j_1, \ldots, j_l)}(t) = \int_0^t \int_0^{s_l} \cdots \int_0^{s_2} dW_{j_1}(s_1) dW_{j_2}(s_2) \cdots dW_{j_l}(s_l),
$$

the multiple Itô integral with multi-index $\alpha = (j_1, \ldots, j_l)$, $j_i \in \{0, 1, \ldots, m\}$, $i = 1, \ldots, l$, $l \geq 1$, and $dW_0(s) := ds$. If coefficients a and σ_r, $r = 1, \ldots, m$, in (2.8) are expanded in the Itô–Taylor series with respect to the initial value x, we can deduce

$$
\begin{aligned}
X_{0,x}(h) ={}& x + a(x)h + \sum_{r=1}^{m} \sigma_r(x)I_{(r)}(h) + \sum_{r=1}^{m} \sum_{i=1}^{m} \Lambda_i \sigma_r(x)I_{(i,r)}(h) \\
&+ \sum_{r=1}^{m} \mathbb{L}\sigma_r(x)I_{(0,r)}(h) + \sum_{r=1}^{m} \Lambda_r a(x)I_{(r,0)}(h) \\
&+ \sum_{r=1}^{m} \sum_{i=1}^{m} \sum_{s=1}^{m} \Lambda_s \Lambda_i \sigma_r(x)I_{(s,i,r)}(h) + \mathbb{L}a(x)\frac{h^2}{2} + \rho.
\end{aligned}
$$

Here, ρ is the remainder (see [197]).

Truncating the Itô–Taylor expansion at a particular point yields a numerical approximation with a certain order. Choosing the first three terms of the above expansion, we arrive at the simplest explicit numerical method, which is called the Euler–Maruyama method

$$X_{k+1} = X_k + a(X_k)h + \sum_{r=1}^{m} \sigma_r(X_k) \Delta_k W_r, \quad k \in \{0, 1, \ldots, N-1\}. \quad (2.9)$$

The increment of the Wiener process $\Delta_k W_r \sim \mathcal{N}(0, h)$ can be simulated by $\xi_r^k \sqrt{h}$, where $\xi_r^k \sim \mathcal{N}(0, 1)$, $r = 1, \ldots, m$, are independent 1-dimensional random variables. Milstein and Tretyakov in [197] prove that, under the condition (1.2) regarding the coefficients of (2.1), the Euler–Maruyama method is of mean-square order $\frac{1}{2}$ via Theorem 2.1.1 with $p_1 = 2$ and $p_2 = 1$. Especially, for the case that σ_r, $r = 1, \ldots, m$, are constants, the value of p_2 could be raised to $\frac{3}{2}$, and it implies that the mean-square order of the Euler–Maruyama method for systems with additive noises is 1. However, when the coefficients of (2.5) are non-globally Lipschitz continuous, the Euler–Maruyama method may diverge (see [197]). By adding more stochastic integral terms from the Itô–Taylor expansion, more accurate methods are obtained. For example, [197] proposes the Milstein method as follows

$$X_{k+1} = X_k + \sum_{r=1}^{m} \sigma_r(X_k) \Delta_k W_r + a(X_k)h$$

$$+ \sum_{i=1}^{m} \sum_{r=1}^{m} (\Lambda_i \sigma_r)(X_k) \int_{t_k}^{t_{k+1}} (W_i(\theta) - W_i(t_k)) dW_r(\theta),$$

where $k = 0, 1, \ldots, N-1$. This method has the first order convergence in the mean-square sense if the condition (1.2) holds. Moreover, it coincides with the Euler–Maruyama method when being used to numerically solve the stochastic differential equations with additive noises. To get a higher mean-square order numerical method based on the Itô–Taylor expansion, now we define $\mathcal{A}_\gamma := \{\alpha : l(\alpha) + n(\alpha) \le 2\gamma, \text{ or }, l(\alpha) = n(\alpha) = \gamma + 1/2\}$ with $l(\alpha)$ and $n(\alpha)$ being the length of the multi-index α and the number of zero components of α, respectively, where $r \in \mathbb{R}_+$. The explicit method that has mean-square convergence of order $\gamma \in \mathbb{R}_+$ can be constructed by truncating the series expansion according to the set $\mathcal{A}_\gamma$, when the coefficients in (2.5) and their partial derivatives are globally Lipschitz continuous. Accordingly, more stochastic integrals are supposed to be simulated. Two random variables are usually used in the formulation of the method together. These are $I_{(1)}(h)$ and $\frac{I_{(1,0)}(h)}{h}$ defined on an interval $[0, h]$. By sampling two independent random variables $\xi, \eta \sim \mathcal{N}(0, 1)$, then they can be easily calculated:

$$I_{(1)}(h) \approx \sqrt{h}\xi, \qquad \frac{I_{(1,0)}(h)}{h} \approx \frac{\sqrt{h}}{2}\left(\xi + \frac{\eta}{\sqrt{3}}\right). \qquad (2.10)$$

Now we turn to the weak convergence of explicit methods generated by the Itô–Taylor expansion. It can be verified that the Euler–Maruyama method has weak order 1. In many applications, it is of vital importance to approximate moments of the solution of a stochastic differential equation with better accuracy. The construction of higher weak order methods has been pursued by many researchers. For instance, [151] shows that explicit numerical methods that have weak convergence of order γ can be constructed via the truncations of the Itô–Taylor expansion according to the set $\mathcal{B}_\gamma := \{\alpha : l(\alpha) \leq \gamma\}$, when coefficients and their partial derivatives in (2.5) are globally Lipschitz continuous.

Further, we can also make use of the Itô–Taylor expansion to propose a class of implicit numerical methods including the increment of Wiener process $\Delta_k W_r = W_r(t_k + h) - W_r(t_k)$ for $r \in \{1, \ldots, m\}$ and $k \in \{0, 1, \ldots, N-1\}$. To overcome the difficulty brought by the unboundedness of $\Delta_k W_r$ for $k \in \{0, 1, \ldots, N-1\}$ and $r \in \{1, \ldots, m\}$, which is simulated by $\sqrt{h}\xi_r^k$ with $\xi_r^k \sim \mathcal{N}(0, 1)$, the truncated random variable $\Delta_k \widehat{W}_r$ is introduced (see [197]). One can define $\Delta_k \widehat{W}_r = \sqrt{h}\zeta_{rh}^k$ as follows

$$\zeta_{rh}^k = \begin{cases} \xi_r^k, & \text{if } |\xi_r^k| \leq A_h; \\ A_h, & \text{if } \xi_r^k > A_h; \\ -A_h, & \text{if } \xi_r^k < -A_h \end{cases} \tag{2.11}$$

with $A_h := \sqrt{2l|\log h|}$, where l is an arbitrary positive integer.

Lemma 2.1.1 (See [200]) *Let* $A_h := \sqrt{2l|\log h|}$, $l \geq 1$, $r \in \{0, 1, \ldots, m\}$, *and*

$$\zeta_{rh} = \begin{cases} \xi_r, & \text{if } |\xi_r| \leq A_h; \\ A_h, & \text{if } \xi_r > A_h; \\ -A_h, & \text{if } \xi_r < -A_h \end{cases}$$

with $\xi_r \sim \mathcal{N}(0, 1)$. *Then it holds that*

$$\mathbb{E}[(\zeta_{rh} - \xi_r)^2] \leq h^l, \tag{2.12}$$

$$0 \leq \mathbb{E}[(\xi_r)^2] - (\zeta_{rh})^2] = 1 - \mathbb{E}[(\zeta_{rh})^2] \leq (1 + 2\sqrt{2l|\log h|})h^l. \tag{2.13}$$

Moreover, it is not difficult to obtain the following properties:

$$\left(\mathbb{E}\left[|\Delta \widehat{W}_r|^{2p}\right]\right)^{1/(2p)} \leq \left(\mathbb{E}\left[|\Delta W_r|^{2p}\right]\right)^{1/(2p)} \leq c_p h^{1/2} \quad \forall\, p \in \mathbb{N}_+,$$

$$\mathbb{E}[(\Delta \widehat{W}_r)^{2p-1}] = \mathbb{E}[(\Delta W_r)^{2p-1}] = 0 \quad \forall\, p \in \mathbb{N}_+,$$

$$\mathbb{E}\left[\Delta \widehat{W}_i \Delta \widehat{W}_j \Delta \widehat{W}_k\right] = \mathbb{E}\left[\Delta W_i \Delta W_j \Delta W_k\right] = 0 \quad \forall\, i, j, k \in \{1, \ldots, m\}, \tag{2.14}$$

where c_p is a constant independent of h.

Taking advantage of above notations, we are able to construct the midpoint method, as a modification of (2.2):

$$X_{k+1} = X_k + \sigma_0 \left(\frac{X_k + X_{k+1}}{2} \right) h + \sum_{r=1}^{m} \sigma_r \left(\frac{X_k + X_{k+1}}{2} \right) \zeta_{rh}^k \sqrt{h}, \qquad (2.15)$$

where $k \in \{0, 1, \ldots, N - 1\}$. When $\sigma_r = 0$, $r \in \{1, \ldots, m\}$, this method coincides with the well-known deterministic midpoint method with the second order of convergence. In the general case, the midpoint method (2.15) is mean-square convergent with order $\frac{1}{2}$. In particular, if $\Lambda_i \sigma_r = \Lambda_r \sigma_i$ for $i, r = 1, \ldots, m$, or $m = 1$, the midpoint method (2.15) has the first mean-square order of convergence, which is stated in the next theorem (see [197]).

Theorem 2.1.5 *Suppose that $\Lambda_i \sigma_r = \Lambda_r \sigma_i$ for $i, r = 1, \ldots, m$. Let ζ_{rh} be defined as (2.11) with $A_h = \sqrt{4|\log h|}$. Then the method (2.15) for (2.1) has the first mean-square order of convergence.*

Notice that the midpoint method is a fully implicit method by introducing implicitness in both drift and diffusion coefficients. Now we present an approach of constructing a class of drift-implicit methods for (2.5) by means of the Itô–Taylor expansion in [197], which introduces implicitness only in drift coefficients. More precisely, according to

$$\mathbb{L}f(x) = \mathbb{L}f(X(h)) - \sum_{r=1}^{m} \int_0^h \Lambda_r \mathbb{L}f(X(\theta)) dW_r(\theta) - \int_0^h \mathbb{L}^2 f(X(\theta)) d\theta,$$

we deduce that the Itô–Taylor expansion for $X_{0,x}(h)$ becomes

$$X_{0,x}(h) = x + a(X_{0,x}(h))h - \sum_{r=1}^{m} h \int_0^h \Lambda_r a(X_{0,x}(\theta)) dW_r(\theta) - h \int_0^h \mathbb{L}a(X_{0,x}(\theta)) d\theta$$

$$+ \sum_{r=1}^{m} \sigma_r(x) I_{(r)}(h) + \sum_{r=1}^{m} \sum_{i=1}^{m} \Lambda_i \sigma_r(x) I_{(i,r)}(h)$$

$$+ \sum_{r=1}^{m} \mathbb{L}\sigma_r(x) I_{(0,r)}(h) + \sum_{r=1}^{m} \Lambda_r a(x) I_{(r,0)}(h)$$

$$+ \sum_{r=1}^{m} \sum_{i=1}^{m} \sum_{s=1}^{m} \Lambda_s \Lambda_i \sigma_r(x) I_{(s,i,r)}(h) + \frac{h^2}{2} \mathbb{L}a(X_{0,x}(h))$$

$$- \frac{h^2}{2} \sum_{r=1}^{m} \int_0^h \Lambda_r \mathbb{L}a(X_{0,x}(\theta)) dW_r(\theta) - \frac{h^2}{2} \int_0^h \mathbb{L}^2 a(X_{0,x}(\theta)) d\theta + \rho.$$

Coefficients in ρ can be treated in the same manner (e.g., the above reasoning is immediately applicable to $\mathbb{L}^2 a$ and $\mathbb{L}^3 a$). When the explicitness is introduced owing to expressions occurring in stochastic integrals, abundant numerical methods which may possess certain stability can be obtained. One of the simplest and the most popular drift-implicit methods based on the above equation has the form

$$X_{k+1} = X_k + a(X_{k+1})h + \sum_{r=1}^{m} \sigma_r(X_k)\zeta_{rh}^k \sqrt{h}, \quad k = 0, 1, \ldots, N-1,$$

which is of mean-square order $\frac{1}{2}$ for general equations and of order 1 for the stochastic differential equation with additive noises. For more accurate drift-implicit methods, they can be generated by utilizing truncations of the above series including more terms and approximating stochastic integrals by bounded random variables (see [151]).

2.1.3 Stochastic Runge–Kutta Methods

It can be observed that the higher order numerical method constructed by the Itô–Taylor expansion requires more higher order partial derivatives of coefficients. Consequently, a great deal of attention has been recently paid to developing derivative-free methods, especially stochastic Runge–Kutta methods. In recent years, derivative-free Runge–Kutta methods have been investigated for the strong approximation of stochastic differential equations (see e.g., [39, 40, 151, 197, 206] and references therein). Stochastic Runge–Kutta methods for the weak approximation have been studied in [151, 153, 155, 226, 227, 254] and references therein. In this subsection, we assume that $\mathbb{E}[\|x\|^2] < \infty$, and the condition (1.2) is fulfilled. Below we shall give a brief introduction to stochastic Runge–Kutta methods.

Consider the general family of s-stage stochastic Runge–Kutta methods given by

$$x_i = X_k + \sum_{l=0}^{m} \sum_{j=1}^{s} \tilde{A}_{ij}^{(l)} \sigma_l(x_j), \quad i = 1, \ldots, s, \ s \in \mathbb{N}_+,$$

$$X_{k+1} = X_k + \sum_{l=0}^{m} \sum_{j=1}^{s} \tilde{b}_j^{(l)} \sigma_l(x_j), \quad k = 0, 1, \ldots, N-1,$$

(2.16)

where $\tilde{A}^{(0)}$ and $\tilde{b}^{(0)}$ are a matrix and a column vector of coefficients, respectively, scaled by the time-step size h, while $\tilde{A}^{(l)}$ and $\tilde{b}^{(l)}$ contain elements which are arbitrary random variables for $l \in \{1, \ldots, m\}$. For instance, a widely used stochastic

Runge–Kutta method of the family (2.16) with $\Delta_k \widehat{W}_r$ is

$$x_i = X_k + \sum_{j=1}^{s} A_{ij}^{(0)} \sigma_0 \left(x_j\right) h + \sum_{r=1}^{m} \sum_{j=1}^{s} A_{ij}^{(r)} \sigma_r \left(x_j\right) \Delta_k \widehat{W}_r, \quad i = 1, \ldots, s, \ s \in \mathbb{N}_+,$$

$$X_{k+1} = X_k + \sum_{i=1}^{s} b_i^{(0)} \sigma_0 \left(x_i\right) h + \sum_{r=1}^{m} \sum_{i=1}^{s} b_i^{(r)} \sigma_r \left(x_i\right) \Delta_k \widehat{W}_r, \quad k = 0, 1, \ldots, N - 1,$$

i.e., in the Butcher tableau

$$\begin{array}{c|cccc} & A^{(0)} & A^{(1)} & \cdots & A^{(m)} \\ \hline & (b^{(0)})^\top & (b^{(1)})^\top & \cdots & (b^{(m)})^\top \end{array}.$$

Denote $e = (1, \ldots, 1)^\top \in \mathbb{R}^s$. If formulae

$$\left(b^{(0)}\right)^\top e = 1, \quad \left(b^{(l)}\right)^\top e = 1, \quad \left(b^{(l)}\right)^\top A^{(l)} e = \frac{1}{2} \tag{2.17}$$

for $l = 1, \ldots, m$, hold, then the above stochastic Runge–Kutta method has strong order $\frac{1}{2}$ when $m > 1$, and is of strong order 1 when $m = 1$ (see e.g., [40, 126] and references therein).

Hong et al. in [126] also study another kind of stochastic Runge–Kutta methods, as a special case in [226], where the number of matrices in the Butcher tableau is independent of the number m of driving Wiener processes. This class of stochastic Runge–Kutta methods for (2.1) is

$$x_i^{(0)} = X_k + \sum_{j=1}^{s} A_{ij}^{(0)} \sigma_0 \left(x_j^{(0)}\right) h + \sum_{r=1}^{m} \sum_{j=1}^{s} B_{ij}^{(0)} \sigma_r \left(x_j^{(r)}\right) \hat{I}_{(r,k)},$$

$$x_i^{(l)} = X_k + \sum_{j=1}^{s} A_{ij}^{(1)} \sigma_0 \left(x_j^{(0)}\right) h + \sum_{r=1}^{m} \sum_{j=1}^{s} B_{ij}^{(1)} \sigma_r \left(x_j^{(r)}\right) \hat{I}_{(r,k)}, \tag{2.18}$$

$$X_{k+1} = X_k + \sum_{i=1}^{s} \alpha_i \sigma_0 \left(x_i^{(0)}\right) h + \sum_{r=1}^{m} \sum_{i=1}^{s} \beta_i \sigma_r \left(x_i^{(r)}\right) \hat{I}_{(r,k)}$$

for $k = 0, 1, \ldots, N - 1$, $i = 1, \ldots, s$, and $l = 1, \ldots, m$. Associated coefficients can be described by the following Butcher tableau

$$\begin{array}{c|c} A^{(0)} & B^{(0)} \\ \hline A^{(1)} & B^{(1)} \\ \hline \alpha^\top & \beta^\top \end{array}.$$

Moreover, random variables $\hat{I}_{(r,k)}, r = 1, \ldots, m, \ k = 0, 1, \ldots, N - 1$, have moments

$$\mathbb{E}\big[\hat{I}_{(r,k)}^q\big] = \begin{cases} 0, & \text{if} \quad q \in \{1, 3, 5\}; \\ (q-1)h^{\frac{q}{2}}, & \text{if} \quad q \in \{2, 4\}; \\ O\big(h^{\frac{q}{2}}\big), & \text{if} \quad q \geq 6. \end{cases}$$

The stochastic Runge–Kutta method (2.18) has weak order 1 if it satisfies

$$\alpha^\top e = 1, \quad \beta^\top e = 1, \quad \beta^\top B^{(1)} e = \frac{1}{2}.$$

Example 2.1.1 For the stage $s = 1$, the midpoint method is

$$\begin{array}{c|cccc} \frac{1}{2} & \frac{1}{2} & \cdots & \frac{1}{2} \\ \hline & 1 & 1 & \cdots & 1 \end{array}.$$

Example 2.1.2 For the stage $s = 1$, we give a stochastic Runge–Kutta method (2.18) of weak convergence order 1 as follows

$$\begin{array}{c|c} \frac{1}{2} & 1 - b \\ \hline b & \frac{1}{2} \\ \hline 1 & 1 \end{array}$$

with $b \in \mathbb{R}$ being an arbitrary parameter.

Example 2.1.3 For $s = 4$, we present a numerical method of weak convergence order 2 when $m = 1$:

$$\left[\begin{array}{cccc|cccc} \frac{1}{8} & 0 & 0 & 0 & \frac{5}{6} - \frac{\sqrt{3}}{3} & -\frac{1}{2} & 0 & 0 \\ \frac{1}{4} & \frac{1}{8} & 0 & 0 & -\frac{1}{6} + \frac{\sqrt{3}}{3} & \frac{1}{2} & 0 & 0 \\ \frac{1}{4} & \frac{1}{4} & \frac{1}{8} & 0 & \frac{1}{2} & \frac{1}{2} & 0 & 0 \\ \frac{1}{4} & \frac{1}{4} & \frac{1}{4} & \frac{1}{8} & -\frac{1}{6} & \frac{1}{2} & 0 & 0 \\ \hline -\frac{1}{6} + \frac{\sqrt{3}}{6} & \frac{1}{3} - \frac{\sqrt{3}}{6} & 0 & \frac{1}{3} & \frac{1}{4} & \frac{1}{4} - \frac{\sqrt{3}}{6} & 0 & 0 \\ \frac{1}{2} & 0 & 0 & 0 & \frac{1}{4} + \frac{\sqrt{3}}{6} & \frac{1}{4} & 0 & 0 \\ 0 & 0 & 0 & 0 & b_1 & b_2 & 0 & b_3 \\ 0 & 0 & 0 & 0 & 0 & -\frac{1}{2} & 0 & 0 \\ \hline \frac{1}{4} & \frac{1}{4} & \frac{1}{4} & \frac{1}{4} & \frac{1}{2} & \frac{1}{2} & 0 & 0 \end{array}\right],$$

where b_1, b_2 and b_3 are arbitrary parameters.

2.2 Stochastic Symplectic Methods

Deterministic Hamiltonian systems have been thoroughly studied and their symplectic methods possess superior performance in long-time simulations, compared with nonsymplectic ones. In response to demand for tools to characterize and analyze continuous and discrete mechanical systems with uncertainty, a prominent line of investigation is the study of constructing stochastic symplectic methods for the $2n$-dimensional stochastic Hamiltonian system

$$
\begin{cases}
dP(t) = -\dfrac{\partial H_0(P(t), Q(t))}{\partial Q}dt - \displaystyle\sum_{r=1}^{m} \dfrac{\partial H_r(P(t), Q(t))}{\partial Q} \circ dW_r(t), \quad P(0) = p, \\[4mm]
dQ(t) = \dfrac{\partial H_0(P(t), Q(t))}{\partial P}dt + \displaystyle\sum_{r=1}^{m} \dfrac{\partial H_r(P(t), Q(t))}{\partial P} \circ dW_r(t), \quad Q(0) = q
\end{cases}
$$

$$(2.19)$$

with $t \in (0, T]$ and $W_r(\cdot)$, $r = 1, \ldots, m$, being independent standard Wiener processes. The systematical study of stochastic symplectic methods that could inherit the geometric feature of the original system started around 2002 (see e.g., [199, 200] and references therein). And fruitful works on stochastic symplectic (partitioned) Runge–Kutta methods, symplectic methods based on the stochastic generating function, stochastic variational integrators and stochastic symplectic integrators via composition methods are presented in [13, 41, 93, 203, 260, 263, 265, 272] and references therein. These stochastic symplectic methods allow us to simulate stochastic Hamiltonian systems on a very long time interval with good numerical stability and probabilistic superiority.

A numerical method $\{P_k, Q_k\}_{k \geq 0}$ with $\left(P_0^\top, Q_0^\top\right)^\top = (p^\top, q^\top)^\top$ applied to (2.19) preserves the stochastic symplectic structure if

$$
dP_{k+1} \wedge dQ_{k+1} = dP_k \wedge dQ_k \quad \forall k \geq 0, \tag{2.20}
$$

or equivalently, $\left(\frac{\partial(P_{k+1}, Q_{k+1})}{\partial(P_k, Q_k)}\right) \in Sp(2n)$, i.e.,

$$
\left(\frac{\partial(P_{k+1}, Q_{k+1})}{\partial(P_k, Q_k)}\right)^\top J_{2n} \left(\frac{\partial(P_{k+1}, Q_{k+1})}{\partial(P_k, Q_k)}\right) = J_{2n} \quad \forall k \geq 0.
$$

Theorem 2.2.1 *Let $X_k = (P_k^\top, Q_k^\top)^\top$ for $k \in \{0, 1, \ldots, N\}$. The implicit midpoint method for (2.19)*

$$
X_{k+1} = X_k + J_{2n}^{-1} \nabla H_0 \left(\frac{X_{k+1} + X_k}{2}\right) h + \sum_{r=1}^{m} J_{2n}^{-1} \nabla H_r \left(\frac{X_k + X_{k+1}}{2}\right) \zeta_{rh}^k \sqrt{h}
$$

$$(2.21)$$

with $A_h = \sqrt{4|\log h|}$ and $k \in \{0, 1, \ldots, N-1\}$ preserves the stochastic symplectic structure.

Proof Since

$$\left(I_{2n} - \frac{h}{2}J_{2n}^{-1}\nabla^2 H_0 - \sum_{r=1}^{m} \zeta_{rh}^{k} \frac{\sqrt{h}}{2} J_{2n}^{-1}\nabla^2 H_r\right)\left(\frac{\partial X_{k+1}}{\partial X_k}\right)$$

$$= \left(I_{2n} + \frac{h}{2}J_{2n}^{-1}\nabla^2 H_0 + \sum_{r=1}^{m} \zeta_{rh}^{k} \frac{\sqrt{h}}{2} J_{2n}^{-1}\nabla^2 H_r\right)$$

for $k \in \{0, 1, \ldots, N-1\}$, a straight computation leads that $\left(\frac{\partial X_{k+1}}{\partial X_k}\right)^{\top} J_{2n}\left(\frac{\partial X_{k+1}}{\partial X_k}\right) = J_{2n}$. $\qquad\square$

A different approach of proving the symplecticity of the midpoint method (using stochastic generating functions) will be given in Sect. 2.2.2. To clarify that both the Euler–Maruyama method and Milstein method from Sect. 2.1 are not symplectic, we apply them to the linear stochastic oscillator

$$\begin{cases} dP(t) = -Q(t)dt + \alpha dW(t), & P(0) = p, \\ dQ(t) = P(t)dt, & Q(0) = q, \end{cases} \tag{2.22}$$

where $W(\cdot)$ is a standard Wiener process. We compare behaviors of several proposed numerical approximations to solve (2.22). The following list gathers the numerical methods together with some relevant references.

1. Exponential method (see [253]):

$$P_{k+1} = \cos(h)P_k - \sin(h)Q_k + \alpha \Delta_k W,$$

$$Q_{k+1} = \sin(h)P_k + \cos(h)Q_k.$$

2. Integral method (see [63]):

$$P_{k+1} = \cos(h)P_k - \sin(h)Q_k + \alpha\cos(h)\Delta_k W,$$

$$Q_{k+1} = \sin(h)P_k + \cos(h)Q_k + \alpha\sin(h)\Delta_k W.$$

3. Optimal method (see [200]):

$$P_{k+1} = \cos(h)P_k - \sin(h)Q_k + \alpha\frac{\sin(h)}{h}\Delta_k W,$$

$$Q_{k+1} = \sin(h)P_k + \cos(h)Q_k + 2\alpha\frac{\sin^2(\frac{h}{2})}{h}\Delta_k W.$$

4. Euler–Maruyama (EM) method (or equivalently, the Milstein method for (2.22)) (see [242]):

$$P_{k+1} = P_k - Q_k h + \alpha \Delta_k W,$$

$$Q_{k+1} = Q_k + P_k h.$$

5. Backward Euler–Maruyama (BEM) method (see [242]):

$$P_{k+1} = P_k - Q_{k+1} h + \alpha \Delta_k W,$$

$$Q_{k+1} = Q_k + P_{k+1} h.$$

6. Predictor–Corrector (EM-BEM) method using EM method as predictor and BEM method as corrector (see [123]):

$$P_{k+1} = (1 - h^2) P_k - Q_k h + \alpha \Delta_k W,$$

$$Q_{k+1} = P_k h + (1 - h^2) Q_k + \alpha h \Delta_k W.$$

Table 2.1 shows good performance of the exponential method as well as the integral method for (2.22). It can be easily checked that the optimal method is symplectic but its second moment does not grow linearly. Moreover, neither symplecticity nor the linear growth property can be preserved by the Euler–Maruyama method, the backward Euler–Maruyama method or the Predictor–Corrector method. One can clearly observe that the Euler–Maruyama method is not symplectic, since the radius of images increases in Fig. 2.1, which represents the evolution of domains in the phase plane of the system (2.22), the symplectic method and the Euler–Maruyama method. The initial domain is the circle with center at the origin and with the unit radius. In the case of the symplectic method, the image of initial circle becomes an ellipse and the area does not change any more. This means that the stochastic symplectic method can preserve geometric features of the original system.

Table 2.1 Fulfillments of symplecticity and linear growth of the second moments by numerical methods for (2.22)

Numerical method	Symplecticity	Linear growth property
Exponential method	Yes	Yes
Integral method	Yes	Yes
Optimal method	Yes	No
EM method	No	No
BEM method	No	No
EM-BEM method	No	No

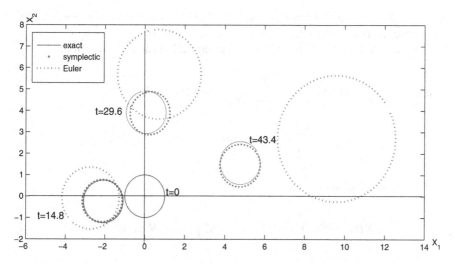

Fig. 2.1 Evolution of domains in the phase plane with $\alpha = 1$ and $h = 0.05$

In the following we introduce some methodologies of constructing stochastic symplectic methods systematically for the stochastic Hamiltonian system (2.19).

2.2.1 Symplectic Methods via Padé Approximation

We start with the Padé approximation approach for constructing symplectic methods for two kinds of linear stochastic Hamiltonian systems.

Linear Stochastic Hamiltonian Systems with Multiplicative Noises Consider the $2n$-dimensional linear stochastic Hamiltonian systems with non-degenerate noises

$$
\begin{cases}
dP(t) = (A_1^0 P(t) + A_2^0 Q(t))dt + \displaystyle\sum_{r=1}^{m}(A_1^r P(t) + A_2^r Q(t)) \circ dW_r(t), & P(0) = p, \\[2mm]
dQ(t) = (A_3^0 P(t) + A_4^0 Q(t))dt + \displaystyle\sum_{r=1}^{m}(A_3^r P(t) + A_4^r Q(t)) \circ dW_r(t), & Q(0) = q,
\end{cases}
$$

(2.23)

where $t \in (0, T]$, A_j^i, $i = 0, 1, \ldots, m$, $j = 1, 2, 3, 4$, are $n \times n$ constant matrices satisfying $A_1^i = -(A_4^i)^\top$, $A_2^i = (A_2^i)^\top$, $A_3^i = (A_3^i)^\top$. Let

$$
A^r = \begin{bmatrix} A_1^r & A_2^r \\ A_3^r & A_4^r \end{bmatrix}, \quad r = 0, 1, \ldots, m,
$$

and $C_r = J_{2n} A^r$. A direct computation leads that $C_r, r = 0, 1, \ldots, m$, are $2n \times 2n$ symmetric matrices. Denoting $X = (P^\top, Q^\top)^\top$, and $H_r(X) = \frac{1}{2} X^\top C_r X, r = 0, 1, \ldots, m$, we deduce that the canonical system (2.23) can be rewritten as

$$dX(t) = J_{2n}^{-1} \nabla H_0(X(t)) dt + \sum_{r=1}^{m} J_{2n}^{-1} \nabla H_r(X(t)) \circ dW_r(t)$$

$$= A^0 X(t) dt + \sum_{r=1}^{m} A^r X(t) \circ dW_r(t), \quad X(0) = (p^\top, q^\top)^\top, \tag{2.24}$$

whose unique solution is

$$X(t) = \exp\left(t A^0 + \sum_{r=1}^{m} W_r(t) A^r\right) X(0) \quad \forall t \in [0, T].$$

Remark 2.2.1 The matrices $A^r, r = 0, 1, \ldots, m$, in (2.24) are infinitesimal symplectic matrices, i.e.,

$$J_{2n} A^r + (A^r)^\top J_{2n} = 0.$$

All infinitesimal symplectic matrices endowed with commutation operation $[A, B] = AB - BA$ form a Lie algebra $sp(2n)$ of the Lie group $Sp(2n)$.

Lemma 2.2.1 (See [100]) *If f is an even polynomial and $B \in sp(2n)$, then*

$$f(B^\top) J_{2n} = J_{2n} f(B).$$

Lemma 2.2.2 (See [100]) *If g is an odd polynomial and $B \in sp(2n)$, then $g(B) \in sp(2n)$, i.e.,*

$$g(B^\top) J_{2n} + J_{2n} g(B) = 0.$$

Lemma 2.2.3 (See [100]) *$S = M^{-1} N \in Sp(2n)$ if and only if*

$$M J_{2n} M^\top = N J_{2n} N^\top.$$

Lemma 2.2.4 (See [100]) *If $B \in sp(2n)$, then $\exp(B) \in Sp(2n)$.*

As is well known, the matrix exponential $\exp(M)$ for an $n \times n$ matrix M has the Taylor expansion

$$\exp(M) = I_n + \sum_{i=1}^{\infty} \frac{M^i}{i!}.$$

A simple way to approximate the exponential function $\exp(x)$ for $x \in \mathbb{R}$ is an application of the rational Padé approximation

$$\exp(x) \approx P_{(r,s)}(x) = D_{(r,s)}^{-1}(x)N_{(r,s)}(x), \qquad (2.25)$$

where $r, s \in \mathbb{N}_+$,

$$D_{(r,s)}(x) = 1 + \sum_{i=1}^{s} \frac{(r+s-i)!s!}{(r+s)!i!(s-i)!}(-x)^i = 1 + \sum_{i=1}^{s} b_i(-x)^i,$$

$$N_{(r,s)}(x) = 1 + \sum_{i=1}^{r} \frac{(r+s-i)!r!}{(r+s)!i!(r-i)!}x^i = 1 + \sum_{i=1}^{r} a_i x^i$$

with $a_i = \frac{(r+s-i)!r!}{(r+s)!i!(r-i)!}$ and $b_j = \frac{(r+s-j)!s!}{(r+s)!j!(s-j)!}$ for $i \in \{1, \ldots, r\}$ and $j \in \{1, \ldots, s\}$. The difference between the exponential function and the Padé approximation $P_{(r,s)}$, $r, s \in \mathbb{N}_+$, satisfies

$$\exp(x) - P_{(r,s)}(x) = o(|x|^{r+s+1})$$

for sufficiently small $|x|$ (see [100]). One can take advantage of the Padé approximation to approach the matrix exponential, and then

$$X(t) \approx \left(I_{2n} + \sum_{j=1}^{s} b_j(-M)^j\right)^{-1}\left(I_{2n} + \sum_{j=1}^{r} a_j M^j\right)X(0) \qquad (2.26)$$

with $M = tA^0 + \sum_{r=1}^{m} W_r(t)A^r$, when the matrix $I_{2n} + \sum_{j=1}^{s} b_j(-M)^j$ is invertible.

Now we investigate whether the Padé approximation can be utilized to construct stochastic symplectic methods for (2.23), similar to the deterministic case in [100]. Based on (2.26), we propose the following numerical method

$$\hat{X}_{k+1}^{(r,s)} = \left(I_{2n} + \sum_{j=1}^{s} b_j(-B_k)^j\right)^{-1}\left(I_{2n} + \sum_{j=1}^{r} a_j B_k^j\right)\hat{X}_k^{(r,s)} \qquad (2.27)$$

where $k = 0, 1, \ldots, N-1$, $B_k = hA^0 + \sum_{r=1}^{m} \Delta_k W_r A^r$, and $\Delta_k W_r = W_r(t_{k+1}) - W_r(t_k)$ is simulated by $\sqrt{h}\xi_r^k$ with $\xi_r^k \sim \mathcal{N}(0, 1)$. To ensure the well-posedness of (2.27), two aspects should be taken into consideration. On one hand, the invertibility of the matrix $I_{2n} + \sum_{j=1}^{s} b_j(-B_k)^j$ can be guaranteed by setting sufficiently small time-step size h such that the matrix is near the identity matrix for $k \in \{0, 1, \ldots, N - 1\}$. On the other hand, to deal with the problem caused by the unboundedness of $\Delta_k W_r = \sqrt{h} \xi_r^k$, we follow (2.11) to truncate the random variable ξ_r^k for obtaining another bounded random variable ζ_{rh}^k for $r \in \{1, \ldots, m\}$ and $k \in \{0, 1, \ldots, N - 1\}$. Milstein et al. in [199] indicate that, the truncation error can be merged into the error of the numerical approximation by choosing a sufficiently large parameter l in $A_h = \sqrt{2l|\log h|}$, which should be at least 2γ if the numerical method containing such a truncation is expected to possess mean-square convergence order $\gamma \in \mathbb{R}_+$. Now we let $\bar{B}_k = hA^0 + \sum_{r=1}^{m} \sqrt{h} \zeta_{rh}^k A^r$ for $k \in \{0, 1, \ldots, N-1\}$. In the construction and implementation of (2.27), we multiply the matrix $I_{2n} + \sum_{j=1}^{s} b_j(-\bar{B}_k)^j$ on both sides of (2.27) instead of taking inverse, and get the following equivalent numerical method of the implicit form

$$X_{k+1}^{(r,s)} = \left(-\sum_{j=1}^{s} b_j(-\bar{B}_k)^j \right) X_{k+1}^{(r,s)} + \left(I_{2n} + \sum_{j=1}^{r} a_j \bar{B}_k^j \right) X_k^{(r,s)}, \qquad (2.28)$$

where $k = 0, 1, \ldots, N - 1$. It is interesting to observe that if $r = s = 1$, we obtain the midpoint method

$$X_{k+1}^{(1,1)} = X_k^{(1,1)} + \frac{1}{2} \bar{B}_k (X_k^{(1,1)} + X_{k+1}^{(1,1)}). \qquad (2.29)$$

If both r and s take the value 2, we attain the method

$$X_{k+1}^{(2,2)} = X_k^{(2,2)} + \frac{1}{2} \bar{B}_k (X_k^{(2,2)} + X_{k+1}^{(2,2)}) + \frac{1}{12} \bar{B}_k^2 (X_k^{(2,2)} - X_{k+1}^{(2,2)}). \qquad (2.30)$$

Similarly, if $r = s = 3$ or 4, two numerical methods are

$$X_{k+1}^{(3,3)} = X_k^{(3,3)} + (\frac{1}{2} \bar{B}_k + \frac{1}{120} \bar{B}_k^3)(X_k^{(3,3)} + X_{k+1}^{(3,3)}) + \frac{1}{10} \bar{B}_k^2 (X_k^{(3,3)} - X_{k+1}^{(3,3)}), \qquad (2.31)$$

$$X_{k+1}^{(4,4)} = X_k^{(4,4)} + (\frac{1}{2} \bar{B}_k + \frac{1}{84} \bar{B}_k^3)(X_k^{(4,4)} + X_{k+1}^{(4,4)}) + (\frac{3}{28} \bar{B}_k^2 + \frac{1}{1680} \bar{B}_k^4)(X_k^{(4,4)} - X_{k+1}^{(4,4)}). \qquad (2.32)$$

Theorem 2.2.2 (See [245]) *Numerical method (2.28) based on the Padé approximation $P_{(l,l)}$, with $A_h = \sqrt{4l|\log h|}$ and $l \geq 1$, preserves stochastic symplectic structure.*

Proof The proof of the theorem is a straight forward extension of its counterpart in deterministic case (see [100]). For the sake of simplicity, we denote $\bar{B}_k := hA^0 + \sum_{r=1}^{m} \sqrt{h} \zeta_{rh}^k A^r$ as above and consider the one-step approximation (2.28) based on

$P_{(l,l)}(\bar{B}_k)$ as follows

$$X_{k+1}^{(l,l)} = \left(D_{(l,l)}(\bar{B}_k)\right)^{-1}\left(N_{(l,l)}(\bar{B}_k)\right)X_k^{(l,l)}, \tag{2.33}$$

where $l \in \mathbb{N}_+$, $k \in \{0, 1, \ldots, N-1\}$,

$$D_{(l,l)}(\bar{B}_k) = I_{2n} + \sum_{j=1}^{l} a_j\left(-hA^0 - \sum_{r=1}^{m}\sqrt{h}\zeta_{rh}^k A^r\right)^j,$$

$$N_{(l,l)}(\bar{B}_k) = I_{2n} + \sum_{j=1}^{l} a_j\left(hA^0 + \sum_{r=1}^{m}\sqrt{h}\zeta_{rh}^k A^r\right)^j.$$

To verify that (2.33) is symplectic, we need to prove $(D_{(l,l)}(\bar{B}_k))^{-1}N_{(l,l)}(\bar{B}_k) \in Sp(2n)$. It can be observed that for $k \in \{0, 1, \ldots, N-1\}$, $N_{(l,l)}(\bar{B}_k) = F(\bar{B}_k) + G(\bar{B}_k)$, $D_{(l,l)}(\bar{B}_k) = F(\bar{B}_k) - G(\bar{B}_k)$, where $F(\bar{B}_k)$ and $G(\bar{B}_k)$ are even and odd polynomials of $\bar{B}_k$, respectively. Because $\bar{B}_k \in sp(2n)$ with $k = 0, 1, \ldots, N-1$, we obtain $F(\bar{B}_k^\top)J_{2n} = J_{2n}F(\bar{B}_k)$ and $G(\bar{B}_k^\top)J_{2n} + J_{2n}G(\bar{B}_k) = 0$ by Lemmas 2.2.1 and 2.2.2. Since

$$\begin{aligned}
&(N_{(l,l)}(\bar{B}_k))^\top J_{2n}N_{(l,l)}(\bar{B}_k) \\
&= (F(\bar{B}_k^\top) + G(\bar{B}_k^\top))J_{2n}(F(\bar{B}_k) + G(\bar{B}_k)) \\
&= J_{2n}(F(\bar{B}_k) - G(\bar{B}_k))(F(\bar{B}_k) + G(\bar{B}_k)) \\
&= J_{2n}(F(\bar{B}_k) + G(\bar{B}_k))(F(\bar{B}_k) - G(\bar{B}_k)) \\
&= (F(\bar{B}_k^\top) - G(\bar{B}_k^\top))J_{2n}(F(\bar{B}_k) - G(\bar{B}_k)) \\
&= (D_{(l,l)}(\bar{B}_k))^\top J_{2n}D_{(l,l)}(\bar{B}_k),
\end{aligned}$$

from Lemma 2.2.3 it follows that $(D_{(l,l)}(\bar{B}_k))^{-1}N_{(l,l)}(\bar{B}_k) \in Sp(2n)$, which ends the proof. $\qquad\square$

Our result regarding the mean-square convergence order of (2.28) is as follows.

Theorem 2.2.3 (See [245]) *Numerical method (2.28) based on the Padé approximation $P_{(r,s)}$, with $A_h = \sqrt{2l|\log h|}$, $l \geq r + s$ and $r, s \geq 1$, is of mean-square convergence order $\frac{r+s}{2}$.*

To examine properties of proposed methods, we apply (2.29)–(2.32) to approximating the Kubo oscillator, which is the simplest case of (2.23),

$$\begin{cases} dP(t) = -aQ(t)dt - \sigma Q(t) \circ dW(t), & P(0) = p, \\ dQ(t) = aP(t)dt + \sigma P(t) \circ dW(t), & Q(0) = q \end{cases} \tag{2.34}$$

with a, σ being constants and $W(\cdot)$ being a 1-dimensional standard Wiener process. Moreover, we carry out numerical experiments from three aspects:

(1) mean-square convergence rates of numerical methods;
(2) sample trajectories produced by numerical methods and those of exact solutions;
(3) evolution of Hamiltonian by numerical methods.

The Euler–Maruyama method applied to (2.34) reads

$$
\begin{aligned}
P_{k+1} &= P_k - a Q_k h - \frac{\sigma^2}{2} P_k h - \sigma Q_k \Delta_k W, \\
Q_{k+1} &= Q_k + a P_k h - \frac{\sigma^2}{2} Q_k h + \sigma P_k \Delta_k W
\end{aligned}
\tag{2.35}
$$

with $k = 0, 1, \ldots, N - 1$, which is of mean-square order $\frac{1}{2}$, but not symplectic. Denote $\bar{B}_{k,i} = (ah + \sigma \zeta_h^{k,i} \sqrt{h}) J_2^{-1}$, where the bound corresponding to truncated random variables $\zeta_h^{k,i}$ is $A_h^i = \sqrt{4i|\log h|}$ for $i = 1, 2, 3, 4$ and $k \in \{0, 1, \ldots, N - 1\}$. Four symplectic methods (2.29)–(2.32) based on four kinds of Padé approximations, respectively, take the following forms for (2.34) accordingly:

$$
X_{k+1}^{(1,1)} = X_k^{(1,1)} + \frac{1}{2} \bar{B}_{k,1} (X_k^{(1,1)} + X_{k+1}^{(1,1)}),
\tag{2.36}
$$

$$
X_{k+1}^{(2,2)} = X_k^{(2,2)} + \frac{1}{2} \bar{B}_{k,2} (X_k^{(2,2)} + X_{k+1}^{(2,2)}) + \frac{1}{12} \bar{B}_{k,2}^2 (X_k^{(2,2)} - X_{k+1}^{(2,2)}),
\tag{2.37}
$$

$$
X_{k+1}^{(3,3)} = X_k^{(3,3)} + \left(\frac{1}{2} \bar{B}_{k,3} + \frac{1}{120} \bar{B}_{k,3}^3 \right) (X_k^{(3,3)} + X_{k+1}^{(3,3)}) + \frac{1}{10} \bar{B}_{k,3}^2 (X_k^{(3,3)} - X_{k+1}^{(3,3)}),
\tag{2.38}
$$

and

$$
\begin{aligned}
X_{k+1}^{(4,4)} &= X_k^{(4,4)} + \left(\frac{1}{2} \bar{B}_{k,4} + \frac{1}{84} \bar{B}_{k,4}^3 \right) (X_k^{(4,4)} + X_{k+1}^{(4,4)}) \\
&\quad + \left(\frac{3}{28} \bar{B}_{k,4}^2 + \frac{1}{1680} \bar{B}_{k,4}^4 \right) (X_k^{(4,4)} - X_{k+1}^{(4,4)}).
\end{aligned}
\tag{2.39}
$$

Figure 2.2 shows that numerical methods (2.36)–(2.39) are of the first, second, third and forth mean-square orders, respectively, which validates the theorem regarding the mean-square convergence order of proposed methods. In numerical experiments, we take $T = 5$, $p = 1$, $q = 0$ and $h \in \{0.01, 0.02, 0.025, 0.05, 0.1\}$, and the expectation is approximated by taking average over 1000 sample paths. Figure 2.3 gives approximations of a sample phase trajectory of the Kubo oscillator (2.34) simulated by symplectic methods (2.36)–(2.39), as well as the Euler–

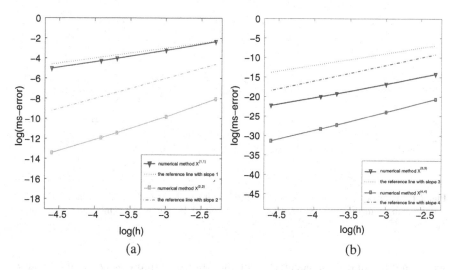

Fig. 2.2 Mean-square convergence orders of methods (**a**) (2.36)–(2.37), (**b**) (2.38)–(2.39)

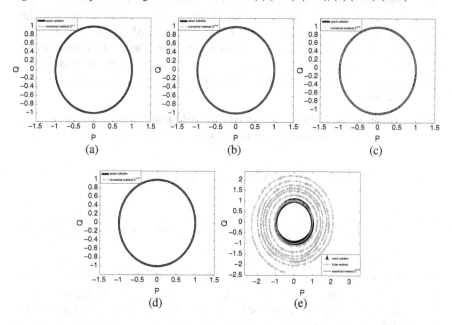

Fig. 2.3 Sample phase trajectories of (**a**) method (2.36), (**b**) method (2.37), (**c**) method (2.38), (**d**) method (2.39), (**e**) Euler–Maruyama method (2.35)

Maruyama method (2.35). It can be seen that when t goes from 0 to 5, the Euler–Maruyama method produces spiral dispersing outside, instead of a closed circle, due to its nonsymplecticity, while the coincidence between the trajectory of the exact solution and those of symplectic methods (2.36)–(2.39) is obvious,

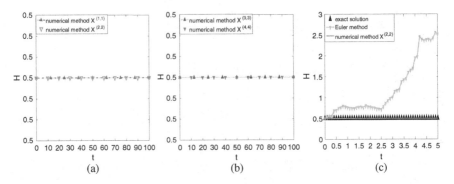

Fig. 2.4 (a) Preservation of the Hamiltonian of methods (2.36)–(2.37), (b) Preservation of the Hamiltonian of methods (2.38)–(2.39), (c) Evolution of Hamiltonian by the Euler–Maruyama method (2.35)

which implies the high accuracy and the ability of preserving circular phase trajectory of the exact solution. Figure 2.4 describes that the Hamiltonian of (2.34) is preserved by symplectic methods (2.36)–(2.39). More precisely, the Hamiltonian $H(P_k, Q_k) = \frac{1}{2}(P_k^2 + Q_k^2)$ with $k = 0, 1, \ldots, N$, produced by (2.36)–(2.39) does not change with the evolution of time. On the contrary, the Euler–Maruyama method (2.35) makes the Hamiltonian to be a nonconservative quantity. We summarize that Figs. 2.2, 2.3 and 2.4 demonstrate the efficiency and superiority of the proposed methods (2.36)–(2.39).

Linear Stochastic Hamiltonian System with Additive Noises Consider the $2n$-dimensional linear stochastic Hamiltonian system with additive noises

$$dX(t) = J_{2n}^{-1} \nabla H_0(X(t))dt + \sum_{r=1}^{m} J_{2n}^{-1} \nabla H_r(X(t)) \circ dW_r(t)$$

$$= J_{2n}^{-1} C_0 X(t)dt + \sum_{r=1}^{m} J_{2n}^{-1} R_r dW_r(t), \quad X(0) = (p^\top, q^\top)^\top, \qquad (2.40)$$

where $H_0(X) = \frac{1}{2} X^\top C_0 X$ with nonnegative and symmetric matrix C_0, and $H_r(X) = \langle R_r, X \rangle$ with $R_r \in \mathbb{R}^{2n}$ for $r \in \{1, \ldots, m\}$. Approximating the matrix exponential $\exp(-h J_{2n} C_0)$ by the Padé approximation $P_{(\hat{r}, \hat{s})}(-h J_{2n} C_0)$ and the Itô integral of the exact solution

$$X(t) = \exp(-t J_{2n} C_0)X(0) + \sum_{r=1}^{m} \int_0^t \exp(-(t-\theta) J_{2n} C_0) J_{2n}^{-1} R_r dW_r(\theta)$$

by the left rectangular formula and the Padé approximation $P_{(1,1)}(-hJ_{2n}C_0)$, we obtain the explicit numerical method

$$
\begin{aligned}
X_{k+1} =& \left(I_{2n} + \sum_{j=1}^{\hat{s}} b_j (hJ_{2n}C_0)^j\right)^{-1} \left(I_{2n} + \sum_{j=1}^{\hat{r}} a_j (-hJ_{2n}C_0)^j\right) X_k \\
&+ \sum_{r=1}^{m} \left(I_{2n} + \frac{1}{2}hJ_{2n}C_0\right)^{-1} \left(I_{2n} - \frac{1}{2}hJ_{2n}C_0\right) J_{2n}^{-1} R_r \Delta_k W_r
\end{aligned}
\tag{2.41}
$$

with $\hat{s}, \hat{r} \geq 1$ and $k = 0, 1, \ldots, N-1$. One can check that (2.41) preserves symplecticity if $\hat{r} = \hat{s}$, and is of mean-square convergence order 1.

Now we perform computations for stochastic linear oscillator (2.22), as a widely used example of (2.40), with initial value $P(0) = 0$ and $Q(0) = 1$. Letting $P_{(\hat{r},\hat{s})}$ in (2.41) be $P_{(1,1)}$, we arrive at

$$
X_{k+1} = \left(I_2 + \frac{h}{2}J_2\right)^{-1} \left(I_2 - \frac{h}{2}J_2\right) X_k + \left(I_2 + \frac{h}{2}J_2\right)^{-1} \left(I_2 - \frac{h}{2}J_2\right) \binom{\alpha}{0} \Delta_k W,
\tag{2.42}
$$

where $k = 0, 1, \ldots, N-1$. It can be seen that the mean-square order of the method (2.42) is 1, as indicated by the reference line of slope 1 in Fig. 2.5a. Figure 2.5b shows the quantity $\mathbb{E}[X_k^\top X_k]$ arising from (2.42) for $k = 0, 1, \ldots, 5000$, where $\mathbb{E}$ is simulated through taking sample average over 500 numerical sample paths. The reference straight line (green-dashed) has slope α^2 with $\alpha = 0.3$, which is equal to

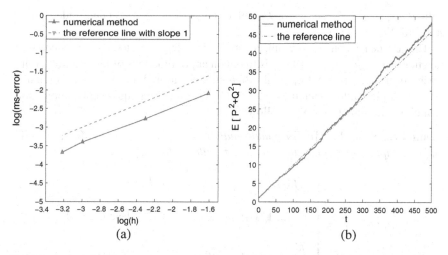

Fig. 2.5 (a) Mean-square convergence order of (2.42), (b) Preservation of the linear growth property of (2.42)

0.09. From Fig. 2.5 it can be observed that (2.42) inherits the linear growth property of the second moment of the solution of (2.22).

Remark 2.2.2 The Padé approximation approaches are proposed here only for the linear stochastic Hamiltonian system. For the nonlinear case, it is also possible to use the Padé approximation to construct efficient numerical methods. To this end, we may employ the Padé approximation to approach the matrix exponential in the solution followed by the variation of constants formula. This provides wider applications of the Padé approximations in the numerical discretizations of the stochastic differential equations.

2.2.2 Symplectic Methods via Generating Function

As mentioned in Sect. 2.1, the symplecticity of the midpoint method can be verified by taking advantage of the stochastic generating function, since the midpoint method can be given by the stochastic type-III generating function. In this section, we adopt the stochastic θ-generating function S_θ, $\theta \in [0, 1]$, to construct stochastic symplectic methods for stochastic Hamiltonian system (2.19)

$$
\begin{cases}
dP(t) = -\dfrac{\partial H_0(P(t), Q(t))}{\partial Q}dt - \displaystyle\sum_{r=1}^{m}\dfrac{\partial H_r(P(t), Q(t))}{\partial Q}\circ dW_r(t), \quad P(0) = p, \\[4mm]
dQ(t) = \dfrac{\partial H_0(P(t), Q(t))}{\partial P}dt + \displaystyle\sum_{r=1}^{m}\dfrac{\partial H_r(P(t), Q(t))}{\partial P}\circ dW_r(t), \qquad Q(0) = q
\end{cases}
$$

with $P, Q, p, q \in \mathbb{R}^n$, $t \in (0, T]$ and $W_r(\cdot)$, $r = 1, \ldots, m$, being independent standard Wiener processes.

The key idea to construct symplectic methods through the stochastic θ-generating function $S_\theta(\hat{P}, \hat{Q}, h)$, $\theta \in [0, 1]$, is to obtain approximations of the solution of the stochastic Hamilton–Jacobi partial differential equation. First, we need to replace every multiple Stratonovich integral in $S_\theta(\hat{P}, \hat{Q}, h)$ with its equivalent combination of multiple Itô integrals via the following lemma.

Lemma 2.2.5 (See [151]) *For a multi-index $\alpha = (j_1, \ldots, j_l)$, $j_i \in \{0, 1, \ldots, m\}$, and an adapted right continuous process f with left-limits,*

$$
I_\alpha[f(\cdot)]_{0,t} := \begin{cases}
f(t), & \text{if } l = 0; \\[3mm]
\displaystyle\int_0^t I_{\alpha-}[f(\cdot)]_{0,s}\,ds, & \text{if } l \geq 1 \text{ and } j_l = 0; \\[3mm]
\displaystyle\int_0^t I_{\alpha-}[f(\cdot)]_{0,s}\,dW_{j_l}(s), & \text{if } l \geq 1 \text{ and } j_l \geq 1.
\end{cases}
$$

Here, $\alpha-$ is the multi-index resulted from discarding the last index of α. Moreover,

$$
J_\alpha := \begin{cases} I_\alpha, & \text{if } l = 1; \\ I_{(j_l)}[J_{\alpha-}] + \chi_{\{j_l = j_{l-1} \neq 0\}} I_{(0)} \left[\dfrac{1}{2} J_{(\alpha-)-} \right], & \text{if } l \geq 2, \end{cases}
$$

where χ_A denotes the indicator function of set A.

Based on the above lemma, we deduce that $I_{(r)}[1]_{0,h} = \int_0^h dW_r(s) = I_{(r)}(h)$, where $r = 1, \ldots, m$, and for any $i \neq j$, $i, j = 1, \ldots, m$,

$$
J_{(0)}(h) = I_{(0)}(h), \quad J_{(i)}(h) = I_{(i)}(h), \quad J_{(0,0)}(h) = I_{(0,0)}(h), \quad J_{(0,i)}(h) = I_{(0,i)}(h),
$$

$$
J_{(i,0)}(h) = I_{(i,0)}(h), \quad J_{(i,j)}(h) = I_{(i,j)}(h), \quad J_{(i,i)}(h) = I_{(i,i)}(h) + \frac{1}{2} I_{(0)}(h).
$$

These relations, together with (1.36), yield

$$
S_\theta(\hat{P}, \hat{Q}, h)
$$

$$
= H_0 I_{(0)}(h) + \sum_{r=1}^{m} H_r I_{(r)}(h) + (2\theta - 1) \sum_{k=1}^{n} \frac{\partial H_0}{\partial \hat{Q}_k} \frac{\partial H_0}{\partial \hat{P}_k} I_{(0,0)}(h)
$$

$$
+ \sum_{r \neq s = 1}^{m} \sum_{k=1}^{n} \left(\theta \frac{\partial H_r}{\partial \hat{Q}_k} \frac{\partial H_s}{\partial \hat{P}_k} + (\theta - 1) \frac{\partial H_r}{\partial \hat{P}_k} \frac{\partial H_s}{\partial \hat{Q}_k} \right) I_{(s,r)}(h)
$$

$$
+ \frac{(2\theta - 1)}{2} \sum_{r=1}^{m} \sum_{k=1}^{n} \frac{\partial H_r}{\partial \hat{Q}_k} \frac{\partial H_r}{\partial \hat{P}_k} I_{(0)}(h) + (2\theta - 1) \sum_{r=1}^{m} \sum_{k=1}^{n} \frac{\partial H_r}{\partial \hat{Q}_k} \frac{\partial H_r}{\partial \hat{P}_k} I_{(r,r)}(h)
$$

$$
+ \sum_{r=1}^{m} \sum_{k=1}^{n} \left(\theta \frac{\partial H_r}{\partial \hat{Q}_k} \frac{\partial H_0}{\partial \hat{P}_k} + (\theta - 1) \frac{\partial H_r}{\partial \hat{P}_k} \frac{\partial H_0}{\partial \hat{Q}_k} \right) I_{(0,r)}(h)
$$

$$
+ \sum_{r=1}^{m} \sum_{k=1}^{n} \left((\theta - 1) \frac{\partial H_0}{\partial \hat{P}_k} \frac{\partial H_r}{\partial \hat{Q}_k} + \theta \frac{\partial H_0}{\partial \hat{Q}_k} \frac{\partial H_r}{\partial \hat{P}_k} \right) I_{(r,0)}(h) + \cdots,
$$

where $(\hat{P}, \hat{Q})$ as the argument of H_r, $r = 0, 1, \ldots, m$, is omitted for notational convenience.

Strongly Convergent Symplectic Methods Using the truncation of $S_\theta(\hat{P}, \hat{Q}, h)$ according to $\mathscr{A}_\gamma := \{\alpha : l(\alpha) + n(\alpha) \leq 2\gamma, \text{ or, } l(\alpha) = n(\alpha) = \gamma + 1/2\}$ and approximating stochastic integrals by bounded random variables, we obtain stochastic symplectic methods that have mean-square order $\gamma \in \mathbb{R}_+$ if coefficients of (2.19) satisfy appropriate conditions. For instance, we deduce an implicit numerical method

$$
P_{k+1} = P_k - \frac{\partial \bar{S}_\theta}{\partial \hat{Q}} \left((1-\theta) P_k + \theta P_{k+1}, (1-\theta) Q_{k+1} + \theta Q_k, h \right),
$$

$$
Q_{k+1} = Q_k + \frac{\partial \bar{S}_\theta}{\partial \hat{P}} \left((1-\theta) P_k + \theta P_{k+1}, (1-\theta) Q_{k+1} + \theta Q_k, h \right) \tag{2.43}
$$

with $k = 0, 1, \ldots, N-1$, when truncating the series S_θ by keeping terms corresponding to multiple Itô integrals I_α with $\alpha \in \mathscr{A}_{\frac{1}{2}}$, i.e.,

$$
\bar{S}_\theta = H_0 h + \sum_{r=1}^{m} H_r \zeta_{rh}^k \sqrt{h} + \frac{(2\theta - 1)}{2} \sum_{r=1}^{m} \sum_{i=1}^{n} \frac{\partial H_r}{\partial \hat{Q}_i} \frac{\partial H_r}{\partial \hat{P}_i} h.
$$

Here, we have used $I_{(0)}(h) = h$, and $I_{(r)}(h) \approx \zeta_{rh} \sqrt{h}$ for $r \in \{1, \ldots, m\}$.

Theorem 2.2.4 (See [199]) *The implicit method (2.43) for stochastic Hamiltonian system (2.19) with condition (1.2) is symplectic and of mean-square order $\frac{1}{2}$.*

Proof The assertion about the convergence order of (2.43) follows from the result in [197], and it suffices to prove the symplecticity. It can be derived that for $k \in \{0, 1, \ldots, N-1\}$,

$$
\theta P_{k+1} + (1-\theta) P_k = P_k - \theta \frac{\partial \bar{S}_\theta}{\partial \hat{Q}} \left((1-\theta) P_k + \theta P_{k+1}, (1-\theta) Q_{k+1} + \theta Q_k, h \right),
$$

$$
Q_{k+1} = (\theta Q_k + (1-\theta) Q_{k+1}) + \theta \frac{\partial \bar{S}_\theta}{\partial \hat{P}} \left((1-\theta) P_k + \theta P_{k+1}, (1-\theta) Q_{k+1} + \theta Q_k, h \right),
$$

which means

$$
P_k^\top d(\theta Q_k + (1-\theta) Q_{k+1}) + Q_{k+1}^\top d(\theta P_{k+1} + (1-\theta) P_k)
$$

$$
= d \left((\theta P_{k+1} + (1-\theta) P_k)^\top (\theta Q_k + (1-\theta) Q_{k+1}) \right)
$$

$$
+ \theta d \bar{S}_\theta \left((1-\theta) P_k + \theta P_{k+1}, (1-\theta) Q_{k+1} + \theta Q_k \right).
$$

Based on Lemma 1.2.1, we deduce that the mapping $(P_k, Q_k) \mapsto (P_{k+1}, Q_{k+1})$ is symplectic for $k \in \{0, 1, \ldots, N-1\}$, which completes the proof.   □

Remark 2.2.3 If $\theta = \frac{1}{2}$, the truncated series

$$
\bar{S}_{\frac{1}{2}} = H_0 \left(\frac{P_k + P_{k+1}}{2}, \frac{Q_{k+1} + Q_k}{2} \right) h + \sum_{r=1}^{m} H_r \left(\frac{P_k + P_{k+1}}{2}, \frac{Q_{k+1} + Q_k}{2} \right) \zeta_{rh}^k \sqrt{h}
$$

with $k = 0, 1, \ldots, N-1$, generates the midpoint method (2.21).

Remark 2.2.4 For the case that $\theta = 1$ and $m = 1$, we obtain

$$\bar{S}_1 = H_1(P_{k+1}, Q_k)\varsigma_{1h}^k\sqrt{h} + \Big(H_0(P_{k+1}, Q_k) + \frac{1}{2}\sum_{i=1}^{n} \frac{\partial H_1(P_{k+1}, Q_k)}{\partial \hat{Q}_i} \frac{\partial H_1(P_{k+1}, Q_k)}{\partial \hat{P}_i}\Big)h,$$

which produces the symplectic Euler method in [199].

Since the complexity of computation increases with the growth of the mean-square order of stochastic symplectic methods, we clarify the construction of higher order methods via the stochastic θ-generating function S_θ, $\theta \in [0, 1]$, in the case of $n = m = 1$. Choosing $\gamma = 1$, the truncated stochastic generating function becomes

$$\tilde{S}_\theta = H_0 h + H_1 \varsigma_{1h}^k \sqrt{h} + \frac{(2\theta - 1)}{2} \frac{\partial H_1}{\partial \hat{Q}} \frac{\partial H_1}{\partial \hat{P}}(\varsigma_{1h}^k)^2 h, \quad k = 0, 1, \ldots, N - 1,$$

which engenders the following symplectic method of mean-square order 1:

$$P_{k+1} = P_k - \frac{\partial \tilde{S}_\theta}{\partial \hat{Q}}\left((1 - \theta)P_k + \theta P_{k+1}, (1 - \theta)Q_{k+1} + \theta Q_k, h\right),$$

$$Q_{k+1} = Q_k + \frac{\partial \tilde{S}_\theta}{\partial \hat{P}}\left((1 - \theta)P_k + \theta P_{k+1}, (1 - \theta)Q_{k+1} + \theta Q_k, h\right).$$

(2.44)

Taking advantage of the following relations

$$J_{(1,1,0)}(h) = I_{(1,1,0)}(h) + \frac{1}{2}I_{(0,0)}(h), \quad J_{(0,1,1)}(h) = I_{(0,1,1)}(h) + \frac{1}{2}I_{(0,0)}(h),$$

$$J_{(1,1,1)}(h) = I_{(1,1,1)}(h) + \frac{1}{2}\left(I_{(0,1)}(h) + I_{(1,0)}(h)\right),$$

$$J_{(1,1,1,1)}(h) = I_{(1,1,1,1)}(h) + \frac{1}{2}\left(I_{(0,1,1)}(h) + I_{(1,0,1)}(h) + I_{(1,1,0)}(h)\right) + \frac{1}{4}I_{(0,0)}(h),$$

we have the truncation of S_θ corresponding to $\mathscr{A}_{\frac{3}{2}}$ which takes the form

$$\hat{S}_\theta\left((1 - \theta)P_k + \theta P_{k+1}, (1 - \theta)Q_{k+1} + \theta Q_k, h\right)$$

$$= G_{(1)}^\theta I_{(1)}(h) + \Big(G_{(0)}^\theta + \frac{1}{2}G_{(1,1)}^\theta\Big)I_{(0)}(h)$$

$$+ \Big(G_{(0,1)}^\theta + \frac{1}{2}G_{(1,1,1)}^\theta\Big)I_{(0,1)}(h) + \Big(G_{(1,0)}^\theta + \frac{1}{2}G_{(1,1,1)}^\theta\Big)I_{(1,0)}(h)$$

$$+ G_{(1,1)}^\theta I_{(1,1)}(h) + \Big(G_{(0,0)}^\theta + \frac{1}{2}\Big(G_{(0,1,1)}^\theta + G_{(1,1,0)}^\theta\Big) + \frac{1}{4}G_{(1,1,1,1)}^\theta\Big)I_{(0,0)}(h),$$

where the coefficients G^θ_α, $\theta \in [0, 1]$, are given by the formula (1.34). The symplectic method of mean-square order $\frac{3}{2}$ can be produced by $\hat{S}_\theta$ with multiple Itô integrals $I_{(0,1)}(h)$, $I_{(1,0)}(h)$ and $I_{(1,1)}(h)$ approximated by bounded random variables as in (2.10).

Remark 2.2.5 For special types of stochastic Hamiltonian systems, such as stochastic Hamiltonian systems with additive noises, stochastic Hamiltonian systems with separable Hamiltonian functions, and stochastic Hamiltonian systems preserving the Hamiltonian functions, the mean-square convergence order of above proposed methods will be higher.

Weakly Convergent Symplectic Methods Now we present a systematic approach of constructing stochastic symplectic methods in the weak convergent sense for the stochastic Hamiltonian system (2.19) based on the stochastic θ-generating function S_θ with $\theta \in [0, 1]$. Analogous to the previous case, the multiple Stratonovich integrals in S_θ should be substituted by the equivalent combination of multiple Itô integrals first. By truncating the series to include only Itô integrals with multi-index α satisfying $l(\alpha) \le \gamma$, we could obtain a weakly convergent symplectic method of order γ.

For the case of $\gamma = 1$, it can be verified that the stochastic θ-generating function S_θ, $\theta \in [0, 1]$, can be approached by

$$\bar{S}_\theta \left((1 - \theta) P_k + \theta P_{k+1}, (1 - \theta) Q_{k+1} + \theta Q_k, h \right)$$

$$= H_0 h + \sum_{r=1}^{m} H_r I_{(r)}(h) + \frac{(2\theta - 1)}{2} \sum_{r=1}^{m} \sum_{i=1}^{n} \frac{\partial H_r}{\partial \hat{Q}_i} \frac{\partial H_r}{\partial \hat{P}_i} h, \quad k = 0, 1, \ldots, N - 1,$$

where the argument of H_r, $r = 0, 1, \ldots, m$, is $((1 - \theta) P_k + \theta P_{k+1}, (1 - \theta) Q_{k+1} + \theta Q_k)$. The increment of the Wiener process $I_{(r)}(h)$ can be simulated by the mutually independent two-point distributed random variables $\sqrt{h} \varsigma_r$ with the law

$$\mathbb{P}(\varsigma_r = \pm 1) = \frac{1}{2}, \quad r = 1, \ldots, m,$$

and then we get the symplectic method of weak order 1

$$P_{k+1} = P_k - \frac{\partial \bar{S}_\theta}{\partial \hat{Q}} \left((1 - \theta) P_k + \theta P_{k+1}, (1 - \theta) Q_{k+1} + \theta Q_k, h \right),$$

$$Q_{k+1} = Q_k + \frac{\partial \bar{S}_\theta}{\partial \hat{P}} \left((1 - \theta) P_k + \theta P_{k+1}, (1 - \theta) Q_{k+1} + \theta Q_k, h \right).$$

To obtain a symplectic method of weak order 2, we use the truncation of S_θ as follows

$$\tilde{S}_\theta \left((1-\theta)P_k + \theta P_{k+1}, (1-\theta)Q_{k+1} + \theta Q_k, h\right)$$

$$= \left(G_{(0)}^\theta + \frac{1}{2}\sum_{r=1}^m G_{(r,r)}^\theta\right)h + \sum_{r=1}^m G_{(r)}^\theta I_{(r)}(h) + \sum_{j,r=1}^m G_{(j,r)}^\theta I_{(j,r)}(h)$$

$$+ \left(G_{(0,0)}^\theta + \frac{1}{2}\sum_{r=1}^m \left(G_{(r,r,0)}^\theta + G_{(0,r,r)}^\theta\right) + \frac{1}{4}\sum_{r,j=1}^m G_{(r,r,j,j)}^\theta\right)I_{(0,0)}(h)$$

$$+ \sum_{r=1}^m \left(\left(G_{(0,r)}^\theta + \frac{1}{2}\sum_{j=1}^m G_{(j,j,r)}^\theta\right)I_{(0,r)}(h) + \left(G_{(r,0)}^\theta + \frac{1}{2}\sum_{j=1}^m G_{(r,j,j)}^\theta\right)I_{(r,0)}(h)\right),$$

where $k = 0, 1, \ldots, N-1$. Simulating the above multiple Itô stochastic integrals as described in [151], we get the approximation

$$\tilde{S}_\theta \left((1-\theta)P_k + \theta P_{k+1}, (1-\theta)Q_{k+1} + \theta Q_k, h\right)$$

$$\approx \left(G_{(0)}^\theta + \frac{1}{2}\sum_{r=1}^m G_{(r,r)}^\theta\right)h + \sum_{r=1}^m G_{(r)}^\theta \sqrt{h}\varsigma_r^k + \frac{h}{2}\sum_{j,r=1}^m G_{(j,r)}^\theta \left(\varsigma_j^k \varsigma_r^k + \varsigma_{j,r}^k\right)$$

$$+ \left(G_{(0,0)}^\theta + \frac{1}{2}\sum_{r=1}^m \left(G_{(r,r,0)}^\theta + G_{(0,r,r)}^\theta\right) + \frac{1}{4}\sum_{r,j=1}^m G_{(r,r,j,j)}^\theta\right)\frac{h^2}{2}$$

$$+ \frac{1}{2}\sum_{r=1}^m \left(G_{(0,r)}^\theta + G_{(r,0)}^\theta + \frac{1}{2}\sum_{j=1}^m \left(G_{(r,j,j)}^\theta + G_{(j,j,r)}^\theta\right)\right)h^{\frac{3}{2}}\varsigma_r^k$$

$$\tag{2.45}$$

with $k = 0, 1, \ldots, N-1$. Here, $\varsigma_r^k, \varsigma_{j,r}^k$ for $j, r = 1, \ldots, m$, and $k = 0, 1, \ldots, N-1$, are mutually independent random variables with the following discrete distributions

$$\mathbb{P}\left(\varsigma_r^k = \pm\sqrt{3}\right) = \frac{1}{6}, \quad \mathbb{P}\left(\varsigma_r^k = 0\right) = \frac{2}{3},$$

and $\varsigma_{j_1,j_1}^k = -1, j_1 = 1, \ldots, m,$

$$\mathbb{P}\left(\varsigma_{j_1,j_2}^k = \pm 1\right) = \frac{1}{2}, j_2 = 1, \ldots, j_1 - 1, \quad \varsigma_{j_1,j_2}^k = -\varsigma_{j_2,j_1}^k, j_2 = j_1 + 1, \ldots, m.$$

Theorem 2.2.5 *The implicit method*

$$P_{k+1} = P_k - \frac{\partial \tilde{S}_\theta}{\partial \hat{Q}} \left((1-\theta)P_k + \theta P_{k+1}, (1-\theta)Q_{k+1} + \theta Q_k, h \right),$$

$$Q_{k+1} = Q_k + \frac{\partial \tilde{S}_\theta}{\partial \hat{P}} \left((1-\theta)P_k + \theta P_{k+1}, (1-\theta)Q_{k+1} + \theta Q_k, h \right),$$

based on $\tilde{S}_\theta$ given in (2.45) for the stochastic Hamiltonian system (2.19) under condition (1.2) is symplectic and of weak order 2.

Proof The proof can be finished by utilizing arguments similar to those in [13, Theorem 4.4].   □

Remark 2.2.6 The authors in [12] also take advantage of special properties of the stochastic Hamiltonian system preserving Hamiltonians to propose efficient symplectic methods via generating functions.

2.2.3 Stochastic Galerkin Variational Integrators

Instead of approaching the solution to the stochastic Hamilton–Jacobi partial differential equation, [122] investigates the generalized Galerkin framework (see e.g., [171, 187, 208, 209] and references therein) for constructing approximations of the stochastic generating function by exploiting the variational characterization, and provides a general class of stochastic symplectic methods called stochastic Galerkin variational integrators.

The stochastic Galerkin variational integrator is constructed by choosing a finite dimensional subspace of $\mathbf{C}([0,T])$ and a quadrature rule for approximating the integrals in the action functional

$$\bar{\mathbb{S}}(P(\cdot), Q(\cdot)) = (P(T))^\top Q(T) - \mathbb{S}(P(\cdot), Q(\cdot))$$

$$= (P(T))^\top Q(T) - \int_0^T \left((P(t))^\top \circ dQ(t) - H_0(P(t), Q(t))dt \right.$$

$$\left. - \sum_{r=1}^m H_r(P(t), Q(t)) \circ dW_r(t) \right).$$

In order to determine the discrete curve $\{(P_k, Q_k)\}_{k=0,1,\ldots,N}$ that approximates the exact solution, we need to construct an approximation on each interval $[t_k, t_{k+1}]$ for $k \in \{0, 1, \ldots, N-1\}$. Let us consider the space defined as

$$\mathbf{C}_s\left([t_k, t_{k+1}]\right) := \{(P(\cdot), Q(\cdot)) \in \mathbf{C}\left([t_k, t_{k+1}]\right) \mid Q(\cdot) \text{ is a polynomial of degree } s\}.$$

For convenience, we express $Q(t)$ by means of Lagrange polynomials. Consider control points $0 = c_0 < c_1 < \cdots < c_s = 1$ and denote the corresponding Lagrange polynomials of degree s by

$$\ell_{u,s}(\tau) := \prod_{\substack{0 \le v \le s \\ v \ne u}} \frac{\tau - c_v}{c_u - c_v},$$

which satisfies $\ell_{u,s}(c_v) = \chi_{\{u=v\}}$. Then a polynomial trajectory $Q_d(t)$ can be given by

$$Q_d(t_k + \tau h) = \sum_{u=0}^{s} Q^u \ell_{u,s}(\tau), \quad \dot{Q}_d(t_k + \tau h) = \frac{1}{h} \sum_{u=0}^{s} Q^u \dot{\ell}_{u,s}(\tau),$$

where $Q^u = Q_d(t_k + c_u h)$ for $u = 0, 1, \ldots, s$, are control values, $\dot{Q}_d$ denotes the time derivative of Q_d, and $\dot{\ell}_{u,s}$ is the derivative of the Lagrange polynomial $\ell_{u,s}$. The restriction of the action functional $\bar{\mathbb{S}}(P(\cdot), Q(\cdot))$ to the space $\mathbf{C}_s([t_k, t_{k+1}])$ is

$$\bar{\mathbb{S}}^s(P(\cdot), Q_d(\cdot)) = (P(t_{k+1}))^\top Q^s - \int_{t_k}^{t_{k+1}} \left((P(t))^\top \dot{Q}_d(t) - H_0(P(t), Q_d(t)) \right) dt$$

$$+ \sum_{r=1}^{m} \int_{t_k}^{t_{k+1}} H_r(P(t), Q_d(t)) \circ dW_r(t). \tag{2.46}$$

Inspired by the ideas given in [35, 40, 199, 200], we approximate the integrals in (2.46) via numerical quadrature rules $(\beta_{r,i}, \gamma_i)_{i=1}^{s'}$, $r = 0, 1, \ldots, m$, where $0 \le \gamma_1 < \cdots < \gamma_{s'} \le 1$ are quadrature points, and $\beta_{r,i}$ is the corresponding weight, and we assume that for each i, there exists at least an $r \in \{0, 1, \ldots, m\}$ such that $\beta_{r,i} \ne 0$. These yield the discrete action functional

$$\hat{\mathbb{S}}^s(P(\cdot), Q_d(\cdot))$$

$$= (P(t_{k+1}))^\top Q^s - \sum_{i=1}^{s'} \beta_{0,i} \left((P(t_k + \gamma_i h))^\top \dot{Q}_d(t_k + \gamma_i h) - H_0\left(P(t_k + \gamma_i h), Q_d(t_k + \gamma_i h)\right) \right) h$$

$$+ \sum_{r=1}^{m} \sum_{i=1}^{s'} \beta_{r,i} H_r\left(P(t_k + \gamma_i h), Q_d(t_k + \gamma_i h)\right) \Delta_k W_r,$$

where $\Delta_k W_r = W_r(t_{k+1}) - W_r(t_k)$. Now we define

$$\bar{S}^s(P_{k+1}, Q_k)$$

$$:= \text{ext}\left\{ (P_{k+1})^\top Q^s - \sum_{i=1}^{s'} \beta_{0,i} \left((P^i)^\top \dot{Q}_d(t_k + \gamma_i h) - H_0\left(P^i, Q_d(t_k + \gamma_i h)\right) \right) h \right.$$

$$+ \sum_{r=1}^{m} \sum_{i=1}^{s'} \beta_{r,i} H_r \left(P^i, Q_d(t_k + \gamma_i h) \right) \Delta_k W_r \ \Big| \ Q^1, \dots, Q^s \in \mathbb{R}^n,$$

$$\tag{2.47}$$

$$P^1, \dots, P^{s'} \in \mathbb{R}^n, \ Q^0 = Q_k \Big\}$$

with $P^i := P(t_k + \gamma_i h)$, which generates an implicit numerical method

$$Q_{k+1} = \frac{\partial \bar{S}^s}{\partial P_{k+1}} (P_{k+1}, Q_k), \qquad P_k = \frac{\partial \bar{S}^s}{\partial Q_k} (P_{k+1}, Q_k). \tag{2.48}$$

The formulae (2.47) and (2.48) can be written together as the following system

$$-P_k = \sum_{i=1}^{s'} \beta_{0,i} \left(P^i \ell_{0,s} (\gamma_i) - \frac{\partial H_0}{\partial Q} (t_k + \gamma_i h) \ell_{0,s} (\gamma_i) h \right)$$

$$- \sum_{r=1}^{m} \sum_{i=1}^{s'} \beta_{r,i} \frac{\partial H_r}{\partial Q} (t_k + \gamma_i h) \ell_{0,s} (\gamma_i) \Delta_k W_r,$$

$$0 = \sum_{i=1}^{s'} \beta_{0,i} \left(P^i \ell_{u,s} (\gamma_i) - \frac{\partial H_0}{\partial Q} (t_k + \gamma_i h) \ell_{u,s} (\gamma_i) h \right)$$

$$- \sum_{r=1}^{m} \sum_{i=1}^{s'} \beta_{r,i} \frac{\partial H_r}{\partial Q} (t_k + \gamma_i h) \ell_{u,s} (\gamma_i) \Delta_k W_r, \tag{2.49}$$

$$P_{k+1} = \sum_{i=1}^{s'} \beta_{0,i} \left(P^i \ell_{s,s} (\gamma_i) - \frac{\partial H_0}{\partial Q} (t_k + \gamma_i h) \ell_{s,s} (\gamma_i) h \right)$$

$$- \sum_{r=1}^{m} \sum_{i=1}^{s'} \beta_{r,i} \frac{\partial H_r}{\partial Q} (t_k + \gamma_i h) \ell_{s,s} (\gamma_i) \Delta_k W_r,$$

$$0 = - \beta_{0,i} \dot{Q}_d (t_k + \gamma_i h) + \beta_{0,i} \frac{\partial H_0}{\partial P} (t_k + \gamma_i h) + \sum_{r=1}^{m} \beta_{r,i} \frac{\partial H_r}{\partial P} (t_k + \gamma_i h) \frac{\Delta_k W_r}{h},$$

$$Q_{k+1} = Q^s,$$

where $u = 1, \dots, s-1$, $i = 1, \dots, s'$, and for brevity we introduce the notation

$$H_r (t_k + \gamma_i h) = H_r (P(t_k + \gamma_i h), Q_d(t_k + \gamma_i h)), \quad r = 0, 1, \dots, m.$$

When $H_r = 0$, $r = 1, \dots, m$, then (2.49) reduces to the deterministic Galerkin variational integrator discussed in [209]. Notice that depending on the choice of Hamiltonians and numerical quadrature rules, the system (2.49) may be explicit, but in general it is implicit. One should define $\Delta_k \widehat{W}_r = \sqrt{h} \zeta_{rh}^k$ as follows

$$
\zeta_{rh}^k = \begin{cases}
\xi_r^k, & \text{if } |\xi_r^k| \le A_h; \\
A_h, & \text{if } \xi_r^k > A_h; \\
-A_h, & \text{if } \xi_r^k < -A_h
\end{cases}
$$

with $A_h := \sqrt{2l |\log h|}$, where l is suitably chosen for the considered problem. The main difficulty lies in the choice of parameters s, s', $\beta_{r,i}$, γ_i, $r = 0, 1, \ldots, m$, $i = 1, \ldots, s'$, so that the resulting numerical method converges to the solution of (2.19) in certain sense. The numbers of parameters and order conditions will grow rapidly, if we add the terms approximating multiple Stratonovich integrals. Below provide concrete choices of coefficients that lead to convergent methods when $m = 1$ in [122].

1. Midpoint method:
   Applying the midpoint rule ($s' = 1$, $\gamma_1 = \frac{1}{2}$, $\beta_{0,1} = \beta_{1,1} = 1$) and letting $s = 1$ give the implicit method

$$
P_{k+1} = P_k - \frac{\partial H_0}{\partial Q} \left(\frac{P_k + P_{k+1}}{2}, \frac{Q_k + Q_{k+1}}{2} \right) h - \frac{\partial H_1}{\partial Q} \left(\frac{P_k + P_{k+1}}{2}, \frac{Q_k + Q_{k+1}}{2} \right) \Delta_k \widehat{W},
$$

$$
Q_{k+1} = Q_k + \frac{\partial H_0}{\partial P} \left(\frac{P_k + P_{k+1}}{2}, \frac{Q_k + Q_{k+1}}{2} \right) h + \frac{\partial H_1}{\partial P} \left(\frac{P_k + P_{k+1}}{2}, \frac{Q_k + Q_{k+1}}{2} \right) \Delta_k \widehat{W}.
$$

   If Hamiltonians are separable, that is, $H_0(P, Q) = T_0(P) + U_0(Q)$ and $H_1(P, Q) = T_1(P) + U_1(Q)$, then P_{k+1} from the first equation can be substituted into the second one, and only n nonlinear equations need to be solved for Q_{k+1}.

2. Stochastic Störmer–Verlet method:
   If we adopt the trapezoidal rule ($s' = 2$, $\gamma_1 = 0$, $\gamma_2 = 1$, $\beta_{0,1} = \beta_{0,2} = \frac{1}{2}$, $\beta_{1,1} = \beta_{1,2} = \frac{1}{2}$) and let $s = 2$, we attain a stochastic generalization of the Störmer–Verlet method (see [116]) as follows

$$
P^1 = P_k - \frac{1}{2} \frac{\partial H_0}{\partial Q} (P^1, Q_k) h - \frac{1}{2} \frac{\partial H_1}{\partial Q} (P^1, Q_k) \Delta_k \widehat{W},
$$

$$
Q_{k+1} = Q_k + \frac{1}{2} \frac{\partial H_0}{\partial P} (P^1, Q_k) h + \frac{1}{2} \frac{\partial H_0}{\partial P} (P^1, Q_{k+1}) h
$$

$$
+ \frac{1}{2} \frac{\partial H_1}{\partial P} (P^1, Q_k) \Delta_k \widehat{W} + \frac{1}{2} \frac{\partial H_1}{\partial P} (P^1, Q_{k+1}) \Delta_k \widehat{W},
$$

$$
P_{k+1} = P^1 - \frac{1}{2} \frac{\partial H_0}{\partial Q} (P^1, Q_{k+1}) h - \frac{1}{2} \frac{\partial H_1}{\partial Q} (P^1, Q_{k+1}) \Delta_k \widehat{W}.
$$

It is particularly efficient, since the first equation can be solved separately from the second one, and the last equation is an explicit update. Moreover, when the Hamiltonians are separable, this method is explicit.

3. Stochastic trapezoidal method:
   This method is based on polynomials of degree $s = 1$ with control points $c_0 = 0$ and $c_1 = 1$, and the trapezoidal rule. The system (2.49) reads

$$P_k = \frac{1}{2}\left(P^1 + P^2\right) + \frac{1}{2}\frac{\partial H_0}{\partial Q}(P^1, Q_k)h + \frac{1}{2}\frac{\partial H_1}{\partial Q}(P^1, Q_k)\Delta_k\widehat{W},$$

$$P_{k+1} = \frac{1}{2}\left(P^1 + P^2\right) - \frac{1}{2}\frac{\partial H_0}{\partial Q}(P^2, Q_{k+1})h - \frac{1}{2}\frac{\partial H_1}{\partial Q}(P^2, Q_{k+1})\Delta_k\widehat{W},$$

$$Q_{k+1} = Q_k + \frac{\partial H_0}{\partial P}(P^1, Q_k)h + \frac{\partial H_1}{\partial P}(P^1, Q_k)\Delta_k\widehat{W},$$

$$Q_{k+1} = Q_k + \frac{\partial H_0}{\partial P}(P^2, Q_{k+1})h + \frac{\partial H_1}{\partial P}(P^2, Q_{k+1})\Delta_k\widehat{W},$$

which is a stochastic generalization of the trapezoidal method for deterministic systems (see [187]). One can easily verify that if Hamiltonians are separable, then $P^1 = P^2$ and stochastic trapezoidal method are equivalent to the stochastic Störmer–Verlet method.

4. If we apply Simpson's rule ($s' = 3$, $\gamma_1 = 0$, $\gamma_2 = \frac{1}{2}$, $\gamma_3 = 1$, $\beta_{0,1} = \beta_{1,1} = \frac{1}{6}$, $\beta_{0,2} = \beta_{1,2} = \frac{2}{3}$, $\beta_{0,3} = \beta_{1,3} = \frac{1}{6}$), and let $H_0(P, Q) = T_0(P) + U_0(Q)$ and $H_1(P, Q) = T_1(P) + U_1(Q)$, then $P^1 = P^2 = P^3$ and the numerical method takes the form

$$Q_{k+1} = Q_k + \frac{\partial T_0}{\partial P}(P^1)h + \frac{\partial T_1}{\partial p}(P^1)\Delta_k\widehat{W},$$

$$P_{k+1} = P^1 - \frac{1}{3}\frac{\partial U_0}{\partial Q}\left(\frac{Q_k + Q_{k+1}}{2}\right)h - \frac{1}{6}\frac{\partial U_0}{\partial Q}(Q_{k+1})h$$

$$- \frac{1}{3}\frac{\partial U_1}{\partial Q}\left(\frac{Q_k + Q_{k+1}}{2}\right)\Delta_k\widehat{W} - \frac{1}{6}\frac{\partial U_1}{\partial Q}(Q_{k+1})\Delta_k\widehat{W},$$

where

$$P^1 = P_k - \frac{1}{6}\frac{\partial U_0}{\partial Q}(Q_k)h - \frac{1}{3}\frac{\partial U_0}{\partial Q}\left(\frac{Q_k + Q_{k+1}}{2}\right)h$$

$$- \frac{1}{6}\frac{\partial U_1}{\partial Q}(Q_k)\Delta_k\widehat{W} - \frac{1}{3}\frac{\partial U_1}{\partial Q}\left(\frac{Q_k + Q_{k+1}}{2}\right)\Delta_k\widehat{W}.$$

Theorem 2.2.6 (See [122]) *The numerical method defined by $\bar{S}^s$ as in (2.48) for (2.19) is symplectic, i.e.,*

$$dP_{k+1} \wedge dQ_{k+1} = dP_k \wedge dQ_k, \quad a.s.,$$

where $k = 0, 1, \ldots, N - 1$.

Proof The proof follows immediately by observing that for any $k \in \{0, 1, \ldots, N - 1\}$,

$$0 = \mathrm{d}^2 \bar{S}^s (P_{k+1}, Q_k) = -\mathrm{d}P_{k+1} \wedge \mathrm{d}Q_{k+1} + \mathrm{d}P_k \wedge \mathrm{d}Q_k,$$

based on (2.48). □

2.2.4 Stochastic (Rough) Symplectic Runge–Kutta Methods

Up to now, a lot of stochastic symplectic methods have been discussed, and most of them can be rewritten as the form of stochastic Runge–Kutta methods, which avoid simulating higher order partial derivatives of coefficients. In this subsection, we will introduce the stochastic symplectic Runge–Kutta method and rough symplectic Runge–Kutta method systemically, and refer to [8, 130, 180, 199, 200, 265] for more details.

Stochastic Symplectic Runge–Kutta Methods For the $2n$-dimensional stochastic Hamiltonian system with additive noises

$$\begin{cases} dP(t) = -\dfrac{\partial H(P(t), Q(t))}{\partial Q}dt - \displaystyle\sum_{r=1}^{m} \sigma_r dW_r(t), \quad P(0) = p, \\[4mm] dQ(t) = \dfrac{\partial H(P(t), Q(t))}{\partial P}dt + \displaystyle\sum_{r=1}^{m} \eta_r dW_r(t), \quad Q(0) = q \end{cases}$$

with σ_r, η_r, $r = 1, \ldots, m$, being constant column vectors, [200] proposes the following s-stage stochastic symplectic Runge–Kutta method

$$p_i = P_k - \sum_{j=1}^{s} A_{ij} \frac{\partial H}{\partial Q}(p_j, q_j)h + \varphi_i,$$

$$q_i = Q_k + \sum_{j=1}^{s} A_{ij} \frac{\partial H}{\partial P}(p_j, q_j)h + \psi_i,$$

$$P_{k+1} = P_k - \sum_{i=1}^{s} b_i \frac{\partial H}{\partial Q}(p_i, q_i)h + \eta,$$

$$Q_{k+1} = Q_k + \sum_{i=1}^{s} b_i \frac{\partial H}{\partial P}(p_i, q_i)h + \zeta,$$

where $k \in \{0, 1, \ldots, N-1\}$, φ_i, ψ_i, η, ζ do not depend on P and Q, and parameters A_{ij} and b_i satisfy the conditions

$$b_i A_{ij} + b_j A_{ji} - b_i b_j = 0, \quad i, j = 1, \ldots, s.$$

In fact, stochastic symplectic Runge–Kutta methods can be generalized to the multiplicative noise case (see [180]). For the sake of simplicity, we discuss the stochastic Hamiltonian system with $m = 1$

$$\begin{cases} dP(t) = -\dfrac{\partial H_0(P(t), Q(t))}{\partial Q} dt - \dfrac{\partial H_1(P(t), Q(t))}{\partial Q} \circ dW(t), \quad P(0) = p, \\[3mm] dQ(t) = \dfrac{\partial H_0(P(t), Q(t))}{\partial P} dt + \dfrac{\partial H_1(P(t), Q(t))}{\partial P} \circ dW(t), \quad Q(0) = q, \end{cases}$$

and consider the associated s-stage stochastic Runge–Kutta method

$$p_i = P_k - \sum_{j=1}^{s} A_{ij}^{(0)} \frac{\partial H_0}{\partial Q}(p_j, q_j) h - \sum_{j=1}^{s} A_{ij}^{(1)} \frac{\partial H_1}{\partial Q}(p_j, q_j) \Delta_k \widehat{W},$$

$$q_i = Q_k + \sum_{j=1}^{s} A_{ij}^{(0)} \frac{\partial H_0}{\partial P}(p_j, q_j) h + \sum_{j=1}^{s} A_{ij}^{(1)} \frac{\partial H_1}{\partial P}(p_j, q_j) \Delta_k \widehat{W},$$

$$\tag{2.50}$$

$$P_{k+1} = P_k - \sum_{i=1}^{s} b_i^{(0)} \frac{\partial H_0}{\partial Q}(p_i, q_i) h - \sum_{i=1}^{s} b_i^{(1)} \frac{\partial H_1}{\partial Q}(p_i, q_i) \Delta_k \widehat{W},$$

$$Q_{k+1} = Q_k + \sum_{i=1}^{s} b_i^{(0)} \frac{\partial H_0}{\partial P}(p_i, q_i) h + \sum_{i=1}^{s} b_i^{(1)} \frac{\partial H_1}{\partial P}(p_i, q_i) \Delta_k \widehat{W}$$

for $i = 1, \ldots, s$, and $k = 0, 1, \ldots, N - 1$, where $A^{(0)} = \left(A_{ij}^{(0)}\right)$ and $A^{(1)} = \left(A_{ij}^{(1)}\right)$ are $s \times s$ matrices of real numbers, and $b^{(0)} = \left(b_1^{(0)}, \ldots, b_s^{(0)}\right)^{\top}$ and $b^{(1)} = \left(b_1^{(1)}, \ldots, b_s^{(1)}\right)^{\top}$ are s-dimensional column vectors. It can be seen that the above stochastic Runge–Kutta method is corresponding to the Butcher tableau as follows

$$\begin{array}{c|cc} & A^{(0)} & A^{(1)} \\ \hline & (b^{(0)})^{\top} & (b^{(1)})^{\top} \end{array}.$$

Theorem 2.2.7 (See [180]) *Assume that the coefficients* $A_{ij}^{(0)}$, $A_{ij}^{(1)}$, $b_i^{(0)}$, $b_i^{(1)}$ *satisfy the relations*

$$b_i^{(0)}b_j^{(0)} - b_i^{(0)}A_{ij}^{(0)} - b_j^{(0)}A_{ji}^{(0)} = 0,$$

$$b_i^{(1)}b_j^{(0)} - b_i^{(1)}A_{ij}^{(0)} - b_j^{(0)}A_{ji}^{(1)} = 0, \qquad (2.51)$$

$$b_i^{(1)}b_j^{(1)} - b_i^{(1)}A_{ij}^{(1)} - b_j^{(1)}A_{ji}^{(1)} = 0$$

for $i, j = 1, \ldots, s$. *Then the stochastic Runge–Kutta method* (2.50) *preserves the symplectic structure almost surely, i.e.,*

$$dP_{k+1} \wedge dQ_{k+1} = dP_k \wedge dQ_k, \quad k = 0, 1, \ldots, N - 1.$$

Proof Introducing the temporary notations

$$f_i = \frac{\partial H_0}{\partial Q}(p_i, q_i), \quad g_i = \frac{\partial H_0}{\partial P}(p_i, q_i), \quad \sigma_i = \frac{\partial H_1}{\partial Q}(p_i, q_i), \quad \eta_i = \frac{\partial H_1}{\partial P}(p_i, q_i)$$

for $i = 1, \ldots, s$, we get

$$dP_{k+1} = dP_k - h\sum_{j=1}^{s} b_j^{(0)} df_j - \Delta_k \widehat{W} \sum_{j=1}^{s} b_j^{(1)} d\sigma_j,$$

$$dQ_{k+1} = dQ_k + h\sum_{j=1}^{s} b_j^{(0)} dg_j + \Delta_k \widehat{W} \sum_{j=1}^{s} b_j^{(1)} d\eta_j.$$

Taking the wedge product leads to

$$dP_{k+1} \wedge dQ_{k+1}$$

$$= dP_k \wedge dQ_k + h\sum_{j=1}^{s} b_j^{(0)} dP_k \wedge dg_j + \Delta_k \widehat{W} \sum_{j=1}^{s} b_j^{(1)} dP_k \wedge d\eta_j$$

$$- h\sum_{j=1}^{s} b_j^{(0)} df_j \wedge dQ_k - h^2 \sum_{i,j=1}^{s} b_i^{(0)} b_j^{(0)} df_i \wedge dg_j - \Delta_k \widehat{W} h \sum_{i,j=1}^{s} b_i^{(0)} b_j^{(1)} df_i \wedge d\eta_j$$

$$- \Delta_k \widehat{W} \sum_{j=1}^{s} b_j^{(1)} d\sigma_j \wedge dQ_k - \Delta_k \widehat{W} h \sum_{i,j=1}^{s} b_i^{(1)} b_j^{(0)} d\sigma_i \wedge dg_j$$

$$- (\Delta_k \widehat{W})^2 \sum_{i,j=1}^{s} b_i^{(1)} b_j^{(1)} d\sigma_i \wedge d\eta_j.$$

By means of following relations

$$dp_i = dP_k - h \sum_{j=1}^{s} A_{ij}^{(0)} df_j - \Delta_k \widehat{W} \sum_{j=1}^{s} A_{ij}^{(1)} d\sigma_j,$$

$$dq_i = dQ_k + h \sum_{j=1}^{s} A_{ij}^{(0)} dg_j + \Delta_k \widehat{W} \sum_{j=1}^{s} A_{ij}^{(1)} d\eta_j$$

for $i = 1, \ldots, s$, we deduce

$$dP_k \wedge dg_i = dp_i \wedge dg_i + h \sum_{j=1}^{s} A_{ij}^{(0)} df_j \wedge dg_i + \Delta_k \widehat{W} \sum_{j=1}^{s} A_{ij}^{(1)} d\sigma_j \wedge dg_i,$$

$$dP_k \wedge d\eta_i = dp_i \wedge d\eta_i + h \sum_{j=1}^{s} A_{ij}^{(0)} df_j \wedge d\eta_i + \Delta_k \widehat{W} \sum_{j=1}^{s} A_{ij}^{(1)} d\sigma_j \wedge d\eta_i,$$

$$df_i \wedge dQ_k = df_i \wedge dq_i - h \sum_{j=1}^{s} A_{ij}^{(0)} df_i \wedge dg_j - \Delta_k \widehat{W} \sum_{j=1}^{s} A_{ij}^{(1)} df_i \wedge d\eta_j,$$

$$d\sigma_i \wedge dQ_k = d\sigma_i \wedge dq_i - h \sum_{j=1}^{s} A_{ij}^{(0)} d\sigma_i \wedge dg_j - \Delta_k \widehat{W} \sum_{j=1}^{s} A_{ij}^{(1)} d\sigma_i \wedge d\eta_j.$$

Based on the above relations, we have

$$dP_{k+1} \wedge dQ_{k+1}$$

$$= dP_k \wedge dQ_k - \Delta_k \widehat{W} h \sum_{i,j=1}^{s} \left(b_i^{(1)} b_j^{(0)} - b_i^{(1)} A_{ij}^{(0)} - b_j^{(0)} A_{ji}^{(1)} \right) d\sigma_i \wedge dg_j$$

$$+ \Delta_k \widehat{W} \sum_{i=1}^{s} b_i^{(1)} (dp_i \wedge d\eta_i - d\sigma_i \wedge dq_i) + h \sum_{i=1}^{s} b_i^{(0)} (dp_i \wedge dg_i - df_i \wedge dq_i)$$

$$- \Delta_k \widehat{W} h \sum_{i,j=1}^{s} \left(b_i^{(0)} b_j^{(1)} - b_i^{(0)} A_{ij}^{(1)} - b_j^{(1)} A_{ji}^{(0)} \right) df_i \wedge d\eta_j$$

$$- h^2 \sum_{i,j=1}^{s} \left(b_i^{(0)} b_j^{(0)} - b_i^{(0)} A_{ij}^{(0)} - b_j^{(0)} A_{ji}^{(0)} \right) df_i \wedge dg_j$$

$$- (\Delta_k \widehat{W})^2 \sum_{i,j=1}^{s} \left(b_i^{(1)} b_j^{(1)} - b_i^{(1)} A_{ij}^{(1)} - b_j^{(1)} A_{ji}^{(1)} \right) d\sigma_i \wedge d\eta_j.$$

Let f_i^j, g_i^j, p_i^j and q_i^j, be the jth component of f_i, g_i, p_i and q_i, for $i \in \{1, \ldots, s\}$, and $j \in \{1, \ldots, n\}$, respectively. It can be checked that

$$dp_i \wedge dg_i - df_i \wedge dq_i$$

$$= \sum_{l,j=1}^{n} \frac{\partial g_i^l}{\partial p_i^j} dp_i^l \wedge dp_i^j + \sum_{l,j=1}^{n} \left(\frac{\partial g_i^l}{\partial q_i^j} dp_i^l \wedge dq_i^j - \frac{\partial f_i^l}{\partial p_i^j} dp_i^j \wedge dq_i^l \right) - \sum_{l,j=1}^{n} \frac{\partial f_i^l}{\partial q_i^j} dq_i^j \wedge dq_i^l$$

$$= \sum_{l \le j} \left(\frac{\partial g_i^l}{\partial p_i^j} - \frac{\partial g_i^j}{\partial p_i^l} \right) dp_i^l \wedge dp_i^j + \sum_{l,j=1}^{n} \left(\frac{\partial g_i^l}{\partial q_i^j} - \frac{\partial f_i^j}{\partial p_i^l} \right) dp_i^l \wedge dq_i^j$$

$$- \sum_{j < l} \left(\frac{\partial f_i^l}{\partial q_i^j} - \frac{\partial f_i^j}{\partial q_i^l} \right) dq_i^j \wedge dq_i^l = 0.$$

Similar to the proof of the above formulae, we obtain

$$dp_i \wedge d\eta_i - d\sigma_i \wedge dq_i = 0$$

for $i = 1, \ldots, s$, which yields $dP_{k+1} \wedge dQ_{k+1} = dP_k \wedge dQ_k$. Thus, the proof of the theorem is completed. $\qquad\square$

In [180, 260], the stochastic symplectic Runge–Kutta method (2.50) under condition (2.51) is also given in terms of three kinds of stochastic generating functions S^i, $i = 1, 2, 3$. Now we unify them, and give the associated stochastic θ-generating function of (2.50) satisfying (2.51) in the following theorem.

Theorem 2.2.8 *The stochastic symplectic Runge–Kutta method (2.50) with (2.51) can be rewritten as*

$$P_{k+1} = P_k - \frac{\partial \tilde{S}_\theta}{\partial \hat{Q}} ((1-\theta)P_k + \theta P_{k+1}, \theta Q_k + (1-\theta)Q_{k+1}, h),$$

$$Q_{k+1} = Q_k + \frac{\partial \tilde{S}_\theta}{\partial \hat{P}} ((1-\theta)P_k + \theta P_{k+1}, \theta Q_k + (1-\theta)Q_{k+1}, h),$$

where $k = 0, 1, \ldots, N - 1$,

$$\tilde{S}_\theta ((1-\theta)P_k + \theta P_{k+1}, \theta Q_k + (1-\theta)Q_{k+1}, h)$$

$$= \sum_{i=1}^{s} b_i^{(0)} H_0 (p_i, q_i) h + \sum_{i=1}^{s} b_i^{(1)} H_1 (p_i, q_i) \Delta_k \widehat{W} + \rho$$

for $\theta \in [0, 1]$ with

$$
\rho = -\theta \sum_{i=1}^{s} \sum_{j=1}^{s} b_i^{(0)} A_{ij}^{(0)} \left(\frac{\partial H_0}{\partial Q}(p_i, q_i) \right)^\top \frac{\partial H_0}{\partial P}(p_j, q_j) h^2
$$

$$
+ (1-\theta) \sum_{i=1}^{s} \sum_{j=1}^{s} b_i^{(0)} A_{ij}^{(0)} \left(\frac{\partial H_0}{\partial P}(p_i, q_i) \right)^\top \frac{\partial H_0}{\partial Q}(p_j, q_j) h^2
$$

$$
- \theta \sum_{i=1}^{s} \sum_{j=1}^{s} b_i^{(1)} A_{ij}^{(0)} \left(\frac{\partial H_1}{\partial Q}(p_i, q_i) \right)^\top \frac{\partial H_0}{\partial P}(p_j, q_j) h \Delta_k \widehat{W}
$$

$$
- \theta \sum_{i=1}^{s} \sum_{j=1}^{s} b_i^{(0)} A_{ij}^{(1)} \left(\frac{\partial H_0}{\partial Q}(p_i, q_i) \right)^\top \frac{\partial H_1}{\partial P}(p_j, q_j) h \Delta_k \widehat{W}
$$

$$
+ (1-\theta) \sum_{i=1}^{s} \sum_{j=1}^{s} b_i^{(0)} A_{ij}^{(1)} \left(\frac{\partial H_0}{\partial P}(p_i, q_i) \right)^\top \frac{\partial H_1}{\partial Q}(p_j, q_j) h \Delta_k \widehat{W}
$$

$$
+ (1-\theta) \sum_{i=1}^{s} \sum_{j=1}^{s} b_i^{(1)} A_{ij}^{(0)} \left(\frac{\partial H_1}{\partial P}(p_i, q_i) \right)^\top \frac{\partial H_0}{\partial Q}(p_j, q_j) h \Delta_k \widehat{W}
$$

$$
- \theta \sum_{i=1}^{s} \sum_{j=1}^{s} b_i^{(1)} A_{ij}^{(1)} \left(\frac{\partial H_1}{\partial Q}(p_i, q_i) \right)^\top \frac{\partial H_1}{\partial P}(p_j, q_j) (\Delta_k \widehat{W})^2
$$

$$
+ (1-\theta) \sum_{i=1}^{s} \sum_{j=1}^{s} b_i^{(1)} A_{ij}^{(1)} \left(\frac{\partial H_1}{\partial P}(p_i, q_i) \right)^\top \frac{\partial H_1}{\partial Q}(p_j, q_j) (\Delta_k \widehat{W})^2.
$$

Proof Based on the abbreviations $H_i = H_0(p_i, q_i)$, $\widetilde{H}_i = H_1(p_i, q_i)$, $f_i = \frac{\partial H_0}{\partial Q}(p_i, q_i)$, $g_i = \frac{\partial H_0}{\partial P}(p_i, q_i)$, $\sigma_i = \frac{\partial H_1}{\partial Q}(p_i, q_i)$ and $\gamma_i = \frac{\partial H_1}{\partial P}(p_i, q_i)$, $i = 1, \ldots, s$, we obtain

$$
\frac{\partial}{\partial \hat{Q}} \left(\sum_{i=1}^{s} b_i^{(0)} H_i \right) = \sum_{i=1}^{s} b_i^{(0)} \left(\frac{\partial P_k}{\partial \hat{Q}} - h \sum_{j=1}^{s} A_{ij}^{(0)} \frac{\partial f_j}{\partial \hat{Q}} - \Delta_k \widehat{W} \sum_{j=1}^{s} A_{ij}^{(1)} \frac{\partial \sigma_j}{\partial \hat{Q}} \right) g_i
$$

$$
+ \sum_{i=1}^{s} b_i^{(0)} \left(\frac{\partial Q_k}{\partial \hat{Q}} + h \sum_{j=1}^{s} A_{ij}^{(0)} \frac{\partial g_j}{\partial \hat{Q}} + \Delta_k \widehat{W} \sum_{j=1}^{s} A_{ij}^{(1)} \frac{\partial \gamma_j}{\partial \hat{Q}} \right) f_i,
$$

$$
\frac{\partial}{\partial \hat{Q}} \left(\sum_{i=1}^{s} b_i^{(1)} \widetilde{H}_i \right) = \sum_{i=1}^{s} b_i^{(1)} \left(\frac{\partial P_k}{\partial \hat{Q}} - h \sum_{j=1}^{s} A_{ij}^{(0)} \frac{\partial f_j}{\partial \hat{Q}} - \Delta_k \widehat{W} \sum_{j=1}^{s} A_{ij}^{(1)} \frac{\partial \sigma_j}{\partial \hat{Q}} \right) \gamma_i
$$

$$
+ \sum_{i=1}^{s} b_i^{(1)} \left(\frac{\partial Q_k}{\partial \hat{Q}} + h \sum_{j=1}^{s} A_{ij}^{(0)} \frac{\partial g_j}{\partial \hat{Q}} + \Delta_k \widehat{W} \sum_{j=1}^{s} A_{ij}^{(1)} \frac{\partial \gamma_j}{\partial \hat{Q}} \right) \sigma_i.
$$

By means of (2.50), we have

$$0 = \frac{\partial P_k}{\partial \hat{Q}} - \theta h \sum_{j=1}^{s} b_j^{(0)} \frac{\partial f_j}{\partial \hat{Q}} - \theta \Delta_k \widehat{W} \sum_{j=1}^{s} b_j^{(1)} \frac{\partial \sigma_j}{\partial \hat{Q}},$$

$$I_n = \frac{\partial Q_k}{\partial \hat{Q}} + (1-\theta) h \sum_{j=1}^{s} b_j^{(0)} \frac{\partial g_j}{\partial \hat{Q}} + (1-\theta) \Delta_k \widehat{W} \sum_{j=1}^{s} b_j^{(1)} \frac{\partial \gamma_j}{\partial \hat{Q}}.$$

Differentiating $\tilde{S}_\theta$ with respect to $\hat{Q}$ leads to

$$\frac{\partial \tilde{S}_\theta}{\partial \hat{Q}}((1-\theta)P_k + \theta P_{k+1}, \theta Q_k + (1-\theta)Q_{k+1}, h) = I + II + III,$$

where

$$I := h \sum_{i=1}^{s} b_i^{(0)} f_i + \Delta_k \widehat{W} \sum_{i=1}^{s} b_i^{(1)} \sigma_i + \theta h^2 \sum_{i,j=1}^{s} (b_i^{(0)} b_j^{(0)} - b_j^{(0)} A_{ji}^{(0)} - b_i^{(0)} A_{ij}^{(0)}) \frac{\partial f_j}{\partial \hat{Q}} g_i$$

$$+ (1-\theta) h^2 \sum_{i,j=1}^{s} (b_j^{(0)} A_{ji}^{(0)} + b_i^{(0)} A_{ij}^{(0)} - b_i^{(0)} b_j^{(0)}) \frac{\partial g_j}{\partial \hat{Q}} f_i,$$

$$II := \theta \Delta_k \widehat{W} h \sum_{i,j=1}^{s} \left(b_i^{(0)} b_j^{(1)} - b_i^{(0)} A_{ij}^{(1)} - b_j^{(1)} A_{ji}^{(0)} \right) \frac{\partial \sigma_j}{\partial \hat{Q}} g_i$$

$$+ (1-\theta) \Delta_k \widehat{W} h \sum_{i,j=1}^{s} \left(b_j^{(0)} A_{ji}^{(1)} + b_i^{(1)} A_{ij}^{(0)} - b_j^{(0)} b_i^{(1)} \right) \frac{\partial g_j}{\partial \hat{Q}} \sigma_i$$

$$+ \theta \Delta_k \widehat{W} h \sum_{i,j=1}^{s} \left(b_j^{(0)} b_i^{(1)} - b_j^{(0)} A_{ji}^{(1)} - b_i^{(1)} A_{ij}^{(0)} \right) \frac{\partial f_j}{\partial \hat{Q}} \gamma_i$$

$$+ (1-\theta) \Delta_k \widehat{W} h \left(b_j^{(1)} A_{ji}^{(0)} + b_i^{(0)} A_{ij}^{(1)} - b_j^{(0)} b_i^{(1)} \right) \frac{\partial \gamma_j}{\partial \hat{Q}} f_i,$$

$$III := \theta (\Delta_k \widehat{W})^2 \sum_{i,j=1}^{s} (b_i^{(1)} b_j^{(1)} - b_j^{(1)} A_{ji}^{(1)} - b_i^{(1)} A_{ij}^{(1)}) \frac{\partial \sigma_j}{\partial \hat{Q}} \gamma_i$$

$$+ (1-\theta)(\Delta_k \widehat{W})^2 \sum_{i,j=1}^{s} (b_j^{(1)} A_{ji}^{(1)} + b_i^{(1)} A_{ij}^{(1)} - b_j^{(1)} b_i^{(1)}) \frac{\partial \gamma_j}{\partial \hat{Q}} \sigma_i.$$

With the help of (2.51),

$$P_{k+1} - P_k = -h \sum_{i=1}^{s} b_i^{(0)} f_i - \Delta_k \widehat{W} \sum_{i=1}^{s} b_i^{(1)} \sigma_i.$$

Taking advantage of similar arguments, we obtain

$$Q_{k+1} - Q_k = h \sum_{i=1}^{s} b_i^{(0)} g_i + \Delta_k \widehat{W} \sum_{i=1}^{s} b_i^{(1)} \gamma_i,$$

which completes the proof.

$\square$

Corollary 2.2.1 (See [180]) *The stochastic symplectic Runge–Kutta method* (2.50) *with* (2.51) *is equivalent to*

$$P_{k+1} = P_k - \frac{\partial \tilde{S}^1}{\partial Q_k}(P_{k+1}, Q_k, h), \quad Q_{k+1} = Q_k + \frac{\partial \tilde{S}^1}{\partial P_{k+1}}(P_{k+1}, Q_k, h),$$

where $k = 0, 1, \ldots, N - 1$, *and*

$$\tilde{S}^1(P_{k+1}, Q_k, h) = \sum_{i=1}^{s} b_i^{(0)} H_0(p_i, q_i) h + \sum_{i=1}^{s} b_i^{(1)} H_1(p_i, q_i) \Delta_k \widehat{W} + \rho$$

with

$$\begin{aligned}
\rho = &- \sum_{i=1}^{s}\sum_{j=1}^{s} b_i^{(0)} A_{ij}^{(0)} \left(\frac{\partial H_0}{\partial Q}(p_i, q_i)\right)^{\top} \frac{\partial H_0}{\partial P}(p_j, q_j) h^2 \\
&- \sum_{i=1}^{s}\sum_{j=1}^{s} b_i^{(1)} A_{ij}^{(0)} \left(\frac{\partial H_1}{\partial Q}(p_i, q_i)\right)^{\top} \frac{\partial H_0}{\partial P}(p_j, q_j) h \Delta_k \widehat{W} \\
&- \sum_{i=1}^{s}\sum_{j=1}^{s} b_i^{(0)} A_{ij}^{(1)} \left(\frac{\partial H_0}{\partial Q}(p_i, q_i)\right)^{\top} \frac{\partial H_1}{\partial P}(p_j, q_j) h \Delta_k \widehat{W} \\
&- \sum_{i=1}^{s}\sum_{j=1}^{s} b_i^{(1)} A_{ij}^{(1)} \left(\frac{\partial H_1}{\partial Q}(p_i, q_i)\right)^{\top} \frac{\partial H_1}{\partial P}(p_j, q_j) (\Delta_k \widehat{W})^2.
\end{aligned}$$

Corollary 2.2.2 (See [180]) *The stochastic symplectic Runge–Kutta method* (2.50) *with* (2.51) *can be given by*

$$P_{k+1} = P_k - \frac{\partial \tilde{S}^2}{\partial Q_{k+1}}(P_k, Q_{k+1}, h), \quad Q_{k+1} = Q_k + \frac{\partial \tilde{S}^2}{\partial P_k}(P_k, Q_{k+1}, h),$$

where $k = 0, 1, \ldots, N - 1$, *and*

$$\tilde{S}^2(P_k, Q_{k+1}, h) = \sum_{i=1}^{s} b_i^{(0)} H_0(p_i, q_i) h + \sum_{i=1}^{s} b_i^{(1)} H_1(p_i, q_i) \Delta_k \widehat{W} + \rho$$

with

$$\rho = \sum_{i=1}^{s}\sum_{j=1}^{s} b_i^{(0)} A_{ij}^{(0)} \left(\frac{\partial H_0}{\partial P}(p_i, q_i)\right)^{\top} \frac{\partial H_0}{\partial Q}(p_j, q_j) h^2$$

$$+ \sum_{i=1}^{s}\sum_{j=1}^{s} b_i^{(0)} A_{ij}^{(1)} \left(\frac{\partial H_0}{\partial P}(p_i, q_i)\right)^{\top} \frac{\partial H_1}{\partial Q}(p_j, q_j) h \Delta_k \widehat{W}$$

$$+ \sum_{i=1}^{s}\sum_{j=1}^{s} b_i^{(1)} A_{ij}^{(0)} \left(\frac{\partial H_1}{\partial P}(p_i, q_i)\right)^{\top} \frac{\partial H_0}{\partial Q}(p_j, q_j) h \Delta_k \widehat{W}$$

$$+ \sum_{i=1}^{s}\sum_{j=1}^{s} b_i^{(1)} A_{ij}^{(1)} \left(\frac{\partial H_1}{\partial P}(p_i, q_i)\right)^{\top} \frac{\partial H_1}{\partial Q}(p_j, q_j) (\Delta_k \widehat{W})^2.$$

Remark 2.2.7 For the stochastic Hamiltonian system driven by multiple noises, the stochastic symplectic Runge–Kutta method and its associated stochastic θ-generating function with $\theta \in [0, 1]$ can be constructed in the same way.

Example 2.2.1 (1-Stage Stochastic Symplectic Runge–Kutta Method) Consider the 1-stage stochastic Runge–Kutta method as follows

$$p_1 = P_k - a_{11}\frac{\partial H_0}{\partial Q}(p_1, q_1)h - b_{11}\frac{\partial H_1}{\partial Q}(p_1, q_1)\Delta_k \widehat{W},$$

$$q_1 = Q_k + a_{11}\frac{\partial H_0}{\partial P}(p_1, q_1)h + b_{11}\frac{\partial H_1}{\partial P}(p_1, q_1)\Delta_k \widehat{W},$$

$$P_{k+1} = P_k - \alpha_1\frac{\partial H_0}{\partial Q}(p_1, q_1)h - \beta_1\frac{\partial H_1}{\partial Q}(p_1, q_1)\Delta_k \widehat{W},$$

$$Q_{k+1} = Q_k + \alpha_1\frac{\partial H_0}{\partial P}(p_1, q_1)h + \beta_1\frac{\partial H_1}{\partial P}(p_1, q_1)\Delta_k \widehat{W},$$

where $k = 0, 1, \ldots, N - 1$. The associated Butcher tableau takes the form of $\dfrac{a_{11}\ b_{11}}{\alpha_1\ \beta_1}$. Using (2.17) and (2.51), we obtain

$$\alpha_1 = 1, \quad \beta_1 = 1, \quad \beta_1 b_{11} = \frac{1}{2}, \quad \alpha_1 = 2a_{11}, \quad \beta_1 = 2b_{11}.$$

Then, we attain a stochastic symplectic Runge–Kutta method by

$$\frac{1/2\ 1/2}{1\ \ \ 1},$$

which is equivalent to the implicit midpoint method and has mean-square order 1.

Example 2.2.2 (2-Stage Stochastic Symplectic Runge–Kutta Method) Now we show a family of 2-stage diagonally implicit stochastic symplectic Runge–Kutta methods of mean-square order 1

$$p_1 = P_k - a_{11}\frac{\partial H_0}{\partial Q}(p_1, q_1)h - b_{11}\frac{\partial H_1}{\partial Q}(p_1, q_1)\Delta_k\widehat{W},$$

$$p_2 = P_k - 2a_{11}\frac{\partial H_0}{\partial Q}(p_1, q_1)h - \left(\frac{1}{2} - a_{11}\right)\frac{\partial H_0}{\partial Q}(p_2, q_2)h$$

$$- 2b_{11}\frac{\partial H_1}{\partial Q}(p_1, q_1)\Delta_k\widehat{W} - \left(\frac{1}{2} - b_{11}\right)\frac{\partial H_1}{\partial Q}(p_2, q_2)\Delta_k\widehat{W},$$

$$q_1 = Q_k + a_{11}\frac{\partial H_0}{\partial P}(p_1, q_1)h + b_{11}\frac{\partial H_1}{\partial P}(p_1, q_1)\Delta_k\widehat{W},$$

$$q_2 = Q_k + 2a_{11}\frac{\partial H_0}{\partial P}(p_1, q_1)h + \left(\frac{1}{2} - a_{11}\right)\frac{\partial H_0}{\partial P}(p_2, q_2)h$$

$$+ 2b_{11}\frac{\partial H_1}{\partial P}(p_1, q_1)\Delta_k\widehat{W} + \left(\frac{1}{2} - b_{11}\right)\frac{\partial H_1}{\partial P}(p_2, q_2)\Delta_k\widehat{W},$$

$$P_{k+1} = P_k - 2a_{11}\frac{\partial H_0}{\partial Q}(p_1, q_1)h - (1 - 2a_{11})\frac{\partial H_0}{\partial Q}(p_2, q_2)h$$

$$- 2b_{11}\frac{\partial H_1}{\partial Q}(p_1, q_1)\Delta_k\widehat{W} - (1 - 2b_{11})\frac{\partial H_1}{\partial Q}(p_2, q_2)\Delta_k\widehat{W},$$

$$Q_{k+1} = Q_k + 2a_{11}\frac{\partial H_0}{\partial P}(p_1, q_1)h + (1 - 2a_{11})\frac{\partial H_0}{\partial P}(p_2, q_2)h$$

$$+ 2b_{11}\frac{\partial H_1}{\partial P}(p_1, q_1)\Delta_k\widehat{W} + (1 - 2b_{11})\frac{\partial H_1}{\partial P}(p_2, q_2)\Delta_k\widehat{W},$$

where $k = 0, 1, \ldots, N - 1$, and $a_{11}, b_{11} \in \left(0, \frac{1}{2}\right)$ are free parameters. If we choose $a_{11} = \frac{1}{8}$ and $b_{11} = \frac{1}{4}$, then the method (2.50) is given by

$$\begin{array}{|cccc}
1/8 & 0 & 1/4 & 0 \\
1/4 & 3/8 & 1/2 & 1/4 \\
\hline
1/4 & 3/4 & 1/2 & 1/2
\end{array}.$$

Rough Symplectic Runge–Kutta Methods Now we turn to the rough Hamiltonian system (1.57)

$$\begin{cases}
dP_t = -\dfrac{\partial H_0(P_t, Q_t)}{\partial Q}dt - \displaystyle\sum_{r=1}^{m}\dfrac{\partial H_r(P_t, Q_t)}{\partial Q}d\mathbb{X}_t^r, & P_0 = p \in \mathbb{R}^n, \\[4mm]
dQ_t = \dfrac{\partial H_0(P_t, Q_t)}{\partial P}dt + \displaystyle\sum_{r=1}^{m}\dfrac{\partial H_r(P_t, Q_t)}{\partial P}d\mathbb{X}_t^r, & Q_0 = q \in \mathbb{R}^n,
\end{cases}$$

whose compact form is

$$\begin{cases} dY_t = \sum_{i=0}^{m} V_i(Y_t)d\mathbb{X}_t^i = V(Y_t)d\mathbb{X}_t, \\ Y_0 = y \end{cases} \tag{2.52}$$

with $\mathbb{X}_t^0 := t$ and $\mathbb{X}_t^i$, $i = 1, \ldots, m$, being Gaussian noises under Assumption 1.4.1. Because of the symplecticity of the phase flow of the rough Hamiltonian system (2.52) (see Sect. 1.4), it is natural to construct numerical methods to inherit this property. However, the simplified step-N Euler schemes studied in [94] cannot inherit this property in general.

Consequently, we introduce rough symplectic Runge–Kutta methods

$$Y_k^h(\alpha) = Y_k^h + \sum_{\beta=1}^{s} a_{\alpha\beta} V\big(Y_k^h(\beta)\big)\mathbb{X}_{t_k,t_{k+1}},$$

$$Y_{k+1}^h = Y_k^h + \sum_{\alpha=1}^{s} b_\alpha V\big(Y_k^h(\alpha)\big)\mathbb{X}_{t_k,t_{k+1}} \tag{2.53}$$

with coefficients $a_{\alpha\beta}$, b_α satisfying

$$a_{\alpha\beta}b_\alpha + a_{\beta\alpha}b_\beta = b_\alpha b_\beta, \quad \alpha, \beta = 1, \ldots, s,$$

for $t_k = kh$, $k = 0, 1, \ldots, N - 1$ and $Y_0^h = y$. The symplecticity of (2.53) can be proved similar to the proof of Theorem 2.2.7. When $s = 1$, $a_{11} = \frac{1}{2}$ and $b_1 = 1$, we have the midpoint method which is the 1-stage symplectic Runge–Kutta method:

$$Y_{k+1/2}^h = Y_k^h + \frac{1}{2}V\big(Y_{k+1/2}^h\big)\mathbb{X}_{t_k,t_{k+1}},$$

$$Y_{k+1}^h = Y_k^h + V\big(Y_{k+1/2}^h\big)\mathbb{X}_{t_k,t_{k+1}}. \tag{2.54}$$

Proposition 2.2.1 (See [130]) *If V in (2.52) is a collection of skew-symmetric linear vector fields of the form $V_i(Y) = A_i Y$ with $i = 0, 1, \ldots, m$, then numerical solutions of the midpoint method (2.54) are uniformly bounded. More precisely, $\|Y_k^h\| = \|y\|$, for any $k = 1, \ldots, N$.*

The above proposition is essential to get the convergence rate of the midpoint method in the linear case. Below, we introduce another two kinds of s-stage Runge–Kutta methods with coefficients expressed in the following Butcher tableaus as follows.

Method I ($s = 2$):

$$
\begin{array}{c|cc}
 & 1/4 & (3 - 2\sqrt{3})/12 \\
 & (3 + 2\sqrt{3})/12 & 1/4 \\
\hline
 & 1/2 & 1/2
\end{array} \quad ;
$$

Method II ($s = 3$):

$$
\begin{array}{c|ccc}
 & a/2 & 0 & 0 \\
 & a & a/2 & 0 \\
 & a & a & 1/2 - a \\
\hline
 & a & a & 1 - 2a
\end{array} \quad ,
$$

where $a = 1.351207$ is the real root of $6x^3 - 12x^2 + 6x - 1 = 0$.

In Definition 1.4.7, the solution of the rough differential equation depends on the information about iterated integrals of the noise, which is difficult to simulate. The Runge–Kutta methods are implementable since they omit nonsymmetric parts of iterated integrals. For example, Lévy's area defined on the 2-dimensional Wiener space $\{\omega \in \mathbf{C}([0, T], \mathbb{R}^2) \mid \omega(0) = 0\}$ by the Itô stochastic integral as follows

$$
S_{Levy}(t) = \frac{1}{2} \int_0^t W_1(s) dW_2(s) - \frac{1}{2} \int_0^t W_2(s) dW_1(s)
$$

is omitted in the standard Brownian motion setting. To analyze the error caused by this, we introduce the piecewise linear approximation, which is a special case of Wong–Zakai approximations in [104]. We define $x_t^h := (x_t^{h,0}, \ldots, x_t^{h,m})$ with

$$
x_t^{h,i} := \mathbb{X}_{t_k}^i + \frac{t - t_k}{h} \mathbb{X}_{t_k,t_{k+1}}^i, \quad t \in (t_k, t_{k+1}], \tag{2.55}
$$

for $k = 0, 1, \ldots, N - 1$, $i = 0, 1, \ldots, m$, and consider the following differential equation in Riemann–Stieltjes integral sense

$$
\begin{cases}
dy_t^h = V(y_t^h) dx_t^h, \\
y_0^h = y.
\end{cases} \tag{2.56}
$$

Theorem 2.2.9 (See [130]) *Suppose that $\mathbb{X}$ and x^h are as in Assumption 1.4.1 and (2.55), respectively. If for some $\gamma > 2\rho$, $V = (V_i)_{0 \le i \le m}$ is a collection of vector fields in Lip^γ or a collection of linear vector fields of the form $V_i(Y) = A_i Y$, then both (2.52) and (2.56) have unique solutions Y and y^h almost surely, respectively.*

Moreover, let $\theta \ge \|y\|$, $v \ge \max\{\|V\|_{Lip^\gamma}, \|A_i\|, i = 0, 1, \ldots, m\}$. Then for any $\eta \in \left[0, \min\{\frac{1}{\rho} - \frac{1}{2}, \frac{1}{2\rho} - \frac{1}{\gamma}\}\right)$, there exist a finite random variable $K(\omega)$ and a null *set M such that*

$$\sup_{t \in [0,T]} \| y_t^h(\omega) - Y_t(\omega) \| \le K(\omega) h^\eta \quad \forall \, \omega \in \Omega \setminus M$$

holds for any $h > 0$. The finite random variable K depends on ρ, η, γ, ν, θ, C and T.

We are in a position to present the convergence rate of the proposed Runge–Kutta methods. The convergence rate of the midpoint method for $\rho \in [1, \frac{3}{2})$ is shown in Theorem 2.2.10 for $V \in Lip^\gamma$ with $\gamma > 2\rho$, and in Theorem 2.2.11 for the case that V is linear.

Theorem 2.2.10 (See [130]) *Suppose that $\rho \in [1, \frac{3}{2})$ and Y_k^h, $k = 0, 1, \ldots, N$, is the numerical solution of the midpoint method (2.54). If $\|V\|_{Lip^\gamma} \le \nu < \infty$ with $\gamma > 2\rho$, then for any $\eta \in \left(0, \min\{\frac{3}{2\rho} - 1, \frac{1}{2\rho} - \frac{1}{\gamma}\} \right)$, there exist a finite random variable K and a null set M such that*

$$\max_{k=1,\ldots,N} \| Y_k^h(\omega) - Y_{t_k}(\omega) \| \le K(\omega) h^\eta \quad \forall \, \omega \in \Omega \setminus M$$

holds for any $0 < h \le 1$. The finite random variable K depends on ρ, η, γ, ν, C and T.

Remark 2.2.8 If components of the noise are independent standard Wiener processes, i.e., $\rho = 1$, the convergence order is (almost) consistent with the mean-square convergence order of the stochastic differential equation in the Stratonovich sense with multiplicative noise. Moreover, we would like to mention that the estimate in Theorem 2.2.10 is valid for the case of $\rho \in [1, \frac{3}{2})$, which is similar to the simplified step-2 Euler scheme in [104].

Theorem 2.2.11 (See [130]) *Suppose that $\gamma > 2\rho$, V is a collection of skew-symmetric linear vector fields of the form $V_i(Y) = A_i Y$ with $i = 0, 1, \ldots, m$, and Y_k^h, $k = 0, 1, \ldots, N$, represents the numerical solution of the midpoint method. Let $\theta \ge \|y\|$ and $\nu \ge \max \{ \|V\|_{Lip^\gamma}, \|A_i\|, i = 0, 1, \ldots, m \}$. For any $\eta \in \left(0, \min\{\frac{1}{\rho} - \frac{1}{2}, \frac{1}{2\rho} - \frac{1}{\gamma}\} \right)$, there exist a finite random variable K and a null set M such that*

$$\max_{k=1,\ldots,N} \| Y_k^h(\omega) - Y_{t_k}(\omega) \| \le K(\omega) h^\eta \quad \forall \, \omega \in \Omega \setminus M$$

holds for any $0 < h \le 1$. The finite random variable K depends on ρ, η, γ, ν, θ, C and T.

Now we show the convergence rate of *Method I* and the *Method II* for $\rho \in [1, 2)$ in the following Theorem.

Theorem 2.2.12 (See [130]) *Suppose that $\rho \in [1, 2)$ and Y_k^h, $k = 0, 1, \ldots, N$, represents the solution of the Method I or the Method II. If $\|V\|_{Lip^\gamma} \leq \nu < \infty$ with $\gamma > \max\{2\rho, 3\}$, then for any $\eta \in \left(0, \min\{\frac{1}{\rho} - \frac{1}{2}, \frac{1}{2\rho} - \frac{1}{\gamma}\}\right)$, there exist a finite random variable K and a null set M such that*

$$\max_{k=1,\ldots,N} \|Y_k^h(\omega) - Y_{t_k}(\omega)\| \leq K(\omega)h^\eta, \quad \forall \, \omega \in \Omega \setminus M.$$

Remark 2.2.9 This convergence rate equals that of the simplified step-3 Euler scheme in [104] for sufficiently large γ. Indeed, there exist order barriers of convergence for the Runge–Kutta method with a higher stage due to the persistence generated by the error of the piecewise linear approximation.

Numerical Experiments Consider a 2-dimensional rough Hamiltonian system

$$\begin{cases} dP_t = \sin(P_t)\sin(Q_t)dt - \cos(Q_t)d\mathbb{X}_t^2, & P_0 = p, \\ dQ_t = \cos(P_t)\cos(Q_t)dt - \sin(P_t)d\mathbb{X}_t^1, & Q_0 = q, \end{cases} \tag{2.57}$$

where $\mathbb{X}^1$ and $\mathbb{X}^2$ are independent fractional Brownian motions with Hurst parameter $\mathbf{H} \in (\frac{1}{4}, \frac{1}{2}]$. Since vector fields are bounded with bounded derivatives, theoretical convergence order of Method I in Theorem 2.2.12 is $(2\mathbf{H} - \frac{1}{2} - \varepsilon)$ for arbitrary small ε. Figure 2.6 shows the maximum error $\max_{k=1,\ldots,N} \|Y_k^h(\omega) - Y_{t_k}(\omega)\|$ in the discretization points of Method I. Pictures are from three sample paths with different $\mathbf{H}$. In addition, the fact that the error becomes larger, as $\mathbf{H}$ decreases, implies the influence of the roughness of the noise.

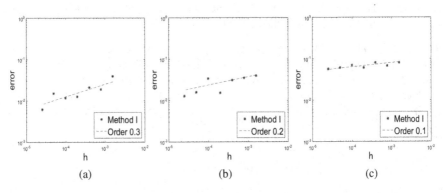

Fig. 2.6 Pathwise maximum error for three sample paths with $p = 1$, $q = 2$ and $T = 0.1$. **(a)** $\mathbf{H} = 0.4$. **(b)** $\mathbf{H} = 0.35$. **(c)** $\mathbf{H} = 0.3$

2.3   Application of Symplectic Methods to Stochastic Poisson Systems

In this section, we concentrate on the stochastic Poisson system and propose a numerical integration methodology which can provide numerical methods preserving both the stochastic Poisson structure and Casimir functions, based on the Darboux–Lie theorem and the stochastic symplectic methods.

As is shown in Sect. 1.3.2, the d-dimensional stochastic Poisson system

$$
\begin{cases}
dX(t) = B(X(t))\left(\nabla K_0(X(t))dt + \displaystyle\sum_{r=1}^{m} \nabla K_r(X(t)) \circ dW_r(t)\right), \\[2mm]
X(0) = x,
\end{cases}
\tag{2.58}
$$

where $t \in (0, T]$, can be transformed to a generalized $2n$-dimensional stochastic Hamiltonian system by the Darboux–Lie theorem. Our strategy is to first utilize stochastic symplectic methods to discretize the stochastic Hamiltonian system, and then transform them back to obtain the stochastic Poisson methods, that is, the numerical method preserving Poisson structure. A numerical method $\{X_k\}_{k\geq 0}$ with $X_0 = x$ applied to (2.58) preserves the Poisson structure if

$$
\frac{\partial X_{k+1}}{\partial X_k} B(X_k) \left(\frac{\partial X_{k+1}}{\partial X_k}\right)^\top = B(X_{k+1}) \quad \forall \, k \geq 0.
$$

The followings are the details of this procedure:

1. Find Casimir functions $\mathscr{C}_i = C_i(X)$ by solving $(\nabla C_i(X))^\top B(X) = 0$, $i = 1, \ldots, l$, where $l = d - 2n$, and $2n$ is the rank of skew-symmetric matrix $B(X)$;
2. Use the coordinate transformation $Y = \Theta(X)$ as described in the proof of Theorem 1.3.4, which has an invertible Jacobian matrix $A(X) = \frac{\partial Y}{\partial X}$, to transform the structure matrix $B(X)$ to $A(X)B(X)(A(X))^\top$;
3. Let $A(X)B(X)(A(X))^\top = B_0$ with B_0 being the constant structure matrix of a generalized stochastic Hamiltonian system, e.g.,

$$
B_0 = \begin{bmatrix} J_{2n}^{-1} & 0 \\ 0 & 0 \end{bmatrix},
$$

and solve the coordinate transformation

$$
Y = \Theta(X) = (P_1, \ldots, P_n, Q_1, \ldots, Q_n, C_1, \ldots, C_l)^\top,
$$

and its inverse $X = \Theta^{-1}(Y)$. Then we obtain the stochastic Hamiltonian system

$$
\begin{cases}
dY(t) = B_0\Big(\nabla H_0(Y(t))dt + \displaystyle\sum_{r=1}^{m} \nabla H_r(Y(t)) \circ dW_r(t)\Big), \\[2mm]
Y(0) = \Theta(x)
\end{cases}
$$

with $H_i(Y) = K_i(X)$, $i = 0, 1, \ldots, m$. Denoting $Y = (Z^\top, \mathscr{C}^\top)^\top$, where

$$
Z = (P^\top, Q^\top)^\top, \quad P = (P_1, \ldots, P_n)^\top, \quad Q = (Q_1, \ldots, Q_n)^\top, \quad \mathscr{C} = (\mathscr{C}_1, \ldots, \mathscr{C}_l)^\top,
$$

we have a stochastic Hamiltonian system

$$
dZ = J_{2n}^{-1}\Big(\nabla_Z H_0(Z, \mathscr{C})dt + \sum_{r=1}^{m} \nabla_Z H_r(Z, \mathscr{C}) \circ dW_r(t)\Big); \tag{2.59}
$$

4. Apply a stochastic symplectic method to the stochastic Hamiltonian system (2.59). Then we have $Y_{k+1} = ((Z^{k+1})^\top, \mathscr{C}^\top)^\top = \psi_h(P^k, Q^k, \mathscr{C})$, where Z^{k+1} and Y_{k+1} are approximations of $Z(t_{k+1})$ and $Y(t_{k+1})$, respectively, for $k \in \{0, 1, \ldots, N-1\}$. Finally, we use the inverse transformation $X = \Theta^{-1}(Y)$ to obtain the numerical method $X_{k+1} = \varphi_h(X_k) = \Theta^{-1}(\psi_h(Y_k))$, $k = 0, 1, \ldots, N-1$, for the stochastic Poisson system (2.58).

Theorem 2.3.1 (See [132]) *The proposed method* $X_{k+1} = \varphi_h(X_k)$, *where* $k = 0, 1, \ldots, N-1$, *preserves both the stochastic Poisson structure and Casimir functions of* (2.58) *almost surely.*

Proof Denoting $Y_k := \Theta(X_k)$, $k = 0, 1, \ldots, N-1$, we get

$$
\frac{\partial X_{k+1}}{\partial X_k} B(X_k) \left(\frac{\partial X_{k+1}}{\partial X_k}\right)^\top
$$

$$
= \frac{\partial X_{k+1}}{\partial \psi_h(Y_k)} \frac{\partial \psi_h(Y_k)}{\partial Y_k} \frac{\partial Y_k}{\partial X_k} B(X_k) \left(\frac{\partial Y_k}{\partial X_k}\right)^\top \left(\frac{\partial \psi_h(Y_k)}{\partial Y_k}\right)^\top \left(\frac{\partial X_{k+1}}{\partial \psi_h(Y_k)}\right)^\top
$$

$$
= \frac{\partial X_{k+1}}{\partial \psi_h(Y_k)} \frac{\partial \psi_h(Y_k)}{\partial Y_k} \begin{bmatrix} J_{2n}^{-1} & 0 \\ 0 & 0 \end{bmatrix} \left(\frac{\partial \psi_h(Y_k)}{\partial Y_k}\right)^\top \left(\frac{\partial X_{k+1}}{\partial \psi_h(Y_k)}\right)^\top,
$$

where we have used $A(X)B(X)(A(X))^\top = B_0$. Since $\{\psi_h(Y_k)\}_{k=0}^{N}$ is given by a symplectic method, we have

$$
\frac{\partial \psi_h(Y_k)}{\partial Y_k} \begin{bmatrix} J_{2n}^{-1} & 0 \\ 0 & 0 \end{bmatrix} \left(\frac{\partial \psi_h(Y_k)}{\partial Y_k}\right)^\top = \begin{bmatrix} J_{2n}^{-1} & 0 \\ 0 & 0 \end{bmatrix}.
$$

Therefore,

$$\frac{\partial X_{k+1}}{\partial X_k} B(X_k) \left(\frac{\partial X_{k+1}}{\partial X_k} \right)^\top = \frac{\partial X_{k+1}}{\partial \psi_h(Y_k)} \begin{bmatrix} J_{2n}^{-1} & 0 \\ 0 & 0 \end{bmatrix} \left(\frac{\partial X_{k+1}}{\partial \psi_h(Y_k)} \right)^\top.$$

From the fact $\psi_h(Y_k) = \Theta(X_{k+1})$ it follows that $\frac{\partial X_{k+1}}{\partial \psi_h(Y_k)} = (A(X_{k+1}))^{-1}$ for $k = 0, 1, \ldots, N-1$. According to

$$A(X_{k+1}) B(X_{k+1}) (A(X_{k+1}))^\top = B_0,$$

we deduce

$$\frac{\partial X_{k+1}}{\partial X_k} B(X_k) \left(\frac{\partial X_{k+1}}{\partial X_k} \right)^\top = B(X_{k+1}) = B(\varphi_h(X_k)), \quad k = 0, 1, \ldots, N-1,$$

which means that the numerical method preserves stochastic Poisson structure. Moreover, it can be checked that the numerical method $\{X_k\}_{k \geq 0}$ preserves Casimir functions.   □

Now we apply the above integration strategy for stochastic Poisson systems to numerically solving the stochastic rigid body system

$$\begin{cases} dX(t) = B(X(t)) \nabla K(X(t))(dt + c \circ dW(t)), \\ X(0) = x = (x_1, x_2, x_3)^\top, \end{cases} \tag{2.60}$$

where $K(X) = \frac{1}{2} \left(\frac{X_1^2}{C_1} + \frac{X_2^2}{C_2} + \frac{X_3^2}{C_3} \right)$, C_1, C_2, C_3, c are non-zero constants, and

$$B(X) = \begin{bmatrix} 0 & -X_3 & X_2 \\ X_3 & 0 & -X_1 \\ -X_2 & X_1 & 0 \end{bmatrix}.$$

It possesses the Casimir function $\mathscr{C}: X \mapsto \frac{1}{2} \|X\|^2$. First we need to find a coordinate transformation $Y = \Theta(X) = (Y_1, Y_2, Y_3)^\top$ with invertible Jacobian matrix

$$A(X) = \left(A_{ij}(X) \right) = \left(\frac{\partial Y_i}{\partial X_j} \right), \quad i, j = 1, 2, 3,$$

such that

$$A(X) B(X) A(X)^\top = B_0 = \begin{bmatrix} 0 & -1 & 0 \\ 1 & 0 & 0 \\ 0 & 0 & 0 \end{bmatrix}. \tag{2.61}$$

Then (2.61) is equivalent to the following relations with respect to the Poisson bracket defined by $B(X)$

$$\{Y_1, Y_1\} = 0, \quad \{Y_1, Y_2\} = -1, \quad \{Y_1, Y_3\} = 0,$$
$$\{Y_2, Y_1\} = 1, \quad \{Y_2, Y_2\} = 0, \quad \{Y_2, Y_3\} = 0,$$
$$\{Y_3, Y_1\} = 0, \quad \{Y_3, Y_2\} = 0, \quad \{Y_3, Y_3\} = 0.$$

Due to the skew-symmetry of the Poisson bracket, nine equations reduce to three relations

$$\{Y_2, Y_1\} = 1, \quad \{Y_3, Y_1\} = 0, \quad \{Y_3, Y_2\} = 0.$$

The last two equations above indicate that we can choose $Y_3 = \mathscr{C}(x)$ with x being the initial value based on the property of the Casimir function. The first equation can be expressed explicitly as

$$(A_{12}A_{23} - A_{13}A_{22})X_1 + (A_{13}A_{21} - A_{11}A_{23})X_2 + (A_{11}A_{22} - A_{12}A_{21})X_3 = 1.$$
$$(2.62)$$

This is actually a partial differential equation which possesses possibly many variants of solutions. For instance, if we let, e.g., $Y_1 = X_2$, then (2.62) becomes

$$A_{23}X_1 - A_{21}X_3 = 1,$$

and it can be verified that $Y_2 = \arctan\left(\frac{X_3}{X_1}\right)$ is a solution. Thus, we find the following coordinate transformation

$$Y_1 = X_2, \quad Y_2 = \arctan\left(\frac{X_3}{X_1}\right), \quad Y_3 = \mathscr{C}(x),$$

and its inverse

$$X_1 = \sqrt{2\mathscr{C} - Y_1^2}\cos(Y_2), \quad X_2 = Y_1, \quad X_3 = \sqrt{2\mathscr{C} - Y_1^2}\sin(Y_2). \quad (2.63)$$

Simultaneously, we obtain a stochastic Hamiltonian system with respect to Y_1 and Y_2

$$d(Y_1, Y_2)^\top = J_2^{-1}\nabla H(Y_1, Y_2)(dt + c \circ dW(t)), \quad (2.64)$$

where

$$H(Y_1, Y_2) = \frac{1}{2C_1}(2\mathscr{C}(x) - Y_1^2)\cos^2(Y_2) + \frac{1}{2C_2}Y_1^2 + \frac{1}{2C_3}(2\mathscr{C}(x) - Y_1^2)\sin^2(Y_2).$$

Substituting derivatives of the Hamiltonian

$$\frac{\partial H}{\partial Y_1} = \left(\frac{1}{C_2} - \frac{\cos^2(Y_2)}{C_1} - \frac{\sin^2(Y_2)}{C_3} \right) Y_1,$$

$$\frac{\partial H}{\partial Y_2} = \left(\frac{1}{2C_3} - \frac{1}{2C_1} \right) (2\mathscr{C}(x) - Y_1^2) \sin(2Y_2),$$

$$\frac{\partial^2 H}{\partial Y_1^2} = \frac{1}{C_2} - \frac{\cos^2(Y_2)}{C_1} - \frac{\sin^2(Y_2)}{C_3},$$

$$\frac{\partial^2 H}{\partial Y_2^2} = \left(\frac{1}{C_3} - \frac{1}{C_1} \right) (2\mathscr{C}(x) - Y_1^2) \cos(2Y_2),$$

$$\frac{\partial^2 H}{\partial Y_1 \partial Y_2} = \left(\frac{1}{C_1} - \frac{1}{C_3} \right) Y_1 \sin(2Y_2)$$

into (2.44), we derive the following symplectic method of mean-square order 1

$$P_{k+1} = P_k - \left(\frac{1}{2C_3} - \frac{1}{2C_1} \right) \left(2\mathscr{C}(x) - \bar{P}_k^2 \right) \sin\left(2\bar{Q}_k \right) (h + c\Delta_k \widehat{W})$$

$$+ c^2 C_\theta (2\mathscr{C}(x) - \bar{P}_k^2) \bar{P}_k \left(\cos(2\bar{Q}_k) \left(\frac{1}{C_2} - \frac{\cos^2(\bar{Q}_k)}{C_1} - \frac{\sin^2(\bar{Q}_k)}{C_3} \right) \right.$$

$$\left. - \sin^2(2\bar{Q}_k) \left(\frac{1}{2C_3} - \frac{1}{2C_1} \right) \right) (\Delta_k \widehat{W})^2,$$

$$Q_{k+1} = Q_k + \left(\frac{1}{C_2} - \frac{\cos^2(\bar{Q}_k)}{C_1} - \frac{\sin^2(\bar{Q}_k)}{C_3} \right) \bar{P}_k (h + c\Delta_k \widehat{W})$$

$$+ c^2 C_\theta \left(\frac{1}{C_2} - \frac{\cos^2(\bar{Q}_k)}{C_1} - \frac{\sin^2(\bar{Q}_k)}{C_3} \right) \left(\frac{3}{2} \bar{P}_k^2 - \mathscr{C}(x) \right) \sin(2\bar{Q}_k)(\Delta_k \widehat{W})^2,$$

where $\bar{P}_k = (1 - \theta) P_k + \theta P_{k+1}$, $\bar{Q}_k = \theta Q_k + (1 - \theta) Q_{k+1}$, $C_\theta = \left(\theta - \frac{1}{2} \right) \left(\frac{1}{C_1} - \frac{1}{C_3} \right)$ with $\theta \in [0, 1]$ and $k = 0, 1, \dots, N - 1$. By means of the inverse coordinate transformation (2.63), we get the following θ-generating function method for the original stochastic rigid body system (2.60) as follows

$$X_{k,1} := \sqrt{2\mathscr{C}(x) - P_k^2} \cos(Q_k),$$

$$X_{k,2} := P_k, \tag{2.65}$$

$$X_{k,3} := \sqrt{2\mathscr{C}(x) - P_k^2} \sin(Q_k), \quad k = 0, 1, \dots, N.$$

It is obvious that (2.65) preserves the Casimir function, since

$$(X_{k,1})^2 + (X_{k,2})^2 + (X_{k,3})^2 = 2\mathscr{C}(x), \quad k = 0, 1, \dots, N.$$

Moreover, (2.65) inherits the stochastic Poisson structure of stochastic rigid body system (2.60) based on Theorem 2.3.1.

On the other hand, the quadratic form of the Casimir function motivates a spherical coordinate transformation $\phi : (\theta_1, \theta_2) \mapsto (X_1, X_2, X_3)$, i.e.,

$$X_1 = R\cos(\theta_1)\cos(\theta_2), \quad X_2 = R\cos(\theta_1)\sin(\theta_2), \quad X_3 = R\sin(\theta_1), \quad (2.66)$$

where $R = \sqrt{2\mathscr{C}(x)}$. Using the inverse mapping of ϕ, we obtain

$$\begin{cases} d\theta_1 = R\left(\dfrac{1}{C_2} - \dfrac{1}{C_1}\right)\cos(\theta_1)\sin(\theta_2)\cos(\theta_2)(dt + c \circ dW(t)), \\[2mm] d\theta_2 = R\sin(\theta_1)\left(\left(\dfrac{1}{C_1} - \dfrac{1}{C_3}\right)\cos^2(\theta_2) - \left(\dfrac{1}{C_3} - \dfrac{1}{C_2}\right)\sin^2(\theta_2)\right)(dt + c \circ dW(t)). \end{cases}$$
$$(2.67)$$

We can apply the stochastic midpoint method to (2.67) and deduce

$$\Theta_{k+1}^1 = \Theta_k^1 + R\left(\frac{1}{C_2} - \frac{1}{C_1}\right)\cos(\bar{\Theta}_k^1)\sin(\bar{\Theta}_k^2)\cos(\bar{\Theta}_k^2)(h + c\Delta_k\widehat{W}),$$

$$\Theta_{k+1}^2 = \Theta_k^2 + R\sin(\bar{\Theta}_k^1)\left(\left(\frac{1}{C_1} - \frac{1}{C_3}\right)\cos^2(\bar{\Theta}_k^2)\right. \qquad\qquad (2.68)$$

$$\left. - \left(\frac{1}{C_3} - \frac{1}{C_2}\right)\sin^2(\bar{\Theta}_k^2)\right)(h + c\Delta_k\widehat{W}),$$

where $\bar{\Theta}_k^1 = \frac{1}{2}(\Theta_k^1 + \Theta_{k+1}^1)$ and $\bar{\Theta}_k^2 = \frac{1}{2}(\Theta_k^2 + \Theta_{k+1}^2)$ with $k = 0, 1, \dots, N-1$. Then by (2.66), we obtain the following numerical method for (2.60)

$$\begin{aligned} X_{k,1} &= R\cos(\Theta_k^1)\cos(\Theta_k^2), \\ X_{k,2} &= R\cos(\Theta_k^1)\sin(\Theta_k^2), \qquad\qquad (2.69) \\ X_{k,3} &= R\sin(\Theta_k^1), \end{aligned}$$

which naturally satisfies

$$(X_{k,1})^2 + (X_{k,2})^2 + (X_{k,3})^2 = 2\mathscr{C}(x), \quad k = 0, 1, \dots, N.$$

It means that (2.69) preserves the Casimir function. In the following, we call the numerical method resulted from the spherical transformation for the stochastic rigid body system (2.60) the 'spherical scheme'. Next we derive the mean-square convergence order of the spherical scheme. Denote a numerical method applied to (2.67) by $\{\Theta_k^1, \Theta_k^2\}_{k=0}^N$, and its spherically transformed method for the original stochastic rigid body system (2.60) by $\{X_{k,1}, X_{k,2}, X_{k,3}\}_{k=0}^N$, respectively.

Theorem 2.3.2 (See [132]) *If $\{\Theta_k^1, \Theta_k^2\}_{k=0}^N$ applied to (2.67) is of mean-square order γ, then $\{X_{k,1}, X_{k,2}, X_{k,3}\}_{k=0}^N$ for (2.60) is also of mean-square order γ, i.e.,*

$$\mathbb{E}\big[\|(X_1(T), X_2(T), X_3(T)) - (X_{N,1}, X_{N,2}, X_{N,3})\|^2 \big] = O(h^{2\gamma}). \tag{2.70}$$

Numerical Experiments Now we demonstrate numerical behaviors of spherical scheme and θ-generating function method with $\theta \in [0, 1]$ for (2.60).

Figure 2.7 shows sample paths of X_1, X_2 and X_3 of stochastic rigid body system (2.60) produced by θ-generating method (2.65) with $\theta \in \{0, 1, 0.5\}$, and by spherical scheme (2.69), respectively. The reference solutions of X_1, X_2 and X_3 in blue color are approximated by the midpoint method with time-step size 10^{-5}. Let $C_1 = \sqrt{2}+\sqrt{\frac{2}{1.51}}, C_2 = \sqrt{2}-0.51\sqrt{\frac{2}{1.51}}, C_3 = 1, c = 0.2$, and $x = \left(\frac{1}{\sqrt{2}}, \frac{1}{\sqrt{2}}, 0\right)^\top$. We take time-step size $h = 0.01$, $err = 10^{-12}$ as the error bound for stopping the inner iterations within each time step by implementing implicit numerical methods. It can be observed that all numerical sample paths coincide well with reference solutions.

Figure 2.8 illustrates evolution of the Casimir function produced by (2.65) with $\theta \in \{0, 0.5, 1\}$, and spherical scheme, each with different initial values giving different Casimir values $\mathscr{C}$. Figure 2.8a shows clearly the exact preservation of Casimir function by θ-generating function method and spherical scheme. Compared with the θ-generating function method and spherical scheme, it can be seen in Fig. 2.8b that the Euler–Maruyama method fails to preserve Casimir function.

Figure 2.9 shows the mean-square convergence order of the θ-generating function method with $\theta \in \{0, 0.5, 1\}$, and that of the spherical scheme. Figure 2.9a

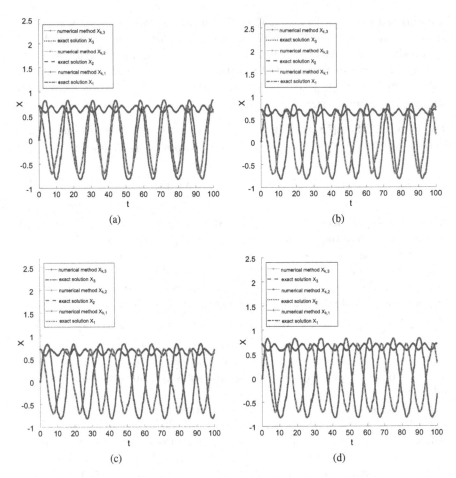

Fig. 2.7 Sample paths of X_1, X_2 and X_3 produced by (2.65) and (2.69). (**a**) 0-generating function method. (**b**) 1-generating function method. (**c**) 0.5-generating function method. (**d**) Spherical scheme

indicates that the θ-generating function method has mean-square order 1 and the 0.5-generating function method is with smaller error than the cases that $\theta = 0, 0.5$. Moreover, we can observe from Fig. 2.9b that the spherical scheme is also of mean-square order 1. In both subfigures we take $h \in \{0.005, 0.01, 0.02, 0.04\}$, $T = 10$, and 500 samples for approximating the expectations.

Beyond the efficiency, we also would like to emphasize the flexibility of proposed numerical algorithms, in allowing different choices of $\theta \in [0, 1]$, and different canonical coordinate transformations.

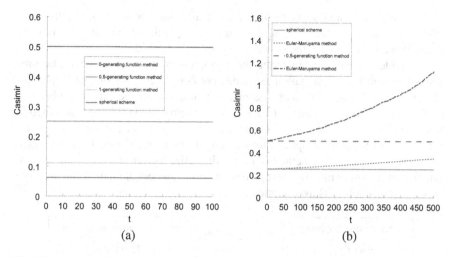

(a)   (b)

Fig. 2.8 Evolution of the Casimir function by (2.65), (2.69) and the Euler–Maruyama method started from different initial values. (a) $T = 100$. (b) $T = 500$

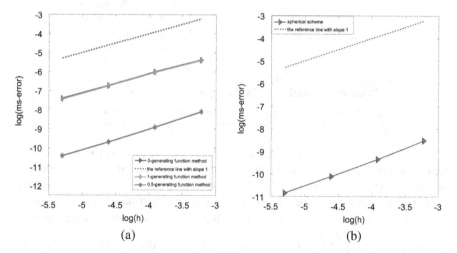

(a)   (b)

Fig. 2.9 Mean-square convergence orders of (2.65) and (2.69). (a) θ-generating function. (b) Spherical scheme

2.4 Superiority of Stochastic Symplectic Methods via Large Deviation Principle (LDP)

Symplectic methods have been rigorously shown to be superior to nonsymplectic ones especially in the long-time simulation, when applied to the deterministic Hamiltonian system. For the stochastic case, large quantities of numerical experiments present that the stochastic symplectic method possesses excellent long-time stability. The existing approach to theoretically explaining the superiority of the stochastic symplectic method is by exploiting the techniques of stochastic modified equations and weak backward error analysis (see e.g., [9, 60, 264] and references therein). In this section, we attempt to apply the LDP to investigating the probabilistic superiority of the stochastic symplectic method.

It is widely used to characterize the asymptotical behavior of stochastic processes for which large deviation estimates are concerned. If a stochastic process $\{X_T\}_{T>0}$ satisfies an LDP with the rate function $\mathbf{I}$, then the hitting probability $\mathbb{P}(X_T \in [a, a+da]) = \exp(-T\mathbf{I}(a))da$ decays exponentially. The rate function describes fluctuations of the stochastic process $\{X_T\}_{T>0}$ in the long-time limit, and has a wide range of applications in engineering and physical sciences (see [101]). When a numerical method is used to discretize a given stochastic differential equation, it is worthwhile to investigate whether the numerical method can preserve the decay rate $\exp(-T\mathbf{I})$ asymptotically.

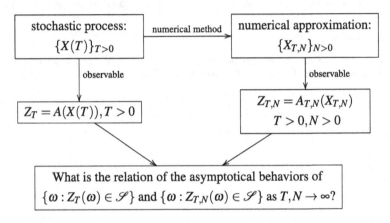

Let $\{Z_T\}_{T>0}$ be an observable of the stochastic Hamiltonian system (2.19), that is, a stochastic process associated with the exact solution of (2.19). Denote a discrete approximation of $\{Z_T\}_{T>0}$ associated with the numerical method $\{(P_k, Q_k)\}_{k\geq 0}$ approximating (2.19) by $\{Z_{T,N}\}_{N\geq 1}$. For instance, one can regard $Z_T = \frac{1}{T}\int_0^T f(P(t), Q(t))dt$ as an observable of (2.19) for some smooth function f, and $Z_{T,N} = \frac{1}{N}\sum_{k=0}^{N-1} f(P_k, Q_k)$ can be taken as a discrete version of Z_T. If $\{Z_T\}_{T>0}$ satisfies an LDP with the rate function $\mathbf{I}$, a natural question is

Question 2.4.1 *Does $\{Z_{T,N}\}_{N\geq 1}$ satisfy an LDP with some rate function $\mathbf{I}^h$ for a fixed time-step size h?*

If the answer is positive, then $\{Z_T\}_{T>0}$ and $\{Z_{T,N}\}_{N\geq 1}$ formally satisfy

$$\mathbb{P}(Z_T \in [a, a+da]) \approx \exp(-T\mathbf{I}(a))da, \quad T \gg 1,$$

$$\mathbb{P}(Z_{T,N} \in [a, a+da]) \approx \exp(-N\mathbf{I}^h(a))da = \exp(-t_N\mathbf{I}^h(a)/h)da, \quad t_N \gg 1.$$

Let $T = t_N$ be the observation scale. It is reasonable to take advantage of $\mathbf{I}^h/h$ to evaluate the ability of the numerical method to preserve the large deviation rate function. Therefore, another meaningful question is

Question 2.4.2 *Is* $\mathbf{I}^h/h$ *a good approximation of* $\mathbf{I}$ *for sufficiently small* h *if* $\{Z_{T,N}\}_{N\geq 1}$ *satisfies an LDP with the rate function* $\mathbf{I}^h$?

Concerning the above questions, we introduce the following definition on the asymptotical or even exact preservation of the LDP.

Definition 2.4.1 (See [54]) Let E be a Polish space, i.e., complete and separable metric space, $\{Z_T\}_{T>0}$ be a stochastic process associated with the exact solution of (2.19), and $\{Z_{T,N}\}_{N\geq 1}$ be a discrete approximation of $\{Z_T\}_{T>0}$, associated with some numerical method $\{(P_k, Q_k)\}_{k\geq 0}$ for (2.19). Assume that $\{Z_T\}_{T>0}$ and $\{Z_{T,N}\}_{N\geq 1}$ satisfy the LDPs on E with rate functions $\mathbf{I}$ and $\mathbf{I}^h$, respectively. We call $\mathbf{I}^h_{mod} := \mathbf{I}^h/h$ the modified rate function of $\mathbf{I}^h$. Moreover, the numerical method $\{(P_k, Q_k)\}_{k\geq 0}$ is said to asymptotically preserve the LDP of $\{Z_T\}_{T>0}$, if

$$\lim_{h\to 0} \mathbf{I}^h_{mod}(y) = \mathbf{I}(y) \quad \forall\, y \in \mathrm{E}. \tag{2.71}$$

In particular, the numerical method $\{(P_k, Q_k)\}_{k\geq 0}$ is said to exactly preserve the LDP of $\{Z_T\}_{T>0}$, if for any sufficiently small time-step size h, $\mathbf{I}^h_{mod}(\cdot) = \mathbf{I}(\cdot)$.

Considering that the linear stochastic oscillator

$$\begin{cases} \ddot{X} + X = \alpha \dot{W}, \\ X(0) = x_0, \ \dot{X}(0) = y_0 \end{cases} \tag{2.72}$$

with $\alpha > 0$ and $W(\cdot)$ being a 1-dimensional standard Wiener process, is one of typical stochastic Hamiltonian systems (see Example 1.2.2), we take it as the test equation to obtain precise results about rate functions of LDP for both exact and numerical solutions. It is known that exact solution can be expressed as

$$X(t) = x_0 \cos(t) + y_0 \sin(t) + \alpha \int_0^t \sin(t - s) dW(s). \tag{2.73}$$

Now we introduce the so-called mean position

$$A_T = \frac{1}{T} \int_0^T X(t) dt \quad \forall\, T > 0. \tag{2.74}$$

A_T is a crucial observable and has many applications in physics. For instance, the Ornstein–Uhlenbeck process is often applied to illustrating the velocity of a particle moving in a random environment (see [212]). In this case, A_T can be interpreted as the mean value of the displacement process $\int_0^T X(t) dt$ (see [101]). By making use of the Gärtner–Ellis theorem (Theorem A.2.1), we prove that the mean position $\{A_T\}_{T>0}$ of the exact solution satisfies an LDP.

Theorem 2.4.1 (See [54]) $\{A_T\}_{T>0}$ *satisfies an LDP with the good rate function* $\mathbf{I}(y) = \frac{y^2}{3\alpha^2}$, *i.e.,*

$$\liminf_{T\to\infty} \frac{1}{T} \log(\mathbb{P}(A_T \in U)) \geq -\inf_{y\in U} \mathbf{I}(y) \quad \textit{for every open set } U \subset \mathbb{R},$$

$$\limsup_{T\to\infty} \frac{1}{T} \log(\mathbb{P}(A_T \in G)) \leq - \inf_{y\in G} \mathbf{I}(y) \quad \textit{for every closed set } G \subset \mathbb{R}.$$

Proof According to (2.73), (2.74) and the stochastic Fubini theorem, we have

$$T A_T = \int_0^T X(t)dt$$

$$= x_0 \sin(T) + y_0(1 - \cos(T)) + \alpha \int_0^T (1 - \cos(T - s)) \, dW(s).$$

Thus, $\mathbb{E}[T A_T] = x_0 \sin(T) + y_0(1 - \cos(T))$, and

$$\mathbb{VAR}[T A_T] = \alpha^2 \int_0^T (1 - \cos(T - s))^2 \, ds = \alpha^2 \left(\frac{3T}{2} - 2\sin(T) + \frac{\sin(2T)}{4} \right).$$

These imply $\lambda T A_T \sim \mathcal{N}\left(\lambda \mathbb{E}[T A_T], \lambda^2 \mathbb{VAR}[T A_T]\right)$ for $\lambda \in \mathbb{R}$. Based on the characteristic function of $\lambda T A_T$, we obtain

$$\mathbb{E}[\exp(\lambda T A_T)] = \exp\left(\lambda \mathbb{E}[T A_T] + \frac{\lambda^2}{2} \mathbb{VAR}[T A_T] \right).$$

In this way, we get the logarithmic moment generating function

$$\Lambda(\lambda) = \lim_{T\to\infty} \frac{1}{T} \log \mathbb{E}\left[\exp(\lambda T A_T) \right] = \frac{3\alpha^2}{4} \lambda^2,$$

which implies that $\Lambda(\cdot)$ is an essentially smooth, lower semi-continuous function. Moreover, we have that the origin belongs to $\mathscr{D}_\Lambda^\circ = \mathbb{R}$, where $\mathscr{D}_\Lambda^\circ$ is the interior of $\mathscr{D}_\Lambda := \{x \in \mathbb{R} : \Lambda(x) < \infty\}$. Due to Theorem A.2.1, we derive that $\{A_T\}_{T>0}$ satisfies an LDP with the good rate function

$$\mathbf{I}(y) = \Lambda^*(y) = \sup_{\lambda \in \mathbb{R}} \{y\lambda - \Lambda(\lambda)\} = \frac{y^2}{3\alpha^2},$$

which completes the proof.   □

Theorem 2.4.1 means that, for any initial value (x_0, y_0), the probability that the mean position $\{A_T\}_{T>0}$ hits the interval $[a, a + da]$ formally satisfies

$$\mathbb{P}(A_T \in [a, a + da]) \approx \exp(-T\mathbf{I}(a))da = \exp\left(-T \frac{a^2}{3\alpha^2} \right)da$$

for sufficiently large T. Moreover, the LDP for $\{A_T\}_{T>0}$ does not depend on the initial value (x_0, y_0) of the stochastic oscillator (2.72).

Let (X_k, Y_k) be the discrete approximation at $t_k = kh$ with $X_k \approx X(t_k)$ and $Y_k \approx \dot{X}(t_k)$, where $k \in \mathbb{N}$. Following [234], we consider the general numerical method in form of

$$\begin{bmatrix} X_{k+1} \\ Y_{k+1} \end{bmatrix} = A \begin{bmatrix} X_k \\ Y_k \end{bmatrix} + \alpha b \Delta_k W := \begin{bmatrix} a_{11} & a_{12} \\ a_{21} & a_{22} \end{bmatrix} \begin{bmatrix} X_k \\ Y_k \end{bmatrix} + \alpha \begin{bmatrix} b_1 \\ b_2 \end{bmatrix} \Delta_k W \quad (2.75)$$

with $\Delta_k W = W(t_{k+1}) - W(t_k)$, $k \in \mathbb{N}$. In fact, the real matrix A and real vector b rely on both the numerical method and time-step size h. In addition, we assume $b_1^2 + b_2^2 \neq 0$, which is reasonable because an effective numerical method for (2.72) must lie on the increment of the Wiener process. Define the discrete mean position as

$$A_N = \frac{1}{N} \sum_{k=0}^{N-1} X_k, \quad N \in \mathbb{N}_+. \quad (2.76)$$

In what follows, we investigate the LDP of $\{A_N\}_{N \geq 1}$ of (2.75) and study how closely the LDP of $\{A_N\}_{N \geq 1}$ approximates the LDP of $\{A_T\}_{T > 0}$, and use the notation $K(c_1, \ldots, c_m)$ to stand for some constant dependent on parameters c_i, $i = 1, \ldots, m$, but independent of N, which may vary from one line to another.

2.4.1 LDP for Numerical Methods

In this part, we consider the LDP of the discrete mean position for the general numerical method. Our idea is to utilize Theorem A.2.1 to study the LDP of $\{A_N\}_{N \geq 1}$. Hence, we first deduce the logarithmic moment generating function

$$\Lambda^h(\lambda) := \lim_{N \to \infty} \frac{1}{N} \log \mathbb{E}\big[\exp(\lambda N A_N) \big]$$

for a fixed appropriate time-step size h. This can be done via the general formula of $\{X_k\}_{k \geq 1}$.

Aiming at giving the general formula of X_k, we denote $M_k = \begin{bmatrix} X_{k+1} \\ X_k \end{bmatrix}$ for $k \geq 1$. The recurrence (2.75) yields that $M_k = B M_{k-1} + r_k$, $k \geq 1$, with

$$B = \begin{bmatrix} \mathrm{Tr}(A) & -\det(A) \\ 1 & 0 \end{bmatrix}, \quad r_k = \begin{bmatrix} \alpha \, (b_1 \Delta_k W + (a_{12}b_2 - a_{22}b_1)\Delta_{k-1}W) \\ 0 \end{bmatrix},$$

where $\mathrm{Tr}(A)$ and $\det(A)$ denote the trace and determinant of A, respectively. In this way, we obtain $M_k = B^k M_0 + \sum_{j=1}^{k} B^{k-j} r_j$, $k \geq 1$.

Assumption 2.4.1 *Suppose that* $\mathrm{Tr}(A)$ *and* $\det(A)$ *satisfy*

$$4\det(A) - (\mathrm{Tr}(A))^2 > 0.$$

Then eigenvalues of B are

$$\lambda_{\pm} = \frac{\mathrm{Tr}(A)}{2} \pm \mathbf{i}\frac{\sqrt{4\det(A) - (\mathrm{Tr}(A))^2}}{2} = \sqrt{\det(A)}\exp(\pm\mathbf{i}\theta), \quad \mathbf{i}^2 = -1$$

for some $\theta \in (0, \pi)$ satisfying

$$\cos(\theta) = \frac{\mathrm{Tr}(A)}{2\sqrt{\det(A)}}, \quad \sin(\theta) = \frac{\sqrt{4\det(A) - (\mathrm{Tr}(A))^2}}{2\sqrt{\det(A)}}. \tag{2.77}$$

Let $\hat{\alpha}_k = (\det(A))^{k/2}\frac{\sin((k+1)\theta)}{\sin(\theta)}$ and $\hat{\beta}_k = -(\det(A))^{(k+1)/2}\frac{\sin(k\theta)}{\sin(\theta)}$ for $k \in \mathbb{N}$. It follows from the expression of M_k (see [234]) that

$$X_{k+1} = \hat{\alpha}_k X_1 + \hat{\beta}_k x_0 + \alpha \sum_{j=1}^{k} \hat{\alpha}_{k-j}\left(b_1\Delta_j W + (a_{12}b_2 - a_{22}b_1)\Delta_{j-1}W\right)$$

for $k \geq 0$. Because $X_1 = a_{11}x_0 + a_{12}y_0 + \alpha b_1\Delta_0 W$, $\hat{\alpha}_{-1} = 0$ and $\hat{\alpha}_0 = 1$, for $k \geq 1$,

$$X_k = \left(a_{11}\hat{\alpha}_{k-1} + \hat{\beta}_{k-1}\right)x_0 + a_{12}\hat{\alpha}_{k-1}y_0$$

$$+ \alpha \sum_{j=0}^{k-1}\left(b_1\hat{\alpha}_{k-1-j} + (a_{12}b_2 - a_{22}b_1)\hat{\alpha}_{k-2-j}\right)\Delta_j W. \tag{2.78}$$

Using (2.76) and (2.78), we arrive at

$$NA_N = x_0 + \sum_{k=1}^{N-1}X_k = \left(1 + a_{11}S_N^{\hat{\alpha}} + S_N^{\hat{\beta}}\right)x_0 + a_{12}S_N^{\hat{\alpha}}y_0 + \alpha\sum_{j=0}^{N-2}c_j\Delta_j W, \tag{2.79}$$

where

$$S_N^{\hat{\alpha}} = \frac{\sin(\theta) - \left(\sqrt{\det(A)}\right)^{N-1}\sin(N\theta) + \left(\sqrt{\det(A)}\right)^N\sin((N-1)\theta)}{\sin(\theta)\left(1 - 2\sqrt{\det(A)}\cos(\theta) + \det(A)\right)},$$

$$S_N^{\hat{\beta}} = -\frac{\det(A)\sin(\theta) - \left(\sqrt{\det(A)}\right)^N\sin((N-1)\theta) + \left(\sqrt{\det(A)}\right)^{N+1}\sin((N-2)\theta)}{\sin(\theta)\left(1 - 2\sqrt{\det(A)}\cos(\theta) + \det(A)\right)},$$

$$c_j = \frac{b_1}{\sin(\theta)}\sin((N-1-j)\theta)\left(\sqrt{\det(A)}\right)^{N-2-j}$$

$$+ \frac{b_1 + a_{12}b_2 - a_{22}b_1}{\sin(\theta)} \left(\frac{\sin(\theta) - \left(\sqrt{\det(A)}\right)^{N-2-j} \sin((N-1-j)\theta)}{1 - 2\sqrt{\det(A)}\cos(\theta) + \det(A)} \right)$$

$$+ \frac{b_1 + a_{12}b_2 - a_{22}b_1}{\sin(\theta)} \left(\frac{\left(\sqrt{\det(A)}\right)^{N-1-j} \sin((N-2-j)\theta)}{1 - 2\sqrt{\det(A)}\cos(\theta) + \det(A)} \right).$$

Next, we investigate the LDP of $\{A_N\}_{N\geq 1}$ for symplectic methods and nonsymplectic ones, respectively. The method (2.75) preserves symplecticity if and only if $\det(A) = 1$. For nonsymplectic methods, we exclude the case of $\det(A) > 1$, where the logarithmic moment generating function Λ^h does not exist. More precisely, we will consider the case of $\det(A) = 1$ and the case of $0 < \det(A) < 1$ separately. Below we introduce assumptions with respect to $\det(A)$.

Assumption 2.4.2 *Assume that* $\det(A) = 1$.

Assumption 2.4.3 *Assume that* $0 < \det(A) < 1$.
LDP of $\{A_N\}_{N\geq 1}$ **for Symplectic Methods** In this part, we deduce the LDP of $\{A_N\}_{N\geq 1}$ for the numerical method (2.75) preserving the symplectic structure. Based on (2.79), NA_N is Gaussian. As a consequence,

$$\Lambda^h(\lambda) = \lim_{N\to\infty} \frac{1}{N} \log \mathbb{E}\big[\exp(\lambda N A_N)\big]$$

$$= \lim_{N\to\infty} \frac{1}{N} \left(\lambda \mathbb{E}[NA_N] + \frac{1}{2}\lambda^2 \mathbb{VAR}[NA_N] \right).$$

To obtain the expression of Λ^h, it suffices to give the estimates of $\mathbb{E}[NA_N]$ and $\mathbb{VAR}[NA_N]$ with respect to N.

Lemma 2.4.1 (See [54]) *Under Assumptions 2.4.1 and 2.4.2, we have*

(1) $b_1^2 + (a_{12}b_2 - a_{22}b_1)^2 \neq 0$;
(2) $(b_1 + a_{12}b_2 - a_{22}b_1)^2(4 + \mathrm{Tr}(A)) - 2b_1(a_{12}b_2 - a_{22}b_1)(2 - \mathrm{Tr}(A)) > 0$.

With the help of the above lemma, we have the following theorem.

Theorem 2.4.2 (See [54]) *If the numerical method (2.75) for approximating (2.72) satisfies Assumptions 2.4.1 and 2.4.2, then its mean position* $\{A_N\}_{N\geq 1}$ *satisfies an LDP with the good rate function*

$$\mathbf{I}^h(y) = \sup_{\lambda\in\mathbb{R}}\{y\lambda - \Lambda^h(\lambda)\}$$

$$= \frac{(2 + \mathrm{Tr}(A))(2 - \mathrm{Tr}(A))^2 y^2}{2\alpha^2 h\left[(b_1 + a_{12}b_2 - a_{22}b_1)^2(4 + \mathrm{Tr}(A)) - 2b_1(a_{12}b_2 - a_{22}b_1)(2 - \mathrm{Tr}(A))\right]}.$$
$$\tag{2.80}$$

Proof Under Assumption 2.4.2, we have $\hat{\alpha}_k = \frac{\sin((k+1)\theta)}{\sin(\theta)}, \hat{\beta}_k = -\frac{\sin(k\theta)}{\sin(\theta)}$ for $k \in \mathbb{N}$, and

$$S_N^{\hat{\alpha}} = \frac{\cos\left(\frac{\theta}{2}\right) - \cos((N - \frac{1}{2})\theta)}{2\sin(\theta)\sin\left(\frac{\theta}{2}\right)}, \quad S_N^{\hat{\beta}} = -\frac{\cos\left(\frac{\theta}{2}\right) - \cos((N - \frac{3}{2})\theta)}{2\sin(\theta)\sin\left(\frac{\theta}{2}\right)},$$

$$c_j = \frac{(b_1 + a_{12}b_2 - a_{22}b_1)\cos\left(\frac{\theta}{2}\right) - b_1\cos((N - \frac{1}{2} - j)\theta)}{2\sin(\theta)\sin\left(\frac{\theta}{2}\right)}$$

$$- \frac{(a_{12}b_2 - a_{22}b_1)\cos((N - \frac{3}{2} - j)\theta)}{2\sin(\theta)\sin\left(\frac{\theta}{2}\right)},$$

which yields that $|S_N^{\hat{\alpha}}| + |S_N^{\hat{\beta}}| \leq K(\theta)$ for $N \geq 2$. According to the independence of increments of Wiener processes, NA_N is Gaussian and satisfies

$$\left|\mathbb{E}[NA_N]\right| = \left|\left(1 + a_{11}S_N^{\hat{\alpha}} + S_N^{\hat{\beta}}\right)x_0 + a_{12}S_N^{\hat{\alpha}}y_0\right| \leq K(x_0, y_0, \theta),$$

$$\mathrm{VAR}[NA_N] = \alpha^2 h \sum_{j=0}^{N-2} c_j^2 = \frac{\alpha^2 h}{4\sin^2(\theta)\sin^2\left(\frac{\theta}{2}\right)} \sum_{j=0}^{N-2} \tilde{c}_j^2$$

with

$$\tilde{c}_j^2 = (b_1 + a_{12}b_2 - a_{22}b_1)^2 \cos^2\left(\frac{\theta}{2}\right) + \frac{1}{2}b_1^2$$

$$+ \frac{1}{2}(a_{12}b_2 - a_{22}b_1)^2 + b_1(a_{12}b_2 - a_{22}b_1)\cos(\theta) + R_j,$$

where

$$R_j = \frac{b_1^2}{2}\cos((2N - 1 - 2j)\theta) + \frac{(a_{12}b_2 - a_{22}b_1)^2}{2}\cos((2N - 3 - 2j)\theta)$$

$$- 2b_1(b_1 + a_{12}b_2 - a_{22}b_1)\cos\left(\frac{\theta}{2}\right)\cos\left(\frac{(2N - 1 - 2j)\theta}{2}\right)$$

$$- 2(b_1 + a_{12}b_2 - a_{22}b_1)(a_{12}b_2 - a_{22}b_1)\cos\left(\frac{\theta}{2}\right)\cos\left(\frac{(2N - 3 - 2j)\theta}{2}\right)$$

$$+ b_1(a_{12}b_2 - a_{22}b_1)\cos((2N - 2 - 2j)\theta).$$

Because $\sum_{n=1}^{N} \cos\left((2n+1)\theta\right) = \frac{\sin((2N+2)\theta)-\sin(2\theta)}{2\sin(\theta)}$, we get

$$\left| \sum_{j=0}^{N-2} \cos((2N-1-2j)\theta) \right| = \left| \sum_{n=1}^{N-1} \cos((2n+1)\theta) \right| = \left| \frac{\sin(2N\theta)-\sin(2\theta)}{2\sin(\theta)} \right| \le K(\theta).$$

Similarly, we have

$$\left| \sum_{j=0}^{N-2} \cos((2N-3-2j)\theta) \right| + \left| \sum_{j=0}^{N-2} \cos\left(\frac{(2N-1-2j)\theta}{2} \right) \right|$$

$$+ \left| \sum_{j=0}^{N-2} \cos\left(\frac{(2N-3-2j)\theta}{2} \right) \right| + \left| \sum_{j=0}^{N-2} \cos((2N-2-2j)\theta) \right| \le K(\theta),$$

which yields $\left| \sum_{j=0}^{N-2} R_j \right| \le K(\theta)$. According to this inequality, we obtain

$$\Lambda^h(\lambda) := \lim_{N\to\infty} \frac{1}{N} \log \mathbb{E}[\exp(\lambda N A_N)]$$

$$= \frac{\alpha^2 h \lambda^2}{8 \sin^2(\theta) \sin^2\left(\frac{\theta}{2}\right)} \left((b_1 + a_{12}b_2 - a_{22}b_1)^2 \cos^2\left(\frac{\theta}{2}\right) + \frac{1}{2}b_1^2 \right.$$

$$\left. + \frac{1}{2}(a_{12}b_2 - a_{22}b_1)^2 + b_1(a_{12}b_2 - a_{22}b_1)\cos(\theta) \right). \qquad (2.81)$$

As a result of (2.77) with $\det(A) = 1$, it holds that

$$\cos(\theta) = \frac{\mathrm{Tr}(A)}{2}, \quad \sin(\theta) = \frac{\sqrt{4 - (\mathrm{Tr}(A))^2}}{2},$$

$$\sin^2\left(\frac{\theta}{2}\right) = \frac{1 - \cos(\theta)}{2} = \frac{2 - \mathrm{Tr}(A)}{4}, \quad \cos^2\left(\frac{\theta}{2}\right) = \frac{1 + \cos(\theta)}{2} = \frac{2 + \mathrm{Tr}(A)}{4},$$

which leads to

$$\Lambda^h(\lambda) = \frac{\alpha^2 h \lambda^2}{2(2 + \mathrm{Tr}(A))(2 - \mathrm{Tr}(A))^2} \left((b_1 + a_{12}b_2 - a_{22}b_1)^2 (4 + \mathrm{Tr}(A)) \right.$$

$$\left. - 2b_1(a_{12}b_2 - a_{22}b_1)(2 - \mathrm{Tr}(A)) \right).$$

Lemma 2.4.1 implies that Λ^h is essentially smooth. By means of Theorem A.2.1, $\{A_N\}_{N\ge1}$ satisfies an LDP with the good rate function (2.80). $\qquad \square$

Remark 2.4.1 Theorem 2.4.2 implies that to make the LDP for $\{A_N\}_{N\geq 1}$ hold, the time-step size h needs to be restricted such that Assumptions 2.4.1 and 2.4.2 are satisfied. Moreover, the rate function $\mathbf{I}^h(y)$ does not rely on the initial value (x_0, y_0). More exactly, for an appropriate time-step size h and arbitrary initial value, $\{A_N\}_{N\geq 1}$ formally satisfies $\mathbb{P}(A_N \in [a, a + da]) \approx \exp(-N\mathbf{I}^h(a))da$ for sufficiently large N.

LDP of $\{A_N\}_{N\geq 1}$ for Nonsymplectic Methods In this part, we study the LDP of $\{A_N\}_{N\geq 1}$ for the numerical method (2.75) which does not inherit the symplectic structure under the following assumption.

Assumption 2.4.4 *Assume that* $b_1 + a_{12}b_2 - a_{22}b_1 \neq 0$.

Theorem 2.4.3 (See [54]) *If the numerical method (2.75) for approximating (2.72) satisfies Assumptions 2.4.1, 2.4.3 and 2.4.4, then its mean position $\{A_N\}_{N\geq 1}$ satisfies an LDP with the good rate function*

$$\widetilde{\mathbf{I}}^h(y) = \frac{y^2}{2\alpha^2 h} \left(\frac{1 - \mathrm{Tr}(A) + \det(A)}{b_1 + a_{12}b_2 - a_{22}b_1} \right)^2.$$

Proof Under Assumption 2.4.3, one may immediately conclude that $\left| S_N^{\hat{\alpha}} \right| + \left| S_N^{\hat{\beta}} \right| \leq K(\theta)$ for all $N \geq 2$, which yields

$$|\mathbb{E}[NA_N]| \leq K(x_0, y_0, \theta), \quad \mathbb{VAR}[NA_N] = \alpha^2 h \sum_{j=0}^{N-2} c_j^2,$$

where

$$c_j^2 = \left(\frac{b_1 + a_{12}b_2 - a_{22}b_1}{1 - 2\sqrt{\det(A)}\cos(\theta) + \det(A)} \right)^2 + \tilde{R}_j$$

with

$$\tilde{R}_j = \frac{b_1^2 \sin^2((N-1-j)\theta)(\det(A))^{N-2-j}}{\sin^2(\theta)} + \frac{(b_1 + a_{12}b_2 - a_{22}b_1)^2}{\sin^2(\theta)\left(1 - 2\sqrt{\det(A)}\cos(\theta) + \det(A)\right)^2}$$

$$\times \left((\det(A))^{N-2-j} \sin^2((N-1-j)\theta) + (\det(A))^{N-1-j} \sin^2((N-2-j)\theta) \right.$$

$$- 2\sin(\theta)\left(\sqrt{\det(A)}\right)^{N-2-j} \sin((N-1-j)\theta)$$

$$+ 2\sin(\theta)\left(\sqrt{\det(A)}\right)^{N-1-j} \sin((N+2-j)\theta)$$

$$\left. - 2\left(\sqrt{\det(A)}\right)^{2N-3-2j} \sin((N-1-j)\theta)\sin((N-2-j)\theta) \right)$$

$$+ \frac{2b_1(b_1 + a_{12}b_2 - a_{22}b_1)}{\sin^2(\theta)\left(1 - 2\sqrt{\det(A)}\cos(\theta) + \det(A)\right)}$$

$$\times \left(\left(\sqrt{\det(A)} \right)^{N-2-j} \sin(\theta) \sin((N-1-j)\theta) - (\det(A))^{N-2-j} \sin^2((N-1-j)\theta) \right.$$
$$\left. + \left(\sqrt{\det(A)} \right)^{2N-3-2j} \sin((N-1-j)\theta) \sin((N-2-j)\theta) \right).$$

Further, we arrive at

$$\sum_{j=0}^{N-2} \left((\det(A))^{N-2-j} + (\det(A))^{N-1-j} \right) \le 2 \sum_{j=0}^{N} (\det(A))^j,$$

$$\sum_{j=0}^{N-2} \left(\left(\sqrt{\det(A)} \right)^{N-2-j} + \left(\sqrt{\det(A)} \right)^{N-1-j} \right) \le 2 \sum_{j=0}^{N} \left(\sqrt{\det(A)} \right)^j,$$

$$\sum_{j=0}^{N-2} \left(\sqrt{\det(A)} \right)^{2N-3-2j} = \sum_{j=1}^{N-1} \left(\sqrt{\det(A)} \right)^{2j-1} \le \frac{1}{\sqrt{\det(A)}} \sum_{j=0}^{N} (\det(A))^j.$$

It follows from the boundedness of $\sin(\cdot)$ and $\det(A) < 1$ that

$$\left| \sum_{j=0}^{N-2} \tilde{R}_j \right| \le K(\theta, A) \sum_{j=0}^{N} \left(\left(\sqrt{\det(A)} \right)^j + (\det(A))^j \right) \le \tilde{K}(\theta, A), \qquad (2.82)$$

which yields

$$\tilde{\Lambda}^h(\lambda) = \lim_{N \to \infty} \frac{1}{N} \log \mathbb{E}[\exp(\lambda N A_N)]$$

$$= \frac{\alpha^2 h \lambda^2}{2} \lim_{N \to \infty} \frac{1}{N} \left(\left(\frac{b_1 + a_{12}b_2 - a_{22}b_1}{1 - 2\sqrt{\det(A)}\cos(\theta) + \det(A)} \right)^2 (N-1) + \sum_{j=0}^{N-2} \tilde{R}_j \right)$$

$$= \frac{\alpha^2 h \lambda^2}{2} \left(\frac{b_1 + a_{12}b_2 - a_{22}b_1}{1 - 2\sqrt{\det(A)}\cos(\theta) + \det(A)} \right)^2.$$

Then, based on Assumption 2.4.4 and Theorem A.2.1, we derive that $\{A_N\}_{N \ge 1}$ satisfies an LDP with the good rate function

$$\tilde{\mathbf{I}}^h(y) = \frac{y^2}{2\alpha^2 h} \left(\frac{1 - 2\sqrt{\det(A)}\cos(\theta) + \det(A)}{b_1 + a_{12}b_2 - a_{22}b_1} \right)^2$$

$$= \frac{y^2}{2\alpha^2 h} \left(\frac{1 - \mathrm{Tr}(A) + \det(A)}{b_1 + a_{12}b_2 - a_{22}b_1} \right)^2,$$

where we have used (2.77) in the second equality. Thus, we finish the proof. $\qquad \square$

2.4.2 Asymptotical Preservation for LDP

Now we introduce the asymptotical preservation for the LDP of $\{A_T\}_{T>0}$ as the time-step size h tends to 0. By Definition 2.4.1, we arrive at the modified rate functions appearing in Theorems 2.4.2 and 2.4.3, respectively, as follows

$$\mathbf{I}^h_{mod}(y) = \frac{(2+\mathrm{Tr}(A))(2-\mathrm{Tr}(A))^2 y^2}{2\alpha^2 h^2 \left[(b_1 + a_{12}b_2 - a_{22}b_1)^2(4+\mathrm{Tr}(A)) - 2b_1(a_{12}b_2 - a_{22}b_1)(2-\mathrm{Tr}(A))\right]},$$

$$(2.83)$$

$$\tilde{\mathbf{I}}^h_{mod}(y) = \frac{y^2}{2\alpha^2 h^2}\left(\frac{1-\mathrm{Tr}(A)+\det(A)}{b_1 + a_{12}b_2 - a_{22}b_1}\right)^2. \qquad (2.84)$$

It may not be possible to obtain the asymptotical convergence for $\mathbf{I}^h_{mod}$ and $\tilde{\mathbf{I}}^h_{mod}$ only by making use of Assumptions 2.4.1–2.4.4 in two aspects: one is that both A and b are some functions of time-step size h, which are unclear unless a specific method is employed; the other is that the numerical approximation with some A and b does not converge to the original system. A solution to this problem is considering the convergence of numerical methods on the finite interval. In what follows, we study the mean-square convergence of (2.75).

For the sake of simplicity, we first introduce some notations.

1. Let $R = O(h^p)$ represent $|R| \le Ch^p$, for sufficiently small time-step size h, where C does not depend on h;
2. $f(h) \sim h^p$ implies that $f(h)$ and h^p are equivalent infinitesimal.

As (2.72) is driven by the additive noise, the convergence order of general numerical methods, which are known for the moment to approximate (2.72), in mean-square sense is no less than 1. Therefore, we present sufficient conditions which make (2.75) have at least first order convergence in the mean-square sense based on the fundamental theorem on the mean-square order of convergence.

Assumption 2.4.5 *Assume that*

$$|a_{11} - 1| + |a_{22} - 1| + |a_{12} - h| + |a_{21} + h| = O(h^2),$$

$$|b_1| + |b_2 - 1| = O(h).$$

The following lemma is adopted to investigate whether (2.75) preserves asymptotically the LDP of $\{A_T\}_{T>0}$ for the exact solution.

Lemma 2.4.2 (See [54]) *Under Assumption 2.4.5, the following properties hold:*

(1) $\mathrm{Tr}(A) \to 2$ *as* $h \to 0$;
(2) $1 - \mathrm{Tr}(A) + \det(A) \sim h^2$;
(3) $b_1 + a_{12}b_2 - a_{22}b_1 \sim h$.

Theorem 2.4.4 (See [54]) *For the numerical method* (2.75) *approximating* (2.72), *if Assumptions 2.4.1, 2.4.2 and 2.4.5 hold, then* (2.75) *asymptotically preserves the LDP of* $\{A_T\}_{T>0}$, *i.e., the modified rate function* $\mathbf{I}^h_{mod}(y) = \mathbf{I}^h(y)/h$ *satisfies*

$$\lim_{h\to 0} \mathbf{I}^h_{mod}(y) = \mathbf{I}(y) \quad \forall\, y \in \mathbb{R},$$

where $\mathbf{I}$ *is the rate function of the LDP for* $\{A_T\}_{T>0}$.

Proof Noting that $\det(A) = 1$, Lemma 2.4.2 means $(2 - \mathrm{Tr}(A)) \sim h^2$. Thus,

$$\lim_{h\to 0} \frac{b_1(a_{12}b_2 - a_{22}b_1)\,(2 - \mathrm{Tr}(A))}{h^2} = 0. \tag{2.85}$$

Based on Lemma 2.4.2, (2.83) and (2.85), we obtain

$$\lim_{h\to 0} \mathbf{I}^h_{mod}(y) = \frac{y^2}{2\alpha^2} \frac{\lim\limits_{h\to 0}(2 + \mathrm{Tr}(A))}{\lim\limits_{h\to 0}\frac{(4+\mathrm{Tr}(A))(b_1+a_{12}b_2-a_{22}b_1)^2}{h^2} - 2\lim\limits_{h\to 0}\frac{b_1(a_{12}b_2-a_{22}b_1)(2-\mathrm{Tr}(A))}{h^2}}$$

$$= \frac{y^2}{3\alpha^2},$$

which completes the proof. □

Theorem 2.4.5 (See [54]) *For the numerical method* (2.75) *of* (2.72), *if Assumptions 2.4.1, 2.4.3, 2.4.4 and 2.4.5 hold, then* (2.75) *does not asymptotically preserve the LDP of* $\{A_T\}_{T>0}$, *i.e., for* $y \neq 0$, $\lim\limits_{h\to 0} \widetilde{\mathbf{I}}^h_{mod}(y) \neq \mathbf{I}(y)$, *where* $\widetilde{\mathbf{I}}^h_{mod}(y) = \widetilde{\mathbf{I}}^h(y)/h$, *and* $\mathbf{I}$ *is the rate function of the LDP for* $\{A_T\}_{T>0}$.

Proof Due to (2.84) and Lemma 2.4.2, we get $\lim\limits_{h\to 0} \widetilde{\mathbf{I}}^h_{mod}(y) = \frac{y^2}{2\alpha^2}$. As a result, we finish the proof. □

Remark 2.4.2 Theorems 2.4.4 and 2.4.5 imply that under appropriate conditions, symplectic methods asymptotically preserve the LDP of the mean position $\{A_T\}_{T>0}$ of the original system (2.72), while nonsymplectic ones do not. This indicates that, in comparison with nonsymplectic methods, symplectic methods have long-time stability in the aspect of the LDP of the mean position.

Now we turn to showing and comparing the LDPs of the mean position for some concrete numerical methods to verify theoretical results (see Table 2.2).

Table 2.2 Fulfillments of symplecticity and asymptotical preservation for the LDP of A_T by numerical methods for (2.72)

Numerical methods	Symplecticity	Asymptotical preservation for the LDP
Symplectic β-method	Yes	Yes
Exponential method	Yes	Yes
Integral method	Yes	Yes
Optimal method	Yes	Yes
Stochastic θ-method	No	No
PEM-MID method	No	No
EM-BEM method	No	No

Symplectic Methods

1. Symplectic β-method ($\beta \in [0, 1]$):

$$A^\beta = \frac{1}{1 + \beta(1-\beta)h^2} \begin{bmatrix} 1 - (1-\beta)^2 h^2 & h \\ -h & 1 - \beta^2 h^2 \end{bmatrix},$$

$$b^\beta = \frac{1}{1 + \beta(1-\beta)h^2} \begin{bmatrix} (1-\beta)h \\ 1 \end{bmatrix}.$$

A straightforward calculation yields

$$\det(A^\beta) = 1, \quad \mathrm{Tr}(A^\beta) = \frac{2 - (2\beta^2 - 2\beta + 1)h^2}{1 + \beta(1-\beta)h^2},$$

$$a_{12}^\beta b_2^\beta - a_{22}^\beta b_1^\beta = \frac{\beta h}{1 + \beta(1-\beta)h^2},$$

$$b_1^\beta + a_{12}^\beta b_2^\beta - a_{22}^\beta b_1^\beta = \frac{h}{1 + \beta(1-\beta)h^2}.$$

It can be verified that if $h \in (0, 2)$, then $\mathbf{I}^h(y) = \frac{hy^2}{3\alpha^2}\left(\frac{3}{2} - \frac{3}{6-(2\beta-1)^2 h^2}\right)$ is the good rate function of the LDP of $\{A_N\}_{N \geq 1}$ for the symplectic β-method by means of Theorem 2.4.2. Further, we obtain $\lim_{h \to 0} \mathbf{I}_{mod}^h(y) = \mathbf{I}(y) = \frac{y^2}{3\alpha^2}$ for $y \in \mathbb{R}$, which is consistent with Theorem 2.4.4. Moreover, for every $h > 0$, the modified rate function of the mean position for the midpoint method, i.e., $\beta = \frac{1}{2}$, is the same as that for the exact solution. It means that the midpoint method exactly preserves the LDP of $\{A_T\}_{T>0}$. In the case of $\beta \neq \frac{1}{2}$, $\mathbf{I}_{mod}^h(y) < \mathbf{I}(y)$ if $y \neq 0$. That is to say, as $T = t_N$ tends to infinity simultaneously, the exponential decay speed of $\mathbb{P}(A_N \in [a, a+da])$ is slower than that of $\mathbb{P}(A_T \in [a, a+da])$ provided $a \neq 0$.

2. Exponential method:

$$A^{EX} = \begin{bmatrix} \cos(h) & \sin(h) \\ -\sin(h) & \cos(h) \end{bmatrix}, \quad b^{EX} = \begin{bmatrix} 0 \\ 1 \end{bmatrix}.$$

It holds that

$$\det(A^{EX}) = 1, \quad \mathrm{Tr}(A^{EX}) = 2\cos(h),$$

$$a_{12}^{EX} b_2^{EX} - a_{22}^{EX} b_1^{EX} = \sin(h),$$

$$b_1^{EX} + a_{12}^{EX} b_2^{EX} - a_{22}^{EX} b_1^{EX} = \sin(h).$$

If $h \in (0, \pi)$, then $\{A_N\}_{N \geq 1}$ satisfies an LDP with the modified rate function $\mathbf{I}_{mod}^h(y) = \frac{2y^2}{\alpha^2} \frac{1-\cos(h)}{h^2(2+\cos(h))}$. Therefore, we deduce that $\lim_{h \to 0} \mathbf{I}_{mod}^h(y) = \frac{y^2}{3\alpha^2} = \mathbf{I}(y)$ and $\mathbf{I}_{mod}^h(y) > \mathbf{I}(y)$ provided that $h \in (0, \pi/6)$ and $y \neq 0$. These mean that if $h \in (0, \pi/6)$, then the mean position $\{A_N\}_{N \geq 1}$ of the exponential method satisfies an LDP, which asymptotically preserves the LDP of $\{A_T\}_{T>0}$. Moreover, as $T = t_N$ tends to infinity simultaneously, the exponential decay speed of $\mathbb{P}(A_N \in [a, a+da])$ is faster than that of $\mathbb{P}(A_T \in [a, a+da])$ if $a \neq 0$.

3. Integral method:

$$A^{INT} = \begin{bmatrix} \cos(h) & \sin(h) \\ -\sin(h) & \cos(h) \end{bmatrix}, \quad b^{INT} = \begin{bmatrix} \sin(h) \\ \cos(h) \end{bmatrix}.$$

A straight calculation yields $\det(A^{INT}) = 1$, $\mathrm{Tr}(A^{INT}) = 2\cos(h)$, $a_{12}^{INT} b_2^{INT} - a_{22}^{INT} b_1^{INT} = 0$, and $b_1^{INT} + a_{12}^{INT} b_2^{INT} - a_{22}^{INT} b_1^{INT} = \sin(h)$. It is obvious that the modified rate function of $\{A_N\}_{N \geq 1}$ is $\mathbf{I}_{mod}^h(y) = \frac{2y^2}{\alpha^2} \frac{1-\cos(h)}{h^2(2+\cos(h))}$. This case is the same as that of the exponential method.

4. Optimal method:

$$A^{OPT} = \begin{bmatrix} \cos(h) & \sin(h) \\ -\sin(h) & \cos(h) \end{bmatrix}, \quad b^{OPT} = \frac{1}{h}\begin{bmatrix} 2\sin^2(\frac{h}{2}) \\ \sin(h) \end{bmatrix}.$$

According to the above two formulae, one gets

$$\det(A^{OPT}) = 1, \quad \mathrm{Tr}(A^{OPT}) = 2\cos(h),$$

$$b_1^{OPT} = a_{12}^{OPT} b_2^{OPT} - a_{22}^{OPT} b_1^{OPT} = \frac{1-\cos(h)}{h}.$$

If $h \in (0, \pi)$, then $\{A_N\}_{N \geq 1}$ of the optimal method satisfies an LDP with the modified rate function $\mathbf{I}_{mod}^h(y) = \frac{y^2}{3\alpha^2} = \mathbf{I}(y)$. Therefore, we conclude that the LDP of the mean position $\{A_N\}_{N \geq 1}$ for the optimal method preserves the LDP of $\{A_T\}_{T>0}$ exactly.

Nonsymplectic Methods

1. Stochastic θ-method ($\theta \in [0, 1/2) \cup (1/2, 1]$):

$$A^\theta = \frac{1}{1 + \theta^2 h^2} \begin{bmatrix} 1 - (1 - \theta)\theta h^2 & h \\ -h & 1 - (1 - \theta)\theta h^2 \end{bmatrix}, \quad b^\theta = \frac{1}{1 + \theta^2 h^2} \begin{bmatrix} \theta h \\ 1 \end{bmatrix}.$$

For this method, we derive

$$\det(A^\theta) = \frac{1 + (1 - \theta)^2 h^2}{1 + \theta^2 h^2}, \quad 1 - \text{Tr}(A^\theta) + \det(A^\theta) = \frac{h^2}{1 + \theta^2 h^2},$$

$$b_1^\theta + a_{12}^\theta b_2^\theta - a_{22}^\theta b_1^\theta = \frac{h}{1 + \theta^2 h^2}.$$

Notice that $0 < \det(A^\theta) < 1$ is equivalent to $\theta \in (1/2, 1]$. One can verify that, for $\theta \in (1/2, 1]$ and $h > 0$, the mean position $\{A_N\}_{N \geq 1}$ satisfies an LDP with the modified rate function $\widetilde{\mathbf{I}}_{mod}^h(y) = \frac{y^2}{2\alpha^2}$, which coincides with Theorem 2.4.5.

2. Predictor–Corrector (PEM-MID) method via the partitioned Euler–Maruyama (PEM) method as the predictor and the midpoint (MID) method as the corrector:

$$A^{(1)} = \begin{bmatrix} 1 - h^2/2 & h(1 - h^2/2) \\ -h & 1 - h^2/2 \end{bmatrix}, \quad b^{(1)} = \begin{bmatrix} h/2 \\ 1 \end{bmatrix}.$$

One gets $1 - \text{Tr}(A^{(1)}) + \det(A^{(1)}) = h^2 - \frac{h^4}{4}$ and $b_1^{(1)} + a_{12}^{(1)} b_2^{(1)} - a_{22}^{(1)} b_1^{(1)} = h - \frac{h^3}{4}$. Then we can obtain that if $h \in (0, \sqrt{2})$, $\{A_N\}_{N \geq 1}$ of this method satisfies an LDP with the modified rate function $\widetilde{\mathbf{I}}_{mod}^h(y) = \frac{y^2}{2\alpha^2}$.

3. Predictor–Corrector (EM-BEM) method via the Euler–Maruyama (EM) method as the predictor and the backward Euler–Maruyama (BEM) method as the corrector:

$$A^{(2)} = \begin{bmatrix} 1 - h^2 & h \\ -h & 1 - h^2 \end{bmatrix}, \quad b^{(2)} = \begin{bmatrix} h \\ 1 \end{bmatrix}.$$

These relations lead to $1 - \text{Tr}(A^{(2)}) + \det(A^{(2)}) = h^2 + h^4$ and $b_1^{(2)} + a_{12}^{(2)} b_2^{(2)} - a_{22}^{(2)} b_1^{(2)} = h + h^3$. Thus, by Theorem 2.4.5, $\{A_N\}_{N \geq 1}$ of this method satisfies an LDP with the modified rate function $\widetilde{\mathbf{I}}_{mod}^h(y) = \frac{y^2}{2\alpha^2}$ provided $h \in (0, 1)$.

Numerical Experiments In this part, we perform numerical experiments to verify our theoretical results. We exploit the algorithm in [225] to numerically approximate the large deviation rate function of $\{A_T\}_{T > 0}$, where the key point is to simulate the logarithmic moment generating function via the Monte–Carlo method.

More precisely, for a numerical method $\{(X_k, Y_k)\}_{k \geq 1}$ approaching (2.72), we first arrive at M samplings of $\{X_k\}_{k=0}^{N_0 - 1}$ for a given N_0, which immediately generates M samplings $A_{N_0}^{(i)}$, $i = 1, \ldots, M$ (recall $A_{N_0} = \frac{1}{N_0} \sum_{k=0}^{N_0-1} X_k$).

Then we consider $G_{M,N_0}(\lambda) = \frac{1}{M} \sum_{i=1}^{M} \exp\left(\lambda N_0 A_{N_0}^{(i)}\right)$ as the approximation of $\mathbb{E}\left[\exp(\lambda N_0 A_{N_0})\right]$. Further, for sufficiently large N_0, we use $\Lambda_{M,N_0}^h(\lambda) = \frac{1}{N_0} \log G_{M,N_0}(\lambda)$ to approximate $\Lambda^h(\lambda) = \lim_{N \to \infty} \frac{1}{N} \log \mathbb{E}\left[\exp(\lambda N A_N)\right]$. Since $(G_{M,N_0})'(\lambda) = \frac{1}{M} \sum_{i=1}^{M} \exp\left(\lambda N_0 A_{N_0}^{(i)}\right) N_0 A_{N_0}^{(i)}$, we can simulate the value of the rate function $\mathbf{I}^h(y) = \sup_{\lambda \in \mathbb{R}} \left\{\lambda y - \Lambda^h(\lambda)\right\}$ at $y(\lambda) := \left(\Lambda_{M,N_0}^h\right)'(\lambda) = \frac{(G_{M,N_0})'(\lambda)}{N_0 G_{M,N_0}(\lambda)}$ by $\mathbf{I}_{M,N_0}^h(y(\lambda)) = \lambda y(\lambda) - \Lambda_{M,N_0}^h(\lambda)$. Thus, we have the following algorithm:

Algorithm 1:

---

1. Choose a proper time-step size h, number M of samples and number N_0 of steps. Compute numerical method $X_k^{(i)}$, $i = 1, \ldots, M$, $k = 0, 1, \ldots, N_0 - 1$.

2. Set $S_{N_0}(i) = \sum_{k=0}^{N_0-1} X_k^{(i)}$, $i = 1, \ldots, M$.

3. For a given $K > 0$, compute $G_{M,N_0}(\lambda) = \frac{1}{M} \sum_{i=1}^{M} \exp\left(\lambda S_{N_0}^{(i)}\right)$ and

$$\left(G_{M,N_0}\right)'(\lambda) = \frac{1}{M} \sum_{i=1}^{M} S_{N_0}^{(i)} \exp\left(\lambda S_{N_0}^{(i)}\right)$$ for sufficiently many $\lambda \in [-K, K]$.

4. Compute $\Lambda_{M,N_0}^h(\lambda) = \log\left(G_{M,N_0}(\lambda)\right)^{1/N_0}$ and $y(\lambda) = \frac{(G_{M,N_0})'(\lambda)}{N_0 G_{M,N_0}(\lambda)}$.

5. Compute $\mathbf{I}_{M,N_0}^h(y(\lambda)) = \lambda y(\lambda) - \Lambda_{M,N_0}^h(\lambda)$ and $\mathbf{I}_{mod}^{h,M,N_0}(y(\lambda)) = \frac{\mathbf{I}_{M,N_0}^h(y(\lambda))}{h}$.

Based on Algorithm 1, we numerically simulate modified rate functions of $\{A_N\}_{N \geq 1}$ of the midpoint method and the PEM-MID method. Set $(x_0, y_0) = (0.5, 0)$, $M = 2000$, $N_0 = 600$ and $\alpha = 1.5$. In the third step, G_{M,N_0} is calculated at $\lambda(j) = -K + 0.001(j-1)$, $j = 1, 2, \ldots, 2000K + 1$ with $K = 1.5$. It is obtained from Fig. 2.10 that as time-step size h decreases, the modified rate function $\mathbf{I}_{mod}^h$ of the midpoint method is closer to the rate function $\mathbf{I} : y \mapsto \frac{y^2}{3\alpha^2}$ of $\{A_T\}_{T>0}$, while the modified rate function of the PEM-MID method gets closer to $\widetilde{\mathbf{I}} : y \mapsto \frac{y^2}{2\alpha^2}$.

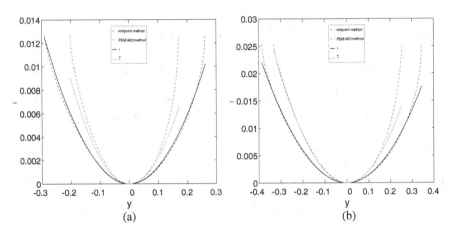

Fig. 2.10 Modified rate functions of the LDP of the mean position of the midpoint method and the PEM-MID method under different time-step sizes. (**a**) $h = 1$. (**b**) $h = 0.5$

Before ending Sect. 2.4, we would like to mention that if one wants to simulate the rate functions of observables associated with stochastic Hamiltonian systems by means of computing the logarithmic generating moment function, making use of the symplectic method is a prime choice, as is shown in Fig. 2.10.

2.5 Stochastic Pseudo-Symplectic Methods and Stochastic Methods Preserving Invariants

This section is devoted to another two kinds of stochastic structure-preserving numerical methods, i.e., the stochastic pseudo-symplectic methods and the stochastic numerical methods preserving invariants.

2.5.1 Stochastic Pseudo-Symplectic Methods

It is known that stochastic symplectic methods are implicit in general, and thus more computation cost will arise. Only for special stochastic Hamiltonian systems such as the separable system, explicit stochastic symplectic methods can be constructed. A nature problem arises.

Question 2.5.1 *Are there nearly symplecticity-preserving numerical methods that can be solved explicitly to fulfill the needs of the application?*

To address this issue, below we shall introduce explicit stochastic pseudo-symplectic methods which preserve the symplecticity in relatively long time frames with certain accuracy for the stochastic Hamiltonian system (2.19).

Definition 2.5.1 (See [207]) If a numerical method based on the one-step approximation $X_1 = \Phi_h(X_0)$ is of mean-square order M for the stochastic Hamiltonian system (2.19) and satisfies

$$\left(\mathbb{E}\left[\left\| \left(\frac{\partial X_1}{\partial X_0} \right)^\top J_{2n} \frac{\partial X_1}{\partial X_0} - J_{2n} \right\|^2 \right] \right)^{\frac{1}{2}} = O(h^{M'+1})$$

with $M' > M$, then this method is called a pseudo-symplectic method of mean-square order (M, M'), and M' is called the pseudo-symplectic order.

Now we introduce several pseudo-symplectic methods in [207] for the stochastic Hamiltonian system with additive noises

$$\begin{cases} dP(t) = -\dfrac{\partial H(P(t), Q(t))}{\partial Q} dt + \displaystyle\sum_{r=1}^m \sigma_r dW_r(t), & P(0) = p, \\[3mm] dQ(t) = \dfrac{\partial H(P(t), Q(t))}{\partial P} dt + \displaystyle\sum_{r=1}^m \gamma_r dW_r(t), & Q(0) = q, \end{cases} \tag{2.86}$$

where the Hamiltonian $H \in \mathbf{C}_b^\eta(\mathbb{R}^{2n}, \mathbb{R})$ for certain $\eta \in \mathbb{N}$, and σ_r, γ_r are constant column vectors for $r \in \{1, \ldots, m\}$. According to equations as follows

$$\begin{aligned} \widetilde{p} &= P_k - \frac{\partial H}{\partial Q}(P_k, Q_k)h + \sum_{r=1}^m \sigma_r \Delta_k W_r, \\[3mm] \widetilde{q} &= Q_k + \frac{\partial H}{\partial P}(P_k, Q_k)h + \sum_{r=1}^m \gamma_r \Delta_k W_r, \end{aligned} \tag{2.87}$$

we introduce the numerical method

$$\begin{aligned} P_{k+1} &= P_k - \alpha \frac{\partial H}{\partial Q}(P_k, Q_k)h - (1-\alpha)\frac{\partial H}{\partial Q}(\tilde{p}, \tilde{q})h + \sum_{r=1}^m \sigma_r \Delta_k W_r, \\[3mm] Q_{k+1} &= Q_k + \alpha \frac{\partial H}{\partial P}(P_k, Q_k)h + (1-\alpha)\frac{\partial H}{\partial P}(\tilde{p}, \tilde{q})h + \sum_{r=1}^m \gamma_r \Delta_k W_r, \end{aligned} \tag{2.88}$$

where $k = 0, 1, \ldots, N-1$, and $\alpha \in [0, 1]$. Moreover, based on (2.87), we construct another numerical method

$$P_{k+1} = P_k - \frac{\partial H}{\partial Q}\left(\alpha\tilde{p} + (1-\alpha)P_k, (1-\alpha)\tilde{q} + \alpha Q_k\right)h + \sum_{r=1}^{m}\sigma_r \Delta_k W_r,$$

$$Q_{k+1} = Q_k + \frac{\partial H}{\partial P}\left(\alpha\tilde{p} + (1-\alpha)P_k, (1-\alpha)\tilde{q} + \alpha Q_k\right)h + \sum_{r=1}^{m}\gamma_r \Delta_k W_r$$

$$(2.89)$$

with $k = 0, 1, \ldots, N-1$, and $\alpha \in [0, 1]$. Comparing the numerical methods (2.88) and (2.89) with the Euler–Maruyama method and then exploiting Theorem 2.1.2, we obtain the convergence of them with mean-square order 1. The following theorem shows that (2.88) and (2.89) are pseudo-symplectic.

Theorem 2.5.1 (See [207]) *Assume that the Hamiltonian $H \in \mathbf{C}_b^2(\mathbb{R}^{2n}, \mathbb{R})$. Then there exists $C_1 := C_1(H, \alpha) > 0$ such that the numerical methods (2.88) and (2.89) for the stochastic Hamiltonian system (2.86) satisfy*

$$\left(\mathbb{E}\left[\left\|\left(\frac{\partial(P_1, Q_1)}{\partial(p, q)}\right)^{\top}J_{2n}\frac{\partial(P_1, Q_1)}{\partial(p, q)} - J_{2n}\right\|^2\right]\right)^{\frac{1}{2}} = C_1 h^2 + O(h^3). \qquad (2.90)$$

Moreover, if $H \in \mathbf{C}_b^4(\mathbb{R}^{2n}, \mathbb{R})$, then there exists $C_2 := C_2(H) > 0$ such that

$$\left(\mathbb{E}\left[\left\|\left(\frac{\partial(P_1, Q_1)}{\partial(p, q)}\right)^{\top}J_{2n}\frac{\partial(P_1, Q_1)}{\partial(p, q)} - J_{2n}\right\|^2\right]\right)^{\frac{1}{2}} = |2\alpha - 1|C_2 h^2 + O(h^3).$$

$$(2.91)$$

Proof We only prove the results for (2.88), and analogous arguments can be applied to (2.89). Denoting $f = -\frac{\partial H}{\partial Q}$, $g = \frac{\partial H}{\partial P}$ and $\Delta W_r = W_r(h)$, $r = 1, \ldots, m$, and using one-step approximations of (2.87) and (2.88), we obtain

$$\tilde{p} := p + f(p, q)h + \sum_{r=1}^{m}\sigma_r \Delta W_r,$$

$$\tilde{q} := q + g(p, q)h + \sum_{r=1}^{m}\gamma_r \Delta W_r,$$

and

$$\overline{F} := P_1 - p - \alpha f(p,q)h - (1-\alpha)f(\tilde{p},\tilde{q})h - \sum_{r=1}^{m}\sigma_r \Delta W_r = 0,$$

$$\overline{G} := Q_1 - q - \alpha g(p,q)h - (1-\alpha)g(\tilde{p},\tilde{q})h - \sum_{r=1}^{m}\gamma_r \Delta W_r = 0.$$

From the above relations it follows that

$$\frac{\partial(\overline{F},\overline{G})}{\partial(P_1,Q_1)}\frac{\partial(P_1,Q_1)}{\partial(p,q)} + \frac{\partial(\overline{F},\overline{G})}{\partial(\tilde{p},\tilde{q})}\frac{\partial(\tilde{p},\tilde{q})}{\partial(p,q)} + \frac{\partial(\overline{F},\overline{G})}{\partial(p,q)} = 0,$$

where

$$\frac{\partial(\overline{F},\overline{G})}{\partial(P_1,Q_1)} = \begin{bmatrix} I_n & 0 \\ 0 & I_n \end{bmatrix}, \quad \frac{\partial(\overline{F},\overline{G})}{\partial(\tilde{p},\tilde{q})} = \begin{bmatrix} (1-\alpha)h\tilde{H}_{pq} & (1-\alpha)h\tilde{H}_{qq} \\ -(1-\alpha)h\tilde{H}_{pp} & -(1-\alpha)h\tilde{H}_{pq} \end{bmatrix},$$

$$\frac{\partial(\tilde{p},\tilde{q})}{\partial(p,q)} = \begin{bmatrix} I_n - hH_{pq} & -hH_{qq} \\ hH_{pp} & I_n + hH_{pq} \end{bmatrix}, \quad \frac{\partial(\overline{F},\overline{G})}{\partial(p,q)} = \begin{bmatrix} -I_n + \alpha h H_{pq} & \alpha h H_{qq} \\ -\alpha h H_{pp} & -I_n - h H_{pq} \end{bmatrix}$$

with $\tilde{H}_{pp} = \frac{\partial^2 H(\tilde{p},\tilde{q})}{\partial P^2}$, $\tilde{H}_{pq} = \frac{\partial^2 H(\tilde{p},\tilde{q})}{\partial P \partial Q}$, $\tilde{H}_{qq} = \frac{\partial^2 H(\tilde{p},\tilde{q})}{\partial Q^2}$, $H_{pp} = \frac{\partial^2 H(p,q)}{\partial P^2}$, $H_{pq} = \frac{\partial^2 H(p,q)}{\partial P \partial Q}$, and $H_{qq} = \frac{\partial^2 H(p,q)}{\partial Q^2}$. Then

$$\frac{\partial(P_1,Q_1)}{\partial(p,q)} = -\frac{\partial(\overline{F},\overline{G})}{\partial(\tilde{p},\tilde{q})}\frac{\partial(\tilde{p},\tilde{q})}{\partial(p,q)} - \frac{\partial(\overline{F},\overline{G})}{\partial(p,q)} =: \begin{bmatrix} \Phi_{11} & \Phi_{12} \\ \Phi_{21} & \Phi_{22} \end{bmatrix}$$

with

$$\Phi_{11} = I_n - (\alpha H_{pq} + (1-\alpha)\tilde{H}_{pq})h + (1-\alpha)(\tilde{H}_{pq}H_{pq} - \tilde{H}_{qq}H_{pp})h^2,$$

$$\Phi_{12} = -(\alpha H_{qq} + (1-\alpha)\tilde{H}_{qq})h + (1-\alpha)(\tilde{H}_{pq}H_{qq} - \tilde{H}_{qq}H_{pq})h^2,$$

$$\Phi_{21} = (\alpha H_{pp} + (1-\alpha)\tilde{H}_{pp})h - (1-\alpha)(\tilde{H}_{pp}H_{pq} - \tilde{H}_{pq}H_{pp})h^2,$$

$$\Phi_{22} = I_n + (\alpha H_{pq} + (1-\alpha)\tilde{H}_{pq})h - (1-\alpha)(\tilde{H}_{pp}H_{qq} - \tilde{H}_{pq}H_{pq})h^2,$$

which implies

$$\left(\frac{\partial(P_1,Q_1)}{\partial(p,q)}\right)^{\top} J_{2n} \frac{\partial(P_1,Q_1)}{\partial(p,q)} = \begin{bmatrix} \Phi_{11}^{\top}\Phi_{21} - \Phi_{21}^{\top}\Phi_{11} & \Phi_{11}^{\top}\Phi_{22} - \Phi_{21}^{\top}\Phi_{12} \\ \Phi_{12}^{\top}\Phi_{21} - \Phi_{22}^{\top}\Phi_{11} & \Phi_{12}^{\top}\Phi_{22} - \Phi_{22}^{\top}\Phi_{12} \end{bmatrix}.$$

In what follows we only prove the estimate for $\Phi_{11}^{\top}\Phi_{21} - \Phi_{21}^{\top}\Phi_{11}$, while similar ideas can be applied to proving the terms $\Phi_{11}^{\top}\Phi_{22} - \Phi_{21}^{\top}\Phi_{12}$, $\Phi_{12}^{\top}\Phi_{21} - \Phi_{22}^{\top}\Phi_{11}$ and

$\Phi_{12}^\top \Phi_{22} - \Phi_{22}^\top \Phi_{12}$. Direct calculations lead to

$$\|\Phi_{11}^\top \Phi_{21} - \Phi_{21}^\top \Phi_{11}\|_{\mathbf{L}^2(\Omega,\mathbb{R}^{n\times n})}$$

$$= \Big\|\alpha^2(H_{pp}H_{pq} - H_{pq}H_{pp}) + (1-\alpha)^2(\tilde{H}_{pp}\tilde{H}_{pq} - \tilde{H}_{pq}\tilde{H}_{pp}) - (1-\alpha)^2$$

$$\times (H_{pp}\tilde{H}_{pq} - H_{pq}\tilde{H}_{pp}) - (1-\alpha)^2(\tilde{H}_{pp}H_{pq} - \tilde{H}_{pq}H_{pp})\Big\|_{\mathbf{L}^2(\Omega,\mathbb{R}^{n\times n})}h^2 + O(h^3).$$

Thus (2.90) holds for certain positive constant C_1.

Now we assume $H \in \mathbf{C}_b^4(\mathbb{R}^{2n}, \mathbb{R})$. Taking Taylor expansions of both $\tilde{H}_{pp}$ and $\tilde{H}_{pq}$ at (p, q), we obtain

$$\tilde{H}_{pp} = H_{pp} + H_{ppp} \otimes (\tilde{p} - p) + H_{ppq} \otimes (\tilde{q} - q) + O(h),$$

$$\tilde{H}_{pq} = H_{pq} + H_{ppq} \otimes (\tilde{p} - p) + H_{pqq} \otimes (\tilde{q} - q) + O(h).$$

Since both $\tilde{p} - p$ and $\tilde{q} - q$ behave like $O(h^{\frac{1}{2}})$, the coefficient of h^2 in the estimation of $\|\Phi_{11}^\top \Phi_{21} - \Phi_{21}^\top \Phi_{11}\|_{\mathbf{L}^2(\Omega,\mathbb{R}^{n\times n})}$ is $|2\alpha - 1|(\mathbb{E}[\|H_{pp}H_{pq} - H_{pq}H_{pp}\|^2])^{\frac{1}{2}} + O(h)$. Thus there exists a positive constant $C_2 := C_2(H)$ such that

$$\|\Phi_{11}^\top \Phi_{21} - \Phi_{21}^\top \Phi_{11}\|_{\mathbf{L}^2(\Omega,\mathbb{R}^{n\times n})} = |2\alpha - 1|C_2 h^2 + O(h^3).$$

It leads to (2.91) and we complete the proof.   □

Remark 2.5.1 The expansion (2.90) yields that (2.88) and (2.89) for the stochastic Hamiltonian system (2.86) are pseudo-symplectic numerical methods of mean-square order $(1, 2)$ if and only if $\alpha = \frac{1}{2}$. In this case, (2.88) and (2.89) are called the stochastic pseudo-symplectic trapezoidal method and the stochastic pseudo-symplectic midpoint method, respectively.

The above pseudo-symplectic methods are proposed for the stochastic Hamiltonian system with additive noises (see [207]). By making use of the stochastic generating function, [10] presents a systematic method to obtain explicit pseudo-symplectic methods for the general stochastic Hamiltonian system

$$\begin{cases} dP(t) = -\dfrac{\partial H_0(P(t), Q(t))}{\partial Q}dt - \displaystyle\sum_{r=1}^m \dfrac{\partial H_r(P(t), Q(t))}{\partial Q} \circ dW_r(t), \quad P(0) = p, \\[4mm] dQ(t) = \dfrac{\partial H_0(P(t), Q(t))}{\partial P}dt + \displaystyle\sum_{r=1}^m \dfrac{\partial H_r(P(t), Q(t))}{\partial P} \circ dW_r(t), \quad Q(0) = q, \end{cases}$$

where the Hamiltonians H_r, $r = 0, 1, \ldots, m$, are assumed to be in $\mathbf{C}_b^k(\mathbb{R}^{2n}, \mathbb{R})$ for certain $k \in \mathbb{N}$. Although the approach in [10] can be applied to constructing stochastic pseudo-symplectic methods of any order, here we only show the numerical methods of order $(\frac{1}{2}, 1)$. Define the truncated generating function as

$$\bar{S}^h_\theta := hH_0 + \sum_{r=1}^m H_r \xi_r \sqrt{h} + \frac{(2\theta - 1)}{2} \sum_{r=1}^m \sum_{k=1}^n \frac{\partial H_r}{\partial \hat{Q}_k} \frac{\partial H_r}{\partial \hat{P}_k} h, \quad \theta = \frac{1}{2}, 1$$

with $\hat{P} = (1 - \theta)p + \theta P$, $\hat{Q} = (1 - \theta)Q + \theta q$, and independent random variables $\xi_r \sim \mathcal{N}(0, 1)$, $r = 1, \ldots, m$. We consider the following one-step approximations with the multi-index $\alpha \in \mathscr{A}_{\frac{1}{2}}$:

$$\tilde{p} = p - \frac{\partial \bar{S}^h_1}{\partial q}(p, q) + \sum_{r_1, r_2=1}^m \xi_{r_1} \xi_{r_2} \frac{\partial^2 H_{r_1}}{\partial P \partial q}(p, q) \frac{\partial H_{r_2}}{\partial Q}(p, q)h,$$

$$P_1 = p - \frac{\partial \bar{S}^h_1}{\partial q}(\tilde{p}, q), \quad Q_1 = q + \frac{\partial \bar{S}^h_1}{\partial P}(\tilde{p}, q),$$

(2.92)

and

$$\tilde{x} = x - J_{2n} \nabla \bar{S}^h_{\frac{1}{2}}(x) + \frac{1}{2} \sum_{r_1, r_2=1}^m \xi_{r_1} \xi_{r_2} J_{2n} \nabla^2 H_{r_1}(x) J_{2n} \nabla H_{r_2}(x)h,$$

$$X_1 = x - J_{2n} \nabla \bar{S}^h_{\frac{1}{2}}\left(\frac{x + \tilde{x}}{2}\right),$$

(2.93)

where $x = (p^\top, q^\top)^\top$.

Theorem 2.5.2 (See [10]) *Assume that* $H_r \in C_b^4(\mathbb{R}^{2n}, \mathbb{R})$, $r = 0, 1, \ldots, m$. *Then the numerical method based on the one-step approximation (2.92) or (2.93) is pseudo-symplectic of order* $(\frac{1}{2}, 1)$.

Remark 2.5.2 For stochastic differential equations, numerical methods with mean-square order higher than 2 are not computationally attractive, since they require generating approximations for multiple stochastic integrals of high order. In addition, stochastic pseudo-symplectic methods of high order will require the values of high order derivatives.

2.5.2 Stochastic Numerical Methods Preserving Invariants

In the fields of mechanics, astronomy, physics, molecular dynamics, etc., physical conservation laws are essential to describe fundamental characters and invariant quantities (see e.g., [201, 250] and references therein). It is a significant issue whether or not the invariants of the stochastic differential equations are preserved in performing reliable numerical methods. In this aspect, a range of numerical methods have been proposed to preserve the invariant in [49, 64, 124, 126, 181, 202] and references therein. Below we shall give some preliminaries about some stochastic numerical methods preserving invariants.

Consider the d-dimensional stochastic differential equation in the Stratonovich sense as follows

$$\begin{cases} dX(t) = \sigma_0(X(t))dt + \sum_{r=1}^{m} \sigma_r(X(t)) \circ dW_r(t), \\ X(0) = x \in \mathbb{R}^d, \end{cases} \qquad (2.94)$$

where $t \in (0, T]$ and $W_r(\cdot)$, $r = 1, \ldots, m$, are independent standard Wiener processes. Assume that coefficients $\sigma_r : \mathbb{R}^d \to \mathbb{R}^d$, $r = 0, 1, \ldots, m$, satisfy (1.2) such that (2.94) has a unique global solution. The stochastic differential Eq. (2.94) with an invariant $\mathbb{I}$ can be rewritten into another form as in the following theorem.

Theorem 2.5.3 (See [124]) *The stochastic differential Eq. (2.94) with an invariant $\mathbb{I}$ has the equivalent skew-gradient form*

$$dX(t) = T_0(X(t))\nabla\mathbb{I}(X(t))dt + \sum_{r=1}^{m} T_r(X(t))\nabla\mathbb{I}(X(t)) \circ dW_r(t), \qquad (2.95)$$

where $T_r(x)$ is a skew-symmetric matrix such that $T_r(x)\nabla\mathbb{I}(x) = \sigma_r(x)$ for $x \in \mathbb{R}^d$ and $r \in \{0, 1, \ldots, m\}$.

The theorem above can be derived by making use of the constructive technique. In general, $T_r(x)$ is not unique for $r \in \{0, 1, \ldots, m\}$. For instance, if $\nabla\mathbb{I}(x) \neq 0$, one simple choice is the default formula

$$T_r(x) = \frac{\sigma_r(x)\nabla\mathbb{I}(x)^\top - \nabla\mathbb{I}(x)\sigma_r(x)^\top}{\|\nabla\mathbb{I}(x)\|^2}, \qquad r = 0, 1, \ldots, m. \qquad (2.96)$$

Single Invariant Preserving Numerical Methods In this part, we take numerical methods preserving a single invariant $\mathbb{I}$ for the stochastic differential Eq. (2.94) into account. First, we consider the s-stage stochastic Runge–Kutta method for (2.94) as follows

$$x_i = X_k + \sum_{j=1}^{s} A_{ij}^{(0)} \sigma_0(x_j) h + \sum_{r=1}^{m} \sum_{j=1}^{s} A_{ij}^{(r)} \sigma_r(x_j) \Delta_k \widehat{W}_r,$$

$$X_{k+1} = X_k + \sum_{i=1}^{s} b_i^{(0)} \sigma_0(x_i) h + \sum_{r=1}^{m} \sum_{i=1}^{s} b_i^{(r)} \sigma_r(x_i) \Delta_k \widehat{W}_r, \qquad (2.97)$$

where $s \in \mathbb{N}_+$, $k = 0, 1, \ldots, N - 1$, and $A^{(r)}$ and $b^{(r)}$ represent a matrix and a vector for $r \in \{0, 1, \ldots, m\}$.

Theorem 2.5.4 (See [126]) *The stochastic Runge–Kutta method* (2.97) *preserves*

1. linear invariants of (2.94);
2. quadratic invariants of (2.94) *if the coefficients satisfy*

$$b_i^{(k)} A_{ij}^{(l)} + b_j^{(l)} A_{ji}^{(k)} = b_i^{(k)} b_j^{(l)} \tag{2.98}$$

for all $i, j = 1, \ldots, s$, *and* $k, l = 0, 1, \ldots, m$.

Corollary 2.5.1 (See [126]) *Under the condition* (2.98), *the stochastic Runge–Kutta method* (2.97) *preserves all invariants of the form* $\mathbb{I}(y) = y^\top M y + \gamma^\top y + c$ *of* (2.94), *where* $M \in \mathbb{R}^{d \times d}$ *is symmetric,* $\gamma \in \mathbb{R}^d$ *and* $c \in \mathbb{R}$.

Now we provide another kind of stochastic Runge–Kutta method

$$x_i^{(0)} = X_k + \sum_{j=1}^{s} A_{ij}^{(0)} \sigma_0 \left(x_j^{(0)} \right) h + \sum_{r=1}^{m} \sum_{j=1}^{s} B_{ij}^{(0)} \sigma_r \left(x_j^{(r)} \right) \hat{I}_{(r,k)},$$

$$x_i^{(k)} = X_k + \sum_{j=1}^{s} A_{ij}^{(1)} \sigma_0 \left(x_j^{(0)} \right) h + \sum_{r=1}^{m} \sum_{j=1}^{s} B_{ij}^{(1)} \sigma_r \left(x_j^{(r)} \right) \hat{I}_{(r,k)}, \tag{2.99}$$

$$X_{k+1} = X_k + \sum_{i=1}^{s} \alpha_i \sigma_0 \left(x_i^{(0)} \right) h + \sum_{r=1}^{m} \sum_{i=1}^{s} \beta_i \sigma_r \left(x_i^{(r)} \right) \hat{I}_{(r,k)}$$

for $i = 1, \ldots, s$, and $k = 1, \ldots, m$, which is equivalent to (2.18).

Theorem 2.5.5 (See [126]) *The stochastic Runge–Kutta method* (2.99) *preserves*

1. linear invariants of (2.94);
2. quadratic invariants of (2.94) *if the coefficients satisfy*

$$\alpha_i A_{ij}^{(0)} + \alpha_j A_{ji}^{(0)} = \alpha_i \alpha_j,$$

$$\alpha_i B_{ij}^{(0)} + \beta_j A_{ji}^{(1)} = \alpha_i \beta_j,$$

$$\beta_i B_{ij}^{(1)} + \beta_j B_{ji}^{(1)} = \beta_i \beta_j$$

for all $i, j = 1, \ldots, s$.

In addition to the above stochastic Runge–Kutta methods, [124] constructs the discrete gradient method preserving the single invariant, which is not restricted to the quadratic invariant of (2.94). The idea of the discrete gradient method is to numerically approximate the skew-gradient system directly. Here, we only present the discrete gradient method for (2.94) in the case of $m = 1$,

$$\begin{cases} dX(t) = \sigma_0(X(t))dt + \sigma_1(X(t)) \circ dW(t), \\ X(0) = x. \end{cases} \tag{2.100}$$

Definition 2.5.2 (See [124]) For a differentiable function $\mathbb{I} \colon \mathbb{R}^d \to \mathbb{R}$, $\bar{\nabla}\mathbb{I} \colon \mathbb{R}^d \times \mathbb{R}^d \to \mathbb{R}^d$ is called a discrete gradient of $\mathbb{I}$ if it satisfies

$$(\bar{\nabla}\mathbb{I}(y, \bar{y}))^\top (\bar{y} - y) = \mathbb{I}(\bar{y}) - \mathbb{I}(y), \quad \bar{\nabla}\mathbb{I}(y, y) = \nabla\mathbb{I}(y), \quad y, \bar{y} \in \mathbb{R}^d.$$

Furthermore, if $\bar{\nabla}\mathbb{I}(y, \bar{y}) = \bar{\nabla}\mathbb{I}(\bar{y}, y)$ holds, we call it the symmetric discrete gradient.

The widely used symmetric discrete gradient is

$$\bar{\nabla}\mathbb{I}(y, \bar{y}) = \frac{1}{2} \left(\bar{\nabla}_1 \mathbb{I}(y, \bar{y}) + \bar{\nabla}_1 \mathbb{I}(\bar{y}, y) \right), \tag{2.101}$$

where

$$\bar{\nabla}_1 \mathbb{I}(y, \bar{y}) := \begin{bmatrix} \frac{\mathbb{I}(\bar{y}_1, y_2, y_3, \dots, y_d) - \mathbb{I}(y_1, y_2, y_3, \dots, y_d)}{\bar{y}_1 - y_1} \\ \frac{\mathbb{I}(\bar{y}_1, \bar{y}_2, y_3, \dots, y_d) - \mathbb{I}(\bar{y}_1, y_2, y_3, \dots, y_d)}{\bar{y}_2 - y_2} \\ \vdots \\ \frac{\mathbb{I}(\bar{y}_1, \bar{y}_2, \bar{y}_3, \dots, \bar{y}_d) - \mathbb{I}(\bar{y}_1, \bar{y}_2, \dots, \bar{y}_{d-1}, y_d)}{\bar{y}_d - y_d} \end{bmatrix}.$$

Let X_k be the numerical approximation of $X(t_k)$ of (2.95). We define

$$X_{k+1} = X_k + T_0(X_k)\, \bar{\nabla}\mathbb{I}(X_k, X_{k+1})\, h + T_1\left(\frac{X_k + X_{k+1}}{2}\right) \bar{\nabla}\mathbb{I}(X_k, X_{k+1})\, \Delta_k \widehat{W}, \tag{2.102}$$

where $k = 0, 1, \dots, N - 1$. Replacing $T_0(X_k)$ by $T_0\left(\frac{X_k + X_{k+1}}{2}\right)$ for $k \in \{0, 1, \dots, N - 1\}$, we have another numerical method

$$X_{k+1} = X_k + T_0\left(\frac{X_k + X_{k+1}}{2}\right) \bar{\nabla}\mathbb{I}(X_k, X_{k+1})\, h$$

$$+ T_1\left(\frac{X_k + X_{k+1}}{2}\right) \bar{\nabla}\mathbb{I}(X_k, X_{k+1})\, \Delta_k \widehat{W}, \quad k = 0, 1, \dots, N - 1. \tag{2.103}$$

In [124], it is proved that both (2.102) and (2.103) for (2.95) are numerically invariant and of mean-square order 1 under some appropriate conditions.

Now we turn to the averaged vector field (AVF) method, which is a special discrete gradient method, for the conservative stochastic differential equation. The AVF method is proposed as follows

$$X_{k+1} = X_k + \left(\int_0^1 \sigma_0(X_k + \tau(X_{k+1} - X_k))d\tau \right) h$$

$$+ \left(\int_0^1 \sigma_1(X_k + \tau(X_{k+1} - X_k))d\tau \right) \Delta_k \widehat{W}, \quad k = 0, 1, \ldots, N - 1.$$

$$(2.104)$$

Since $\mathbb{I} \in \mathbf{C}^1 (\mathbb{R}^d, \mathbb{R})$ is an invariant, there are two skew-symmetric matrices T_i, $i = 0, 1$, such that $\sigma_i(y) = T_i(y)\nabla\mathbb{I}(y)$ (see [124]). If T_0 and T_1 are constant matrices, then (2.104) preserves the invariant $\mathbb{I}$. To preserve the invariant $\mathbb{I}$ for general σ_0 and σ_1, [49] combines the skew-gradient form of (2.100) and the AVF method (2.104) to give a variant of the AVF method as follows

$$X_{k+1} = X_k + \left(T_0 \left(\frac{X_k + X_{k+1}}{2} \right) \int_0^1 \nabla\mathbb{I}\left(X_k + \tau\left(X_{k+1} - X_k\right)\right)d\tau \right) h$$

$$+ \left(T_1 \left(\frac{X_k + X_{k+1}}{2} \right) \int_0^1 \nabla\mathbb{I}\left(X_k + \tau\left(X_{k+1} - X_k\right)\right)d\tau \right) \Delta_k \widehat{W},$$

where $k = 0, 1, \ldots, N - 1$. It can be derived that the method (2.104) and the above one in [49] could preserve an invariant.

Multiple Invariants Preserving Numerical Methods In this part, we illustrate both the modified averaged vector field (MAVF) method and the projection method preserving multiple invariants $\mathbb{I}_i$, $i = 1, \ldots, \nu$, simultaneously. For the convenience, we denote $\mathbb{I}(y) := (\mathbb{I}_1(y), \ldots, \mathbb{I}_\nu(y))^\top$ for $y \in \mathbb{R}^d$.

Motivated by the ideas of line integral methods for deterministic conservative ordinary differential equations in [38], [53] devises and analyzes the MAVF method which keeps multiple invariants of the original system unchanged. The MAVF method preserving multiple invariants is constructed by adding some modification terms to the AVF method. In detail, for (2.100), we define

$$X_{k+1} = X_k + \left(\int_0^1 \sigma_0(\Theta(\tau)) \, d\tau - \int_0^1 (\nabla\mathbb{I}(\Theta(\tau)))^\top d\tau \, \alpha_0 \right) h$$

$$+ \left(\int_0^1 \sigma_1(\Theta(\tau)) \, d\tau - \int_0^1 (\nabla\mathbb{I}(\Theta(\tau)))^\top d\tau \, \alpha_1 \right) \Delta_k \widehat{W}, \quad (2.105)$$

where $\Theta(\tau) = X_k + \tau(X_{k+1} - X_k)$, $k = 0, 1, \ldots, N - 1$, and α_0, α_1 are $\mathbb{R}^\nu$-valued random variables satisfying

$$\int_0^1 \nabla\mathbb{I}(\Theta(\tau))\,d\tau \int_0^1 (\nabla\mathbb{I}(\Theta(\tau)))^\top d\tau\alpha_0 = \int_0^1 \nabla\mathbb{I}(\Theta(\tau))\,d\tau \int_0^1 \sigma_0(\Theta(\tau))\,d\tau,$$

$$\int_0^1 \nabla\mathbb{I}(\Theta(\tau))\,d\tau \int_0^1 (\nabla\mathbb{I}(\Theta(\tau)))^\top d\tau\alpha_1 = \int_0^1 \nabla\mathbb{I}(\Theta(\tau))\,d\tau \int_0^1 \sigma_1(\Theta(\tau))\,d\tau.$$

Here, $\int_0^1 (\nabla\mathbb{I}(\Theta(\tau)))^\top d\tau\, \alpha_0$ and $\int_0^1 (\nabla\mathbb{I}(\Theta(\tau)))^\top d\tau\, \alpha_1$ are the modification terms, and $\boldsymbol{\alpha} = (\alpha_0, \alpha_1)$ is called the modification coefficient. By the Taylor expansion and (2.105), one has

$$\mathbb{I}(X_{k+1}) - \mathbb{I}(X_k) = \int_0^1 \nabla\mathbb{I}(X_k + \tau(X_{k+1} - X_k))\,d\tau(X_{k+1} - X_k)$$

$$= \left(\int_0^1 \nabla\mathbb{I}(\Theta(\tau))\,d\tau \int_0^1 \sigma_0(\Theta(\tau))\,d\tau - \int_0^1 \nabla\mathbb{I}(\Theta(\tau))\,d\tau \int_0^1 (\nabla\mathbb{I}(\Theta(\tau)))^\top d\tau\, \alpha_0 \right) h$$

$$+ \left(\int_0^1 \nabla\mathbb{I}(\Theta(\tau))\,d\tau \int_0^1 \sigma_1(\Theta(\tau))\,d\tau - \int_0^1 \nabla\mathbb{I}(\Theta(\tau))\,d\tau \int_0^1 (\nabla\mathbb{I}(\Theta(\tau)))^\top d\tau\, \alpha_1 \right) \Delta_k \widehat{W},$$

where $k = 0, 1, \ldots, N - 1$. It can be observed that the modification coefficient $\boldsymbol{\alpha}$ ensures the preservation of multiple invariants, i.e., $\mathbb{I}(X_{k+1}) = \mathbb{I}(X_k)$ for $n \in \{0, 1, \ldots, N - 1\}$.

Based on the above ideas, now we study the MAVF method for (2.94) with multiple invariants as follows

$$X_{k+1} = X_k + \left(\int_0^1 \sigma_0(\Theta(\tau))\,d\tau - \int_0^1 (\nabla\mathbb{I}(\Theta(\tau)))^\top d\tau\, \alpha_0 \right) h$$

$$+ \sum_{r=1}^m \left(\int_0^1 \sigma_r(\Theta(\tau))\,d\tau - \int_0^1 [\nabla\mathbb{I}(\Theta(\tau))]^\top d\tau\, \alpha_r \right) \Delta_k \widehat{W}_r, \qquad (2.106)$$

$$\int_0^1 \nabla\mathbb{I}(\Theta(\tau))\,d\tau \int_0^1 (\nabla\mathbb{I}(\Theta(\tau)))^\top d\tau\alpha_r = \int_0^1 \nabla\mathbb{I}(\Theta(\tau))\,d\tau \int_0^1 \sigma_r(\Theta(\tau))\,d\tau$$

for $r \in \{0, 1, \ldots, m\}$, where $\Theta(\tau) = X_k + \tau(X_{k+1} - X_k)$, $k \in \{0, 1, \ldots, N-1\}$, and $\Delta_k \widehat{W}_r = \sqrt{h}\zeta_{rh}^k$ is defined by (2.11) with $l = 2$. In addition, $\boldsymbol{\alpha} = (\alpha_0, \alpha_1, \ldots, \alpha_m)$ is called the modification coefficient, with α_r, $r = 0, 1, \ldots, m$, being $\mathbb{R}^\nu$-valued random variables.

Remark 2.5.3 If the invariant $\mathbb{I}$ of (2.94) is quadratic, and σ_r, $r = 0, 1, \ldots, m$, are linear functions, then

$$\int_0^1 \sigma_r(\Theta(\tau))\, d\tau = \sigma_r\left(\frac{X_{k+1} + X_k}{2}\right),$$

$$\int_0^1 \nabla\mathbb{I}(\Theta(\tau))\, d\tau = \nabla\mathbb{I}\left(\frac{X_{k+1} + X_k}{2}\right)$$

for $k \in \{0, 1, \ldots, N - 1\}$. Noting that $\nabla\mathbb{I}(y)\sigma_r(y) = 0$, $r = 0, 1, \ldots, m$, we have $\alpha = 0$. In this case, (2.106) becomes the midpoint method.

Theorem 2.5.6 (See [53]) *Let $\sigma_0 \in C_b^2(\mathbb{R}^d, \mathbb{R}^d)$, and $\sigma_r \in C_b^3(\mathbb{R}^d, \mathbb{R}^d)$, $r = 1, \ldots, m$. Suppose that $\nabla\mathbb{I}(\cdot)\nabla\mathbb{I}^\top(\cdot)$ is invertible on $\mathscr{M}_x = \{y \in \mathbb{R}^d \mid \mathbb{I}_i(y) = \mathbb{I}_i(x), i = 1, \ldots, \nu\}$, and that for some bounded*

$$\mathscr{M}_x^\delta = \left\{y \in \mathbb{R}^d \mid \mathrm{dist}(y, \mathscr{M}_x) \leq \delta\right\}$$

with $\delta > 0$ and $\mathrm{dist}(y, \mathscr{M}_x) = \inf\limits_{z \in \mathscr{M}_x} \|z - y\|$, $\nabla\mathbb{I} \in C^1(\mathscr{M}_x^\delta, \mathbb{R}^{\nu \times d})$. Then the numerical method (2.106) for (2.94) possesses the following properties:

(1) *it preserves multiple invariants $\mathbb{I}_i$, $i = 1, \ldots, \nu$, i.e., $\mathbb{I}(X_k) = \mathbb{I}(x)$ for $k \in \{1, \ldots, N\}$.*
(2) *if diffusion coefficients of (2.94) satisfy the commutative conditions, i.e., $\Lambda_r \sigma_i = \Lambda_i \sigma_r$, $i, r = 1, \ldots, m$, it is of mean-square order 1.*

Remark 2.5.4 If diffusion coefficients are not commutative, the mean-square order of the method (2.106) is only $\frac{1}{2}$. Moreover, when the integrals contained in the MAVF method can not be obtained directly, they need to be approximated by using the numerical integration.

For stochastic differential Eq. (2.94) with multiple invariants, [272] couples a common supporting method with a projection to construct numerical methods in the stochastic case. The basic idea of the projection method is to combine an arbitrary numerical approximation $\widehat{X}_{k+1}$ starting at X_k, where $k = 0, 1, \ldots, N - 1$, together with a projection onto the invariant submanifold $\mathscr{M}_x$ in every step (see Fig. 2.11). Thus the procedures of the projection methods at each step as follows

1. compute the numerical approximation $\widehat{X}_{k+1}$ for $k = \{0, 1, \ldots, N - 1\}$;
2. compute $\lambda \in \mathbb{R}^\nu$, for $X_{k+1} = \widehat{X}_{k+1} + \Phi(X_k, \widehat{X}_{k+1}, X_{k+1})\lambda$ such that $\mathbb{I}(X_{k+1}) = \mathbb{I}(X_k)$ for $k = \{0, 1, \ldots, N - 1\}$.

The matrix $\Phi \in \mathbb{R}^{d \times \nu}$ defines the direction of the projection, and $\lambda \in \mathbb{R}^\nu$ needs to be chosen such that X_{k+1} belongs to the invariant manifold $\mathscr{M}_x$. In fact, Φ is not unique, and here $\Phi = \left(\nabla\mathbb{I}(\widehat{X}_{k+1})\right)^\top$. To realize the projection method above, we need to solve a ν-dimensional nonlinear system, which can be implemented by

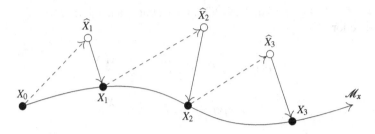

Fig. 2.11 Basic idea of stochastic projection methods preserving multiple invariants

some iterative algorithms such as the Newton method, and we refer to [272] for more details.

2.6   End Notes

The subject of this chapter is stochastic numerical methods that preserve geometric properties and physical quantities of the flow of original systems: stochastic (pseudo-)symplectic methods for stochastic Hamiltonian systems, rough symplectic methods for rough Hamiltonian systems, and invariant-preserving numerical methods for stochastic differential equations. In the last few decades, despite fruitful results on the construction and analysis of stochastic structure-preserving methods, there still exist many unclear parts and directions to be explored in this field.

One significant but challenging problem is that the great majority of numerical methods could not preserve simultaneously mathematical structure, physical property, and dynamics for general stochastic differential equations. A typical example is the failure of symplectic Euler methods in preserving the Hamiltonian of the Kubo oscillator whose solution lies in the isoenergetic surface and admits a lot of invariant measures (see [131]). Although stochastic symplectic methods could not preserve the Hamiltonian in general, they play an important role in the construction of numerical methods preserving several intrinsic properties at the same time for some non-canonical stochastic Hamiltonian systems. For instance, Sect. 2.3 has shown that stochastic symplectic methods can be utilized to design numerical approximations preserving both the Casimir function and stochastic Poisson structure for stochastic Poisson systems. Hong et al. in [133] apply stochastic symplectic methods to construct numerical methods preserving both the positivity and stochastic K-symplectic structure for the stochastic Lotka–Volterra predator-prey model. However, not many results on structure-preserving methods are known for rough differential equations. Especially, dynamical properties, such as the periodicity, the existence and uniqueness of the invariant measure, for rough symplectic methods are still in their infancy.

Although a lot of numerical experiments indicate the long-time superiority of stochastic structure-preserving methods (see e.g., [130, 197, 207, 245, 260] and references therein), less attention has been paid on the probabilistic aspect of numerical methods. Inspired by the fact that some numerical methods have been used to derive the LDP of the original systems (see e.g., [46, 91] and references therein), we investigate probabilities belonging to some intervals of the random variable associated with exact and numerical solutions, hoping to obtain some intrinsic characteristics of stochastic symplectic methods in Sect. 2.4. In this direction, a lot of interesting aspects remain to be solved:

- Whether all stochastic symplectic methods could asymptotically preserve the LDP for all observables associated with the linear stochastic oscillator?
- Can stochastic symplectic methods asymptotically preserve the LDPs for observables associated with the general stochastic Hamiltonian system?

Due to the loss of the explicit expression of large deviation rate functions for general stochastic differential equations, it is difficult to analyze the asymptotical behavior of rate functions. We leave the above problems as open problems.

The study of asymptotical preservation of the LDP for numerical methods is related to the accuracy of approximations in strong or weak convergent sense. For instance, strong convergence of the Euler–Maruyama method is taken into account in Sect. 2.4. The fundamental theorems on strong and weak convergence under the global Lipschitz condition have been presented in Theorems 2.1.1 and 2.1.4. However, special care must be taken to construct and analyze strong and weak convergent stochastic structure-preserving numerical methods for the stochastic differential equation in the non-global Lipschitz setting. In recent years, this interesting and hotspot topic has received extensive attentions (see e.g., [5, 55, 118, 141, 184, 231, 266, 267, 270] and references therein). Notably, the fundamental theorem of weak convergence for stochastic numerical methods still remains unclear and needs to be further developed for non-globally Lipschitz stochastic differential equations and rough differential equations.

In general, stochastic symplectic methods are implicit. In the course of numerical implementation, they have to be solved by means of the fixed point iteration or the Picard iteration. Thus, a lot of computational cost has been paid when solving nonlinear algebraic equations, and the preservation of mathematical structures will be somewhat damaged due to the iteration error. One remaining problem is how to analyze the overall error estimate of the implicit stochastic structure-preserving numerical method. In the deterministic case, one may use explicit symplectic methods (see e.g., [189, 244, 249] and references therein) and explicit adaptive symplectic integrators (see e.g., [31, 96] and references therein) to avoid using implicit solvers. However, there has not been any result on the design and numerical analysis for explicit adaptive stochastic (rough) symplectic methods with high accuracy and less computation cost.

Chapter 3
Stochastic Modified Equations and Applications

The modified equation is extremely successful in understanding numerical behaviors of deterministic ordinary differential equation. For instance, it allows that a symplectic method can be approximated by a perturbed Hamiltonian system. A natural question is whether such theory could be extended to the stochastic differential equation and in which sense. This is an important and subtle question, since, unlike the deterministic case, there exist various different notions of convergence for the stochastic numerical method. This chapter is devoted to investigating the stochastic modified equation with respect to the weak convergence and its applications to constructing high weak order numerical methods.

3.1 Stochastic Modified Equations

Stochastic modified equations associated with various numerical methods for the stochastic differential equation are presented in [1, 60, 90, 157, 158, 215, 235, 264, 274] and references therein. For instance, Shardlow [235] deduces stochastic modified equations of the Euler–Maruyama method for the Itô stochastic differential equation with additive noises. Zygalakis in [274] provides a framework for deriving the stochastic modified equation of numerical methods for general stochastic differential equations. The stochastic modified equation for the Euler–Maruyama method is utilized in [215] to exhibit the fact that the Euler–Maruyama method performs poorly for small random perturbations of Hamiltonian flows.

Now we derive the stochastic modified equation of numerical methods with respect to weak convergence for the d-dimensional autonomous stochastic differential equation in the sense of Itô

$$\begin{cases} dX(t) = a(X(t))dt + \displaystyle\sum_{r=1}^{m} \sigma_r(X(t))dW_r(t), & t \in (0, T], \\ X(0) = x, \end{cases} \qquad (3.1)$$

where $a, \sigma_i \colon \mathbb{R}^d \to \mathbb{R}^d$, $i = 1, \ldots, m$, and $W_r(\cdot)$, $r = 1, \ldots, m$, are independent standard Wiener processes. Suppose that drift and diffusion coefficients are globally Lipschitz continuous and sufficiently smooth, and the numerical method $\{X_k\}_{k=0}^{N}$ with time-step size $h = \frac{T}{N}$ converges weakly of order $\gamma \in \mathbb{N}_+$, i.e.,

$$|\mathbb{E}[\phi(X_N)] - \mathbb{E}[\phi(X(T))]| = O(h^\gamma)$$

for $\phi \in \mathbf{C}_P^{2(\gamma+1)}(\mathbb{R}^d, \mathbb{R})$. Actually, for the test function ϕ, it is usually enough to consider polynomials up to order $2\gamma + 1$ with $\gamma \in \mathbb{N}_+$ (see e.g., [235, 274] and references therein). The simplest example of a numerical method, which has order 1 in the weak convergent sense, is the Euler–Maruyama method

$$X_{k+1} = X_k + a\,(X_k)\,h + \sum_{r=1}^{m} \sigma_r(X_k)\Delta_k W_r, \quad k = 0, 1, \ldots, N-1$$

with $X_0 = x$.

A shift of key ideas comes about by interpreting the numerical solution as the exact solution of a stochastic modified equation. The objective is to modify the stochastic differential equation (3.1) and define a stochastic process $\tilde{X}(\cdot)$, that better describes the numerical approximation X_k, in the sense that

$$\left| \mathbb{E}[\phi(X_k)] - \mathbb{E}[\phi(\tilde{X}(t_k))] \right| = O(h^{\gamma+\gamma'}), \quad k = 1, \ldots, N, \qquad (3.2)$$

with $\gamma' \in \mathbb{N}_+$ being the increase in the order of accuracy. In detail, $\tilde{X}$ is the solution to the stochastic modified equation in the sense of Itô

$$
\begin{cases}
d\tilde{X}(t) = A(\tilde{X}(t))dt + \sum_{r=1}^{m} \Gamma_r(\tilde{X}(t))dW_r(t), & t \in (0, T], \\
\tilde{X}(0) = x
\end{cases}
\tag{3.3}
$$

with

$$
A(\tilde{X}) = a(\tilde{X}) + a_1(\tilde{X})h + a_2(\tilde{X})h^2 + \cdots + a_\iota(\tilde{X})h^\iota,
$$
$$
\Gamma_r(\tilde{X}) = \sigma_r(\tilde{X}) + \sigma_{r,1}(\tilde{X})h + \sigma_{r,2}(\tilde{X})h^2 + \cdots + \sigma_{r,\iota}(\tilde{X})h^\iota,
$$

where functions a_i and $\sigma_{r,i}$, $i = 1, \ldots, \iota$, $r = 1, \ldots, m$, need to be determined for some $\iota \in \mathbb{N}_+$. It can be seen that the stochastic modified equation can fit its corresponding numerical method with high accuracy, that is, the solution of (3.3) is weakly γ'-order 'closer' to the numerical solution than that of the stochastic differential equation (3.1) does, where $\gamma' \in \mathbb{N}_+$.

The main technique utilized to determine unknown functions a_i and $\sigma_{r,i}$, $i = 1, \ldots, \iota$, $r = 1, \ldots, m$, is as follows. From Sect. 2.1 it can be known that the weak convergence analysis of a stochastic numerical method can be given by studying the approximation error over one time step. This is analogous to the idea of local error for the deterministic ordinary differential equation, where the global error of the numerical approximation can essentially be determined by means of only the local error. More precisely, to achieve the γth order weak convergence, one would expect that

$$
|\mathbb{E}[\phi(X_1)] - \mathbb{E}[\phi(X(h))]| = O(h^{\gamma+1})
\tag{3.4}
$$

for $\phi \in \mathbf{C}_p^{2(\gamma+1)}(\mathbb{R}^d, \mathbb{R})$. In order to achieve (3.2), we need

$$
\left| \mathbb{E}[\phi(X_1)] - \mathbb{E}[\phi(\tilde{X}(h))] \right| = O(h^{\gamma+\gamma'+1})
\tag{3.5}
$$

for $\phi \in \mathbf{C}_p^{2(\gamma+\gamma'+1)}(\mathbb{R}^d, \mathbb{R})$.

Below we will illustrate a general framework for deriving the stochastic modified equation of arbitrary order with respect to the weak convergence for stochastic numerical methods based on the backward Kolmogorov equation (see [274]), and that for stochastic symplectic methods via stochastic generating function (see [264]).

3.1.1   Modified Equation via Backward Kolmogorov Equation

As we have previously discussed, to fit a stochastic modified equation of a stochastic numerical method of weak order γ, we need (3.4) to be satisfied. This implies that we need to figure out what happens to the expectation of functionals of $X(h)$ given by (3.1). To address this issue, a natural way is exploiting the backward Kolmogorov equation.

The backward Kolmogorov equation associated with the stochastic differential equation (3.1) reads

$$\begin{cases} \dfrac{\partial u(x,t)}{\partial t} = \mathbb{L}u(x,t), \\[2mm] u(x,0) = \phi(x), \end{cases} \tag{3.6}$$

where

$$\mathbb{L} = a^\top \frac{\partial}{\partial x} + \frac{1}{2} \sum_{r=1}^{m} \sum_{i=1}^{d} \sum_{j=1}^{d} \sigma_r^i \sigma_r^j \frac{\partial^2}{\partial x^i \partial x^j}$$

with σ_r^i being the ith component of σ_r for $r \in \{1, \ldots, m\}$. The probabilistic way of considering the solution of this equation is that

$$u(x,t) = \mathbb{E}[\phi(X(t))|\, X(0) = x].$$

If u is $N+1$ times differentiable with respect to t, we have the following Taylor expansion

$$u(x,s) = u(x,0) + s\frac{\partial u(x,0)}{\partial s} + \cdots + \frac{s^N}{N!}\frac{\partial^N u(x,0)}{\partial s^N} + R,$$

where $s \in [0, T]$ and R is the remainder. Taking advantage of the relation $u(x,h) - u(x,0) = \int_0^h \mathbb{L}u(x,s)ds$, we achieve

$$u(x,h) - u(x,0) = h\mathbb{L}u(x,0) + \sum_{i=1}^{N} \frac{h^{i+1}}{(i+1)!}\mathbb{L}\frac{\partial^i u(x,0)}{\partial s^i} + O(h^{N+2}).$$

Now making use of the fact that $u(x,0) = \phi(x)$ together with $\frac{\partial^i u(x,0)}{\partial s^i} = \mathbb{L}^i \phi(x)$ for $i \in \{1, \ldots, N\}$ leads to

$$u(x,h) - \phi(x) = \sum_{i=0}^{N} \frac{h^{i+1}}{(i+1)!}\mathbb{L}^{i+1}\phi(x) + O(h^{N+2}). \tag{3.7}$$

A weak γth order numerical method for (3.1) should obey this expansion up to terms of order $\gamma \in \mathbb{N}_+$ since the local weak error is limited to $O(h^{\gamma+1})$ while subtracting the numerical method from (3.7), which infers that the global weak error is of the γth order. In detail, denote the expectation of ϕ for a one-step numerical approximation of weak order $\gamma \in \mathbb{N}_+$ by

$$u_{num}(x, h) := \mathbb{E}[\phi(X_1)|\ X_0 = x].$$

The expectation has an expansion coinciding with (3.7) up to terms of order γ which corresponds to a local error of order $\gamma + 1$, i.e.,

$$u_{num}(x, h) - \phi(x) = \sum_{i=0}^{\gamma-1} \frac{h^{i+1}}{(i+1)!} \mathbb{L}^{i+1}\phi(x) + \sum_{i=\gamma}^{N} h^{i+1} \mathscr{A}_i\phi(x) + O\left(h^{N+2}\right),$$

(3.8)

where $\mathscr{A}_i$, $i = \gamma, \dots, N$, are partial differential operators acting on ϕ that depend on the choice of used numerical methods.

Moreover, the backward Kolmogorov equation associated with the modified equation (3.3) reads

$$\begin{cases} \dfrac{\partial u_{mod}(x, t)}{\partial t} = \mathbb{L}^h u_{mod}(x, t), \\[2mm] u_{mod}(x, 0) = \phi(x), \end{cases}$$

where

$$\mathbb{L}^h := A^\top \frac{\partial}{\partial x} + \frac{1}{2} \sum_{r=1}^{m} \sum_{i=1}^{d} \sum_{j=1}^{d} \Gamma_r^i \Gamma_r^j \frac{\partial^2}{\partial x^i \partial x^j}$$

with Γ_r^i being the ith component of Γ_r for $i \in \{1, \dots, d\}$ and $r \in \{1, \dots, m\}$. To obtain the stochastic modified equation (3.3) satisfying (3.2), one also needs to equate the expansion of $u_{mod}(x, h)$ with that of $u_{num}(x, h)$ up to terms of order $\gamma + \gamma'$. We now write the generator $\mathbb{L}^h$ in a more compact form

$$\mathbb{L}^h := \mathbb{L} + \sum_{i=1}^{2\iota} h^i \mathbb{L}_i,$$

where ι is the highest power of h in the modified drift term $A(\tilde{X})$, $\mathbb{L}_i$, $i \in \{1, \ldots, 2\iota\}$ contains unknown coefficients $a_j, \sigma_{r,j}$, with $i = j, \ldots, \iota$, and $r = 1, \ldots, m$. It can be verified that if $d = m = 1$, then

$$\mathbb{L}_0 := \mathbb{L} = a\frac{d}{dx} + \frac{\sigma_1^2}{2}\frac{d^2}{dx^2},$$

$$\mathbb{L}_1 := a_1\frac{d}{dx} + \sigma_1\sigma_{1,1}\frac{d^2}{dx^2},$$

$$\mathbb{L}_2 := a_2\frac{d}{dx} + \left(\frac{\sigma_{1,1}^2}{2} + \sigma_1\sigma_{1,2}\right)\frac{d^2}{dx^2}.$$

Using the same procedure as before, we arrive at

$u_{mod}(x, h) - \phi(x)$

$= h\mathbb{L}\phi(x) + h^2\left(\mathbb{L}_1\phi(x) + \frac{1}{2}\mathbb{L}^2\phi(x)\right)$

$+ h^3\left(\mathbb{L}_2\phi(x) + \frac{1}{2}\mathbb{L}\mathbb{L}_1\phi(x) + \frac{1}{2}\mathbb{L}_1\mathbb{L}\phi(x) + \frac{1}{6}\mathbb{L}^3\phi(x)\right) + \cdots$

$+ h^{\iota+1}\left(\mathbb{L}_\iota\phi(x) + \frac{1}{2}\left(\mathbb{L}\mathbb{L}_{\iota-1}\phi(x) + \mathbb{L}_{\iota-1}\mathbb{L}\phi(x)\right) + \cdots + \frac{1}{(\iota+1)!}\mathbb{L}^{\iota+1}\phi(x)\right) + O\left(h^{\iota+2}\right)$

$=: \sum_{i=0}^{N} h^{i+1}\tilde{L}_{i+1}\phi(x) + O\left(h^{N+2}\right),$

which yields

$$u_{mod}(x, h) - u_{num}(x, h) = \sum_{i=1}^{\gamma} h^i\left(\tilde{L}_i\phi(x) - \frac{1}{i!}\mathbb{L}^i\phi(x)\right)$$

$$+ \sum_{i=\gamma}^{N} h^{i+1}\left(\tilde{L}_{i+1}\phi(x) - \mathscr{A}_i\phi(x)\right) + O\left(h^{N+2}\right).$$

In order to achieve (3.5), we need

$$\tilde{L}_i\phi(x) = \frac{1}{i!}\mathbb{L}^i\phi(x), \quad \tilde{L}_{j+1}\phi(x) = \mathscr{A}_j\phi(x) \tag{3.9}$$

for any $\phi \in \mathbf{C}_P^{(2\gamma+2\gamma'+2)}(\mathbb{R}^d, \mathbb{R})$, $i \in \{1, \ldots, \gamma\}$ and $j \in \{\gamma, \ldots, \gamma + \gamma' - 1\}$. If there exist functions a_i and $\sigma_{r,i}$, $i = 1, \ldots, \iota$, $r = 1, \ldots, m$, such that the above formulae hold simultaneously, then the stochastic modified equation can be obtained.

Now take the case of $d = m = 1$ as an example to investigate the existence of stochastic modified equations for some numerical methods of weak order 1. In this case, Eq. (3.1) becomes

$$\begin{cases} dX(t) = a(X(t))dt + \sigma(X(t))dW(t), \\ X(0) = x, \end{cases} \tag{3.10}$$

and the associated operator $\mathbb{L}$ meets

$$\mathbb{L}\phi(x) = a(x)\frac{d}{dx}\phi(x) + \frac{1}{2}\sigma^2(x)\frac{d^2}{dx^2}\phi(x).$$

It follows that

$$\mathbb{L}^2\phi(x)$$

$$= \left(a(x)a'(x) + \frac{\sigma^2(x)}{2}a''(x)\right)\phi'(x)$$

$$+ \left(a^2(x) + \sigma^2(x)a'(x) + a(x)\sigma(x)\sigma'(x) + \frac{\sigma^2(x)}{2}(\sigma'(x))^2 + \frac{\sigma^3(x)}{2}\sigma''(x)\right)\phi''(x)$$

$$+ \left(\sigma^3(x)\sigma'(x) + \sigma^2(x)a(x)\right)\phi^{(3)}(x) + \frac{\sigma^4(x)}{4}\phi^{(4)}(x),$$

where $\phi^{(i)}$, $i \in \{3, 4\}$, is the ith derivative of ϕ. Moreover, we consider the modified equation

$$\begin{cases} d\tilde{X}(t) = \left(a(\tilde{X}(t)) + \sum_{i=1}^{\iota} a_i(\tilde{X}(t))h^i\right)dt + \left(\sigma(\tilde{X}(t)) + \sum_{i=1}^{\iota} \sigma_i(\tilde{X}(t))h^i\right)dW(t), \\ X(0) = x \end{cases}$$

$$\tag{3.11}$$

for the Euler–Maruyama method and the Milstein method in the following examples.

Example 3.1.1 (Euler–Maruyama Method) The one-step approximation of the Euler–Maruyama method for (3.11) is given by

$$\bar{x} = x + a(x)h + \sigma(x)\xi\sqrt{h},$$

where $\xi \sim \mathcal{N}(0, 1)$. Taking the Taylor expansion of $\phi \in \mathbf{C}^4(\mathbb{R}, \mathbb{R})$ yields that

$\mathbb{E}[\phi(\bar{x})| X(0) = x]$

$$= \phi(x) + h\mathbb{L}\phi(x) + h^2\left(\frac{a^2(x)}{2}\phi''(x) + \frac{\sigma^2(x)a(x)}{2}\phi^{(3)}(x) + \frac{\sigma^4(x)}{8}\phi^{(4)}(x)\right) + O\left(h^3\right),$$

which means

$$\mathscr{A}_1\phi(x) = \frac{a^2(x)}{2}\phi''(x) + \frac{\sigma^2(x)}{2}a(x)\phi^{(3)}(x) + \frac{\sigma^4(x)}{8}\phi^{(4)}(x). \tag{3.12}$$

Based on (3.12), (3.9) becomes

$$\mathbb{L}_1\phi(x) = -\frac{1}{2}\left(a(x)a'(x) + \frac{\sigma^2(x)}{2}a''(x)\right)\phi'(x)$$

$$-\frac{1}{2}\left(\sigma^2(x)a'(x) + a(x)\sigma(x)\sigma'(x) + \frac{\sigma^2(x)}{2}\left(\sigma'(x)\right)^2 + \frac{\sigma^3(x)}{2}\sigma''(x)\right)\phi''(x)$$

$$-\frac{\sigma^3(x)}{2}\sigma'(x)\phi^{(3)}(x).$$

Since $\mathbb{L}_1$ is a second order differential operator and the right hand side of the above equation contains the third order derivative of ϕ, it is impossible to write down a stochastic modified equation in Itô sense which fits the Euler–Maruyama method for stochastic differential equations with multiplicative noises. In the following, it can be found that if we use the Milstein method, such a derivation of the stochastic modified equation is available. Another way of getting around the problem is to consider the additive noise case, that is, $\sigma(X) = \sigma$. Using the expression for $\mathbb{L}_1$, we obtain

$$a_1 = -\left(\frac{1}{2}aa' + \frac{\sigma^2}{4}a''\right), \quad \sigma_1 = -\frac{\sigma}{2}a',$$

which agrees with the coefficients of the stochastic modified equation derived in [235] for the Euler–Maruyama method. However, if the coefficient a is nonlinear, the modified equation is an Itô stochastic differential equation with multiplicative noise and we thus cannot obtain the higher order stochastic modified equation.

Example 3.1.2 (Milstein Method) The one-step approximation of the Milstein method for (3.11) reads

$$\bar{x} = x + a(x)h + \sigma(x)\sqrt{h}\xi + \frac{1}{2}\sigma(x)\sigma'(x)\left(h\xi^2 - h\right)$$

where $\xi \sim \mathcal{N}(0, 1)$. After taking the Taylor expansion of $\phi \in \mathbf{C}^4(\mathbb{R}, \mathbb{R})$, we obtain

$$\mathbb{E}[\phi(\bar{x})| \, X(0) = x]$$

$$= \phi(x) + h\mathbb{L}\phi(x) + \frac{h^2}{2}\left(a^2(x) + \frac{1}{2}\left(\sigma(x)\sigma'(x)\right)^2\right)\phi''(x)$$

$$+ \frac{h^2}{2}\left(\sigma^3(x)\sigma'(x) + \sigma^2(x)a(x)\right)\phi^{(3)}(x) + \frac{h^2}{8}\sigma^4(x)\phi^{(4)}(x) + O\left(h^3\right),$$

which leads to

$$\mathscr{A}_1\phi(x) = \frac{a^2(x)\phi''(x)}{2} + \frac{1}{4}\left(\sigma(x)\sigma'(x)\right)^2\phi''(x)$$

$$+ \frac{1}{2}\left(\sigma^3(x)\sigma'(x) + \sigma^2(x)a(x)\right)\phi^{(3)}(x) + \frac{\sigma^4(x)}{8}\phi^{(4)}(x).$$

Following similar procedures as in the last example, thus for the Milstein method, the modified coefficients are

$$a_1 = -\frac{1}{2}\left(aa' + \frac{\sigma^2}{2}a''\right), \quad \sigma_1 = -\frac{1}{2}\left(\sigma a' + a\sigma' + \frac{\sigma^2}{2}\sigma''\right).$$

It can be checked that the Milstein method agrees with the Euler–Maruyama method in the case that the noise is additive, and the above coefficients of the stochastic modified equation coincide with the ones in Example 3.1.1.

Remark 3.1.1 Actually the nature of the noise is a subtle issue, since it is impossible to write down a stochastic modified equation of the Euler–Maruyama method in the case of multiplicative noise and in the weak convergent sense. For the existence of the stochastic modified equation in case of multiple dimension, we refer to [274] for more details.

3.1.2 Modified Equation via Stochastic Generating Function

Due to the inheritance of the symplectic structure of the stochastic Hamiltonian system, stochastic symplectic methods are superior to nonsymplectic ones in tracking phase trajectories of the original systems in the long-time simulation. In this part, we propose an approach of constructing the stochastic modified equation for the weakly convergent stochastic symplectic method based on stochastic generating function, rather than the backward Kolmogorov equation associated with stochastic Hamiltonian system.

Consider the stochastic Hamiltonian system

$$
\begin{cases}
dP(t) = -\dfrac{\partial H_0(P(t), Q(t))}{\partial Q}dt - \displaystyle\sum_{r=1}^{m} \dfrac{\partial H_r(P(t), Q(t))}{\partial Q} \circ dW_r(t), \quad P(0) = p, \\[3mm]
dQ(t) = \dfrac{\partial H_0(P(t), Q(t))}{\partial P}dt + \displaystyle\sum_{r=1}^{m} \dfrac{\partial H_r(P(t), Q(t))}{\partial P} \circ dW_r(t), \quad Q(0) = q,
\end{cases}
$$

$$(3.13)$$

where $P, Q, p, q \in \mathbb{R}^n$, and $W_r(\cdot), r = 1, \ldots, m$, are independent standard Wiener processes. It is given in Chap. 1 that the phase flow of the stochastic Hamiltonian system (3.13) can be generated by the stochastic θ-generating function

$$
S_\theta(\hat{P}, \hat{Q}, t) = \sum_{\alpha} G_\alpha^\theta(\hat{P}, \hat{Q}) J_\alpha(t), \quad \theta \in [0, 1],
$$

via the relations

$$
P(t) = p - \frac{\partial S_\theta}{\partial q}(\hat{P}(t), \hat{Q}(t), t), \quad Q(t) = q + \frac{\partial S_\theta}{\partial P}(\hat{P}(t), \hat{Q}(t), t)
$$

with $\hat{P}(t) = (1-\theta)p + \theta P(t)$, $\hat{Q}(t) = (1-\theta)Q(t) + \theta q$. The stochastic generating function approach to constructing stochastic symplectic methods is established, based on the fact that each symplectic mapping $(P_k, Q_k) \mapsto (P_{k+1}, Q_{k+1})$ for $k \in \{0, 1, \ldots, N-1\}$ can be associated with a stochastic generating function, e.g., the truncated type-I stochastic generating function $\bar{S}_1(P_{k+1}, Q_k, h)$, such that

$$
P_{k+1} = P_k - \frac{\partial \bar{S}_1(P_{k+1}, Q_k, h)}{\partial Q_k}, \quad Q_{k+1} = Q_k + \frac{\partial \bar{S}_1(P_{k+1}, Q_k, h)}{\partial P_{k+1}}. \qquad (3.14)
$$

Now we show the theorem about the stochastic modified equation of the weakly convergent symplectic method (3.14) for (3.13).

Theorem 3.1.1 (See [264]) *Suppose that the stochastic Hamiltonian system (3.13), for which the noises are additive, or $H_r(P, Q)$, $r \in \{1, \ldots, m\}$, depends only on P or only on Q, has a weakly convergent symplectic method based on the one-step approximation $\psi_h: (p, q) \mapsto (P, Q)$ given by (3.14) whose associated generating function is*

$$
\bar{S}_1(P, q, h) = \sum_{\alpha \in \Lambda_\psi} F_\alpha^1(P, q) \bar{I}_\alpha(h),
$$

where $F_\alpha^1(P, q)$ is a linear combination of some functions $G_\beta^1(P, q)$ for appropriate multi-index β when α is fixed, and $\bar{I}_\alpha(h)$ is an appropriate realization of the multiple *stochastic Itô integral $I_\alpha(h)$. Λ_ψ is the set of multi-indices associated with ψ_h. Then the stochastic modified equation of the considered weakly convergent symplectic method is a perturbed stochastic Hamiltonian system*

$$
\begin{cases}
d\tilde{P}(t) = -\dfrac{\partial \tilde{H}_0(\tilde{P}(t), \tilde{Q}(t))}{\partial \tilde{Q}} dt - \displaystyle\sum_{r=1}^{m} \dfrac{\partial \tilde{H}_r(\tilde{P}(t), \tilde{Q}(t))}{\partial \tilde{Q}} \circ dW_r(t), \ \tilde{P}(0) = p, \\[4mm]
d\tilde{Q}(t) = \dfrac{\partial \tilde{H}_0(\tilde{P}(t), \tilde{Q}(t))}{\partial \tilde{P}} dt + \displaystyle\sum_{r=1}^{m} \dfrac{\partial \tilde{H}_r(\tilde{P}(t), \tilde{Q}(t))}{\partial \tilde{P}} \circ dW_r(t), \ \tilde{Q}(0) = q,
\end{cases}
$$

$$(3.15)$$

where

$$
\tilde{H}_r(\tilde{P}, \tilde{Q}) = H_r(\tilde{P}, \tilde{Q}) + H_r^{[1]}(\tilde{P}, \tilde{Q})h + H_r^{[2]}(\tilde{P}, \tilde{Q})h^2 + \cdots, \quad r = 0, 1, \ldots, m,
$$

$$(3.16)$$

with functions $H_j^{[i]}$, $j = 0, 1, \ldots, m$, $i = 1 \in \mathbb{N}$, to be determined.

Proof For the sake of simplicity, we denote $H_r = H_r^{[0]}$ for $r \in \{0, 1, \ldots, m\}$. The proof is a procedure of finding unknown functions $H_j^{[i]}$ for $j = 0, 1, \ldots, m$, and $i \in \mathbb{N}$. Suppose that the stochastic Hamiltonian system (3.15) is associated with $\tilde{S}_1(\tilde{P}(t), q, t)$ satisfying the stochastic Hamilton–Jacobi partial differential equation

$$
\begin{cases}
d\tilde{S}_1 = \tilde{H}_0\left(\tilde{P}, q + \dfrac{\partial \tilde{S}_1}{\partial \tilde{P}}\right) dt + \displaystyle\sum_{r=1}^{m} \tilde{H}_r\left(\tilde{P}, q + \dfrac{\partial \tilde{S}_1}{\partial \tilde{P}}\right) \circ dW_r(t), \\[4mm]
\tilde{S}_1(\tilde{P}, q, 0) = 0.
\end{cases}
$$

Assume that the solution of the above equation has the form

$$
\tilde{S}_1(\tilde{P}, q, t) = \sum_\alpha \tilde{G}_\alpha^1(\tilde{P}, q, h) J_\alpha(t)
$$

with coefficients

$$
\tilde{G}_\alpha^1 = \sum_{i=1}^{l(\alpha)-1} \frac{1}{i!} \sum_{k_1, \ldots, k_i = 1}^{n} \frac{\partial^i \tilde{H}_r}{\partial q_{k_1} \cdots \partial q_{k_i}} \sum_{\substack{l(\alpha_1) + \cdots + l(\alpha_i) = l(\alpha) - 1 \\ \alpha - \in \Lambda_{\alpha_1, \ldots, \alpha_i}}} \frac{\partial \tilde{G}_{\alpha_1}^1}{\partial \tilde{P}_{k_1}} \cdots \frac{\partial \tilde{G}_{\alpha_i}^1}{\partial \tilde{P}_{k_i}}
$$

$$(3.17)$$

for $\alpha = (i_1, \ldots, i_{l-1}, r)$ with $l \geq 2$, and $\tilde{G}_{(r)}^1 = \tilde{H}_r$ for $\alpha = (r)$, $r = 0, 1, \ldots, m$. Denoting

$$
\tilde{G}_\alpha^1(\tilde{P}, q, h) = G_\alpha^{1[0]}(\tilde{P}, q) + G_\alpha^{1[1]}(\tilde{P}, q)h + G_\alpha^{1[2]}(\tilde{P}, q)h^2 + \cdots,
$$

we obtain that for $\alpha = (i_1, \ldots, i_{l-1}, r)$ with $l \geq 2$,

$$
G_\alpha^{1[k]} = \sum_{i=1}^{l(\alpha)-1} \frac{1}{i!} \sum_{k_1,\ldots,k_i=1}^{n} \sum_{j+j_1+\cdots+j_i=k} \frac{\partial^i H_r^{[j]}}{\partial q_{k_1} \cdots \partial q_{k_i}}
$$

$$
\times \sum_{\substack{l(\alpha_1)+\cdots+l(\alpha_i)=l(\alpha)-1 \\ \alpha- \in \Lambda_{\alpha_1,\ldots,\alpha_i}}} \frac{\partial G_{\alpha_1}^{1[j_1]}}{\partial \tilde{P}_{k_1}} \cdots \frac{\partial G_{\alpha_i}^{1[j_i]}}{\partial \tilde{P}_{k_i}},
$$

and that for $\alpha = (r)$, $r = 0, 1, \ldots, m$,

$$
G_{(r)}^{1[k]} = H_r^{[k]},
$$

where $k \in \mathbb{N}$. Replacing t by h, one gets

$$
\tilde{S}_1(\tilde{P}, q, h) = \sum_\alpha \sum_{k=0}^{\infty} G_\alpha^{1[k]}(\tilde{P}, q) h^k J_\alpha(h)
$$

$$
= \sum_\alpha \sum_{k=0}^{\infty} G_\alpha^{1[k]}(\tilde{P}, q) \sum_{\beta \in \Lambda_{0_k,\alpha}} k! J_\beta(h), \tag{3.18}
$$

where 0_k denotes the index containing k zeros $\underbrace{(0, \ldots, 0)}_{k}$. The equality (3.18) is due

to the fact that $h^k = k! J_{0_k}(h)$ and the relation (see e.g., [13, 93] and references therein)

$$
\prod_{i=1}^{n} J_{\alpha_i} = \sum_{\beta \in \Lambda_{\alpha_1,\ldots,\alpha_n}} J_\beta.
$$

Rearranging the summation terms in (3.18) yields

$$
\tilde{S}_1(\tilde{P}, q, h) = \sum_\beta \left(\sum_{\substack{k=0,\ldots,l(\beta)-1, \\ \beta \in \Lambda_{0_k,\alpha}}} k! \tau_\beta(0_k, \alpha) G_\alpha^{1[k]}(\tilde{P}, q) \right) J_\beta(h)
$$

$$
=: \sum_\beta \tilde{G}_\beta^1(\tilde{P}, q) J_\beta(h),
$$

where $\tau_\beta(0_k, \alpha)$ denotes the number of β appearing in $\Lambda_{0_k,\alpha}$, and

$$\tilde{G}^1_\beta(\tilde{P}, q) := \sum_{\substack{k=0,\ldots,l(\beta)-1, \\ \beta \in \Lambda_{0_k,\alpha}}} k!\tau_\beta(0_k, \alpha)G^{1[k]}_\alpha(\tilde{P}, q).$$

For the sake of simplicity, letting $\tilde{P}(h) = \tilde{P}$ and $\tilde{Q}(h) = \tilde{Q}$, we have

$$\tilde{P} = p - \frac{\partial \tilde{S}_1(\tilde{P}, q, h)}{\partial q}, \quad \tilde{Q} = q + \frac{\partial \tilde{S}_1(\tilde{P}, q, h)}{\partial \tilde{P}},$$

which implies

$$\mathbb{E}[\phi(\tilde{P}, \tilde{Q})] = \phi(p, q) + \sum_{i=1}^n \frac{\partial \phi}{\partial p_i}(p, q)\mathbb{E}\left[-\frac{\partial \tilde{S}_1}{\partial q_i}\right] + \sum_{i=1}^n \frac{\partial \phi}{\partial q_i}(p, q)\mathbb{E}\left[\frac{\partial \tilde{S}_1}{\partial \tilde{P}_i}\right]$$

$$+ \frac{1}{2}\sum_{i,j=1}^n \frac{\partial^2 \phi(p, q)}{\partial p_i \partial p_j}\mathbb{E}\left[\frac{\partial \tilde{S}_1}{\partial q_i}\frac{\partial \tilde{S}_1}{\partial q_j}\right] - \sum_{i,j=1}^n \frac{\partial^2 \phi(p, q)}{\partial q_i \partial p_j}\mathbb{E}\left[\frac{\partial \tilde{S}_1}{\partial \tilde{P}_i}\frac{\partial \tilde{S}_1}{\partial q_j}\right]$$

$$+ \frac{1}{2}\sum_{i,j=1}^n \frac{\partial^2 \phi(p, q)}{\partial q_i \partial q_j}\mathbb{E}\left[\frac{\partial \tilde{S}_1}{\partial \tilde{P}_i}\frac{\partial \tilde{S}_1}{\partial \tilde{P}_j}\right] + \cdots, \tag{3.19}$$

where the function $\tilde{S}_1$ takes value at $(\tilde{P}, q, h)$.

As is known from Chap. 2, the type-I stochastic generating function for weakly convergent symplectic methods can be obtained by transforming $J_\beta(h)$ to their equivalent linear combination of Itô integrals $I_\alpha(h)$, choosing those terms with multi-index α satisfying $l(\alpha) \leq \gamma$ for a symplectic method of weak order γ, and then approximating the chosen $I_\alpha(h)$ by some appropriate $\bar{I}_\alpha(h)$. This can be expressed as follows

$$\bar{S}_1(P, q, h) = \sum_\beta G^1_\beta(P, q) \sum_{l(\alpha)\leq\gamma} C^\beta_\alpha \bar{I}_\alpha(h) = \sum_{l(\alpha)\leq\gamma}\left(\sum_\beta C^\beta_\alpha G^1_\beta(P, q)\right)\bar{I}_\alpha(h)$$

$$=: \sum_{l(\alpha)\leq\gamma} F^1_\alpha(P, q)\bar{I}_\alpha(h), \tag{3.20}$$

where C^β_α are some constants resulted from the relation

$$J_\beta(h) = \sum_\alpha C^\beta_\alpha I_\alpha(h).$$

Meanwhile, the considered stochastic symplectic method is defined by the one-step approximation

$$P = p - \frac{\partial \bar{S}_1(P, q, h)}{\partial q}, \quad Q = q + \frac{\partial \bar{S}_1(P, q, h)}{\partial P}.$$

Thus we have

$$
\mathbb{E}[\phi(P, Q)] = \phi(p, q) + \sum_{i=1}^{n} \frac{\partial \phi}{\partial p_i}(p, q) \mathbb{E}\left[-\frac{\partial \bar{S}_1}{\partial q_i} \right] + \sum_{i=1}^{n} \frac{\partial \phi}{\partial q_i}(p, q) \mathbb{E}\left[\frac{\partial \bar{S}_1}{\partial P_i} \right]
$$
$$
+ \frac{1}{2} \sum_{i,j=1}^{n} \frac{\partial^2 \phi(p, q)}{\partial p_i \partial p_j} \mathbb{E}\left[\frac{\partial \bar{S}_1}{\partial q_i} \frac{\partial \bar{S}_1}{\partial q_j} \right] - \sum_{i,j=1}^{n} \frac{\partial^2 \phi(p, q)}{\partial q_i \partial p_j} \mathbb{E}\left[\frac{\partial \bar{S}_1}{\partial P_i} \frac{\partial \bar{S}_1}{\partial q_j} \right]
$$
$$
+ \frac{1}{2} \sum_{i,j=1}^{n} \frac{\partial^2 \phi(p, q)}{\partial q_i \partial q_j} \mathbb{E}\left[\frac{\partial \bar{S}_1}{\partial P_i} \frac{\partial \bar{S}_1}{\partial P_j} \right] + \cdots,
$$

where the function $\bar{S}_1$ takes value at (P, q, h).

Now we ought to let the modified equation (3.15) be globally weakly γ' order closer to the numerical method based on ψ_h than the original system (3.13) does, which means

$$|\mathbb{E}[\phi(P, Q)] - \mathbb{E}[\phi(\tilde{P}, \tilde{Q})]| = O(h^{\gamma + \gamma' + 1}),$$

where

$$
\mathbb{E}[\phi(P, Q)] - \mathbb{E}[\phi(\tilde{P}, \tilde{Q})]
$$
$$
= \sum_{i=1}^{n} \frac{\partial \phi}{\partial p_i}(p, q) \mathbb{E}\left[\frac{\partial \bar{S}_1}{\partial q_i} - \frac{\partial \tilde{S}_1}{\partial q_i} \right] + \sum_{i=1}^{n} \frac{\partial \phi}{\partial q_i}(p, q) \mathbb{E}\left[\frac{\partial \bar{S}_1}{\partial P_i} - \frac{\partial \tilde{S}_1}{\partial \tilde{P}_i} \right]
$$
$$
+ \frac{1}{2} \sum_{i,j=1}^{n} \frac{\partial^2 \phi(p, q)}{\partial p_i \partial p_j} \mathbb{E}\left[\frac{\partial \bar{S}_1}{\partial q_i} \frac{\partial \bar{S}_1}{\partial q_j} - \frac{\partial \tilde{S}_1}{\partial q_i} \frac{\partial \tilde{S}_1}{\partial q_j} \right]
$$
$$
- \sum_{i,j=1}^{n} \frac{\partial^2 \phi(p, q)}{\partial q_i \partial p_j} \mathbb{E}\left[\frac{\partial \bar{S}_1}{\partial P_i} \frac{\partial \bar{S}_1}{\partial q_j} - \frac{\partial \tilde{S}_1}{\partial \tilde{P}_i} \frac{\partial \tilde{S}_1}{\partial q_j} \right]
$$
$$
+ \frac{1}{2} \sum_{i,j=1}^{n} \frac{\partial^2 \phi(p, q)}{\partial q_i \partial q_j} \mathbb{E}\left[\frac{\partial \bar{S}_1}{\partial P_i} \frac{\partial \bar{S}_1}{\partial P_j} - \frac{\partial \tilde{S}_1}{\partial \tilde{P}_i} \frac{\partial \tilde{S}_1}{\partial \tilde{P}_j} \right] + \cdots. \tag{3.21}
$$

Let every item in the right hand side of (3.21) be behaving like $O(h^{\gamma + \gamma' + 1})$ with $\gamma + \gamma' \geq 2$. Note that, after taking expectations, the terms of mean-square order

$\frac{1}{2}$, $\frac{3}{2}$, etc., will vanish. Thus we only need to consider terms whose mean-square order are integers. Since the lowest mean-square order of the $J_\alpha(h)$ is $\frac{1}{2}$, the highest degree of partial derivatives that can produce h is 2. Similarly, the highest degree of partial derivatives that can generate h^s is $2s$ with $s \in \mathbb{N}_+$.

As a consequence, we need coefficients of h to be equal within each of the following pairs, for $i, j = 1, \ldots, n$,

$$\mathbb{E}\left[\frac{\partial \bar{S}_1}{\partial q_i}\right] \text{ and } \mathbb{E}\left[\frac{\partial \tilde{S}_1}{\partial q_i}\right]; \ \mathbb{E}\left[\frac{\partial \bar{S}_1}{\partial P_i}\right] \text{ and } \mathbb{E}\left[\frac{\partial \tilde{S}_1}{\partial \tilde{P}_i}\right]; \ \mathbb{E}\left[\frac{\partial \bar{S}_1}{\partial q_i}\frac{\partial \bar{S}_1}{\partial q_j}\right] \text{ and } \mathbb{E}\left[\frac{\partial \tilde{S}_1}{\partial q_i}\frac{\partial \tilde{S}_1}{\partial q_j}\right];$$

$$\mathbb{E}\left[\frac{\partial \bar{S}_1}{\partial q_i}\frac{\partial \bar{S}_1}{\partial P_j}\right] \text{ and } \mathbb{E}\left[\frac{\partial \tilde{S}_1}{\partial q_i}\frac{\partial \tilde{S}_1}{\partial \tilde{P}_j}\right]; \ \mathbb{E}\left[\frac{\partial \bar{S}_1}{\partial P_i}\frac{\partial \bar{S}_1}{\partial P_j}\right] \text{ and } \mathbb{E}\left[\frac{\partial \tilde{S}_1}{\partial \tilde{P}_i}\frac{\partial \tilde{S}_1}{\partial \tilde{P}_j}\right],$$

where derivatives of $\bar{S}_1$ take values at (P, q, h) and those of $\tilde{S}_1$ take values at $(\tilde{P}, q, h)$. For instance, to compare the above first pair, we need to perform the Taylor expansion of partial derivatives of $\bar{S}_1(P, q, h)$ and those of $\tilde{S}_1(\tilde{P}, q, h)$ at (p, q, h), recursively. More precisely, we have

$$\mathbb{E}\left[\frac{\partial \bar{S}_1}{\partial q_i}(P, q, h)\right]$$

$$= \sum_{\alpha \in \Lambda_\psi} \mathbb{E}\left[\frac{\partial F_\alpha^1}{\partial q_i}(p, q, h) + \sum_{j=1}^n \frac{\partial^2 F_\alpha^1}{\partial q_i \partial P_j}(p, q, h)\left(-\frac{\partial \bar{S}_1}{\partial q_j}(P, q, h)\right) + \cdots\right] \bar{I}_\alpha(h),$$

and

$$\mathbb{E}\left[\frac{\partial \tilde{S}_1}{\partial q_i}(\tilde{P}, q, h)\right]$$

$$= \sum_\beta \mathbb{E}\left[\frac{\partial \tilde{G}_\beta^1}{\partial q_i}(p, q, h) + \sum_{j=1}^n \frac{\partial^2 \tilde{G}_\beta^1}{\partial q_i \partial \tilde{P}_j}(p, q, h)\left(-\frac{\partial \tilde{S}_1}{\partial q_j}(\tilde{P}, q, h)\right) + \cdots\right] J_\beta(h),$$

where partial derivatives of $\bar{S}_1$ and those of $\tilde{S}_1$ are expanded at (p, q, h) in the same way once again and further on. It is worth mentioning that, $\bar{I}_\alpha(h)$ is an approximation of $I_\alpha(h)$ for some α, and the approximation error could be controlled. For example, choosing an appropriate truncation bound of Gaussian random variables will affect neither the convergence order of numerical methods, nor the finding of the modified equation of desired order by comparing like powers of h within each pair.

Notice that the pairs given above contain all possible h^μ with $\mu \geq 1$, so for coefficients of h^2. Therefore, the above pairs still need to be compared, and in addition, the following pairs are required for $i, j, l, s = 1, \ldots, n$,

$$\mathbb{E}\left[\frac{\partial \bar{S}_1}{\partial q_i}\frac{\partial \bar{S}_1}{\partial q_j}\frac{\partial \bar{S}_1}{\partial q_l}\right] \quad \text{and} \quad \mathbb{E}\left[\frac{\partial \tilde{S}_1}{\partial q_i}\frac{\partial \tilde{S}_1}{\partial q_j}\frac{\partial \tilde{S}_1}{\partial q_l}\right];$$

$$\mathbb{E}\left[\frac{\partial \bar{S}_1}{\partial q_i}\frac{\partial \bar{S}_1}{\partial q_j}\frac{\partial \bar{S}_1}{\partial P_l}\right] \quad \text{and} \quad \mathbb{E}\left[\frac{\partial \tilde{S}_1}{\partial q_i}\frac{\partial \tilde{S}_1}{\partial q_j}\frac{\partial \tilde{S}_1}{\partial \tilde{P}_l}\right];$$

$$\mathbb{E}\left[\frac{\partial \bar{S}_1}{\partial q_i}\frac{\partial \bar{S}_1}{\partial P_j}\frac{\partial \bar{S}_1}{\partial P_l}\right] \quad \text{and} \quad \mathbb{E}\left[\frac{\partial \tilde{S}_1}{\partial q_i}\frac{\partial \tilde{S}_1}{\partial \tilde{P}_j}\frac{\partial \tilde{S}_1}{\partial \tilde{P}_l}\right];$$

$$\mathbb{E}\left[\frac{\partial \bar{S}_1}{\partial P_i}\frac{\partial \bar{S}_1}{\partial P_j}\frac{\partial \bar{S}_1}{\partial P_l}\right] \quad \text{and} \quad \mathbb{E}\left[\frac{\partial \tilde{S}_1}{\partial \tilde{P}_i}\frac{\partial \tilde{S}_1}{\partial \tilde{P}_j}\frac{\partial \tilde{S}_1}{\partial \tilde{P}_l}\right];$$

$$\mathbb{E}\left[\frac{\partial \bar{S}_1}{\partial q_i}\frac{\partial \bar{S}_1}{\partial q_j}\frac{\partial \bar{S}_1}{\partial q_l}\frac{\partial \bar{S}_1}{\partial q_s}\right] \quad \text{and} \quad \mathbb{E}\left[\frac{\partial \tilde{S}_1}{\partial q_i}\frac{\partial \tilde{S}_1}{\partial q_j}\frac{\partial \tilde{S}_1}{\partial q_l}\frac{\partial \tilde{S}_1}{\partial q_s}\right];$$

$$\mathbb{E}\left[\frac{\partial \bar{S}_1}{\partial q_i}\frac{\partial \bar{S}_1}{\partial q_j}\frac{\partial \bar{S}_1}{\partial q_l}\frac{\partial \bar{S}_1}{\partial P_s}\right] \quad \text{and} \quad \mathbb{E}\left[\frac{\partial \tilde{S}_1}{\partial q_i}\frac{\partial \tilde{S}_1}{\partial q_j}\frac{\partial \tilde{S}_1}{\partial q_l}\frac{\partial \tilde{S}_1}{\partial \tilde{P}_s}\right];$$

$$\mathbb{E}\left[\frac{\partial \bar{S}_1}{\partial q_i}\frac{\partial \bar{S}_1}{\partial q_j}\frac{\partial \bar{S}_1}{\partial P_l}\frac{\partial \bar{S}_1}{\partial P_s}\right] \quad \text{and} \quad \mathbb{E}\left[\frac{\partial \tilde{S}_1}{\partial q_i}\frac{\partial \tilde{S}_1}{\partial q_j}\frac{\partial \tilde{S}_1}{\partial \tilde{P}_l}\frac{\partial \tilde{S}_1}{\partial \tilde{P}_s}\right];$$

$$\mathbb{E}\left[\frac{\partial \bar{S}_1}{\partial q_i}\frac{\partial \bar{S}_1}{\partial P_j}\frac{\partial \bar{S}_1}{\partial P_l}\frac{\partial \bar{S}_1}{\partial P_s}\right] \quad \text{and} \quad \mathbb{E}\left[\frac{\partial \tilde{S}_1}{\partial q_i}\frac{\partial \tilde{S}_1}{\partial \tilde{P}_j}\frac{\partial \tilde{S}_1}{\partial \tilde{P}_l}\frac{\partial \tilde{S}_1}{\partial \tilde{P}_s}\right];$$

$$\mathbb{E}\left[\frac{\partial \bar{S}_1}{\partial P_i}\frac{\partial \bar{S}_1}{\partial P_j}\frac{\partial \bar{S}_1}{\partial P_l}\frac{\partial \bar{S}_1}{\partial P_s}\right] \quad \text{and} \quad \mathbb{E}\left[\frac{\partial \tilde{S}_1}{\partial \tilde{P}_i}\frac{\partial \tilde{S}_1}{\partial \tilde{P}_j}\frac{\partial \tilde{S}_1}{\partial \tilde{P}_l}\frac{\partial \tilde{S}_1}{\partial \tilde{P}_s}\right].$$

Further on, for a stochastic modified equation which is of weak order $\gamma + \gamma'$ apart from the numerical method, the coefficients in pairs up to those of $2\left(\gamma + \gamma'\right)$th power of the partial derivatives of both $\bar{S}_1$ and $\tilde{S}_1$ need to be equated. In this process the unknown functions $H_j^{[i]}$ can be determined for $j \in \{0, 1, \ldots, m\}$ and $i \in \mathbb{N}_+$.

It is natural that we hope to have a solution for the basic non-trivial case of $\gamma = \gamma' = 1$. As we compare the coefficients of h^2 in $\mathbb{E}\left[\left(\frac{\partial \bar{S}_1}{\partial q_i}\right)^2\right]$ and $\mathbb{E}\left[\left(\frac{\partial \tilde{S}_1}{\partial q_i}\right)^2\right]$, we should have

$$\left(\frac{\partial}{\partial q_i}\left(G_{(0)}^1 + \frac{1}{2}\sum_{r=1}^m G_{(r,r)}^1\right)\right)^2$$

$$= \left(\frac{\partial}{\partial q_i}\left(\tilde{G}_{(0)}^1 + \frac{1}{2}\sum_{r=1}^m \tilde{G}_{(r,r)}^1\right)\right)^2 + \frac{1}{2}\sum_{r=1}^m\left(\frac{\partial}{\partial q_i}\tilde{G}_{(r,r)}\right)^2, \tag{3.22}$$

where the functions take values at (p, q). Since for $r = 1, \ldots, m$,

$$G^1_{(0)}(p, q) = \tilde{G}^1_{(0)}(p, q), \quad G^1_{(r,r)}(p, q) = \tilde{G}^1_{(r,r)}(p, q),$$

we ought to make

$$\frac{\partial}{\partial q_i} \tilde{G}^1_{(r,r)} = 0, \quad i = 1, \ldots, n.$$

Analogously, it should hold that

$$\frac{\partial}{\partial p_i} \tilde{G}^1_{(r,r)} = 0, \quad r = 1, \ldots, m, \quad i = 1, \ldots, n.$$

These imply that H_r, $r = 1, \ldots, m$, should either be linear functions of p and q, or functions depending only on p or only on q, where the former means that the noises of the stochastic Hamiltonian system are additive, and the latter includes the multiplicative noises case. □

However, for the truncated type-III stochastic generating function $\bar{S}_{\frac{1}{2}}$ which generates the stochastic symplectic method

$$
\begin{aligned}
P_{k+1} &= P_k - \frac{\partial \bar{S}_{\frac{1}{2}}\left(\frac{P_{k+1}+P_k}{2}, \frac{Q_{k+1}+Q_k}{2}, h\right)}{\partial Q_k}, \\[2mm]
Q_{k+1} &= Q_k + \frac{\partial \bar{S}_{\frac{1}{2}}\left(\frac{P_{k+1}+P_k}{2}, \frac{Q_{k+1}+Q_k}{2}, h\right)}{\partial P_{k+1}}
\end{aligned}
\tag{3.23}
$$

with $k = 0, 1, \ldots, N - 1$, it can be verified that $\tilde{G}^{\frac{1}{2}}_{(r,r)} = 0$ for $r = 0, 1, \ldots, m$. Thus the restriction on the diffusion coefficients H_r, $r = 1, \ldots, m$, of the above theorem may be relaxed.

Theorem 3.1.2 *For the stochastic Hamiltonian system* (3.13), *suppose that it has a weakly convergent symplectic method based on* $\psi_h \colon (p, q) \mapsto (P, Q)$ *given by* (3.23) *which has the generating function*

$$\bar{S}_{\frac{1}{2}}\left(\frac{P+p}{2}, \frac{Q+q}{2}, h\right) = \sum_{\alpha \in \Lambda_\psi} F^{\frac{1}{2}}_\alpha\left(\frac{P+p}{2}, \frac{Q+q}{2}\right) \bar{I}_\alpha(h),$$

where $F^{\frac{1}{2}}_\alpha(\frac{P+p}{2}, \frac{Q+q}{2})$ *is a combination of some functions* $G^{\frac{1}{2}}_\beta(\frac{P+p}{2}, \frac{Q+q}{2})$ *for appropriate multi-index* β *when* α *is fixed, and* $\bar{I}_\alpha(h)$ *is an appropriate realization of the multiple stochastic Itô integral* $I_\alpha(h)$ *defined on* $(0, h)$. Λ_ψ *is the set of multi-indices associated with* ψ_h. *Then the stochastic modified equation of the considered method is a perturbed stochastic Hamiltonian system*

$$
\begin{cases}
d\tilde{P}(t) = -\dfrac{\partial \tilde{H}_0(\tilde{P}(t), \tilde{Q}(t))}{\partial \tilde{Q}}dt - \displaystyle\sum_{r=1}^{m} \dfrac{\partial \tilde{H}_r(\tilde{P}(t), \tilde{Q}(t))}{\partial \tilde{Q}} \circ dW_r(t), \ \ \tilde{P}(0) = p, \\[4mm]
d\tilde{Q}(t) = \ \ \dfrac{\partial \tilde{H}_0(\tilde{P}(t), \tilde{Q}(t))}{\partial \tilde{P}}dt + \displaystyle\sum_{r=1}^{m} \dfrac{\partial \tilde{H}_r(\tilde{P}(t), \tilde{Q}(t))}{\partial \tilde{P}} \circ dW_r(t), \ \ \tilde{Q}(0) = q,
\end{cases}
$$

where

$$
\tilde{H}_r(\tilde{P}, \tilde{Q}) = H_r(\tilde{P}, \tilde{Q}) + H_r^{[1]}(\tilde{P}, \tilde{Q})h + H_r^{[2]}(\tilde{P}, \tilde{Q})h^2 + \cdots, \quad r = 0, 1, \ldots, m,
$$

with undetermined functions $H_j^{[i]}$, $j = 0, 1, \ldots, m$, $i \in \mathbb{N}$.

Below we shall provide some examples to test the validity of theoretical results.

Example 3.1.3 (Symplectic Euler Method for Linear Stochastic Oscillator) The linear stochastic oscillator

$$
\begin{cases}
dP(t) = -Q(t)dt + \alpha dW(t), \quad P(0) = p_0, \\
dQ(t) = P(t)dt, \quad Q(0) = q_0,
\end{cases}
$$

is a stochastic Hamiltonian system with $H_0(P, Q) = \frac{1}{2}(P^2 + Q^2)$, $H_1(Q) = -\alpha Q$. It can be deduced that coefficient functions of the generating function S_1 associated with this system are

$$
G_{(0)}^1(P, q) = H_0(P, q) = \frac{1}{2}(P^2 + q^2), \quad G_{(1)}^1(P, q) = H_1(P, q) = -\alpha q,
$$

$$
G_{(1,1)}^1(P, q) = 0, \quad G_{(0,1)}^1(P, q) = -\alpha P, \quad G_{(1,0)}^1(P, q) = 0,
$$

$$
G_{(1,1,1)}^1(P, q) = 0, \quad G_{(0,0)}^1(P, q) = Pq, \ldots
$$

In order to achieve a symplectic method of weak order 1, we need to include the terms of I_β with $l(\beta) \leq 1$ in the series of S_1. Then, we make use of the relations

$$
J_{(0)}(h) = I_{(0)}(h), \quad J_{(1)}(h) = I_{(1)}(h), \quad J_{(1,1)}(h) = I_{(1,1)}(h) + \frac{1}{2}I_{(0)}(h)
$$
(3.24)

to obtain the coefficient of $I_{(0)}$, which is $G_{(0)}^1 + \frac{1}{2}G_{(1,1)}^1$, and that of $I_{(1)}$ which is $G_{(1)}^1$. Thus, the generating function for a symplectic method of weak order 1 is

$$
\bar{S}_1(P, q, h) = G_{(1)}^1(P, q)\sqrt{h}\xi + \left(G_{(0)}^1(P, q) + \frac{1}{2}G_{(1,1)}^1(P, q) \right)h
$$
(3.25)

with $\xi \sim \mathcal{N}(0, 1)$. According to the relation (3.14), the symplectic method of weak order 1 generated by (3.25) is

$$P_{k+1} = P_k - Q_k h + \alpha \sqrt{h} \xi_1^k,$$
$$Q_{k+1} = Q_k + P_{k+1} h, \quad k = 0, 1, \ldots, N - 1, \tag{3.26}$$

which is just the stochastic symplectic Euler method. Next we find $\tilde{S}_1(\tilde{P}, q, h) = \sum_\beta \tilde{G}_\beta^1(\tilde{P}, q) J_\beta(h)$ associated with this numerical method. According to (3.18), we have

$$\tilde{G}_{(0)}^1(\tilde{P}, q) = \frac{1}{2}(\tilde{P}^2 + q^2), \quad \tilde{G}_{(1)}^1(\tilde{P}, q) = -\alpha q, \quad \tilde{G}_{(1,1)}^1(\tilde{P}, q) = 0,$$

$$\tilde{G}_{(0,1)}^1(\tilde{P}, q) = -\alpha \tilde{P} + H_1^{[1]}(\tilde{P}, q), \quad \tilde{G}_{(1,0)}^1(\tilde{P}, q) = H_1^{[1]}(\tilde{P}, q),$$

$$\tilde{G}_{(0,0)}^1(\tilde{P}, q) = \tilde{P} q + 2 H_0^{[1]}(\tilde{P}, q), \quad \tilde{G}_{(1,1,0)}^1(\tilde{P}, q) = -\alpha \frac{\partial H_1^{[1]}}{\partial \tilde{P}}(\tilde{P}, q),$$

$$\tilde{G}_{(0,1,1)}^1(\tilde{P}, q) = \alpha^2 - \alpha \frac{\partial H_1^{[1]}}{\partial \tilde{P}}(\tilde{P}, q), \quad \tilde{G}_{(1,1,1)}^1(\tilde{P}, q) = 0, \quad \tilde{G}_{(1,1,1,1)}^1(\tilde{P}, q) = 0, \ldots$$

It is obvious that coefficients of h are naturally equal. Equating coefficients of h^2 gives

$$\frac{1}{2}\alpha \frac{\partial^2 H_1^{[1]}}{\partial p \partial q} - \frac{\partial H_0^{[1]}}{\partial q} = \frac{1}{2}p, \quad \frac{1}{2}\alpha \frac{\partial^2 H_1^{[1]}}{\partial p^2} - \frac{\partial H_0^{[1]}}{\partial p} = \frac{1}{2}q, \quad \frac{\partial H_1^{[1]}}{\partial q} = 0, \quad \frac{\partial H_1^{[1]}}{\partial p} = \frac{1}{2}\alpha,$$

which leads to

$$\frac{\partial H_0^{[1]}}{\partial p} = -\frac{1}{2}q, \quad \frac{\partial H_0^{[1]}}{\partial q} = -\frac{1}{2}p. \tag{3.27}$$

Thus, the stochastic modified equation of (3.26) is

$$\begin{cases} dP(t) = \left(-Q(t) + \frac{h}{2}P(t)\right)dt + \alpha dW(t), & P(0) = p_0, \\ dQ(t) = \left(P(t) - \frac{h}{2}Q(t)\right)dt + h\frac{\alpha}{2}dW(t), & Q(0) = q_0, \end{cases}$$

which coincides with the result in [274] about the stochastic modified equation of the symplectic Euler method for (1.38) with $F(q) = \frac{1}{2}q^2$ and $v = 0$.

For the stochastic modified equation of weak third order apart from the symplectic method, we need to equate coefficients of h^3 within corresponding pairs to determine more unknown coefficients.

Example 3.1.4 (Weak Order 2 Symplectic Method for Linear Stochastic Oscillator)
To obtain a symplectic method of weak second order, we still need to calculate the
following coefficients

$$G^1_{(1,1,0)}(P,q) = 0, \quad G^1_{(0,1,1)}(P,q) = \alpha^2, \quad G^1_{(1,0,1)}(P,q) = 0, \quad G^1_{(1,1,1,1)}(P,q) = 0.$$

Thus, the corresponding truncated generating function is

$$\bar{S}_1(P,q,h) = -\alpha q \sqrt{h}\xi + \frac{1}{2}(P^2+q^2)h - \alpha P\left(\frac{\xi}{2} - \frac{\sqrt{3}\eta}{6}\right)h^{\frac{3}{2}} + \left(Pq + \frac{1}{2}\alpha^2\right)\frac{h^2}{2},$$

where $\eta \sim \mathcal{N}(0,1)$ is independent of ξ, and $(\frac{\xi}{2} - \frac{\sqrt{3}\eta}{6})h^{\frac{3}{2}}$ is the simulation of
$I_{(0,1)}(h)$. The associated numerical method is

$$P_{k+1} = P_k - Q_k h + \alpha\sqrt{h}\xi_1^k - \frac{P_{k+1}}{2}h^2,$$

$$Q_{k+1} = Q_k + P_{k+1}h - \alpha\left(\frac{\xi_1^k}{2} - \frac{\sqrt{3}\eta_1^k}{6}\right)h^{\frac{3}{2}} + \frac{Q_k}{2}h^2 \tag{3.28}$$

with $k = 0, 1, \ldots, N-1$. In addition to $\tilde{G}_\beta(\tilde{P},q)$ in the previous example, we also
need to have the following relations

$$\tilde{G}^1_{(0,0,1)}(\tilde{P},q) = -\alpha q + 2H^{[2]}_1(\tilde{P},q), \quad \tilde{G}^1_{(1,0,0)}(\tilde{P},q) = 2H^{[2]}_1(\tilde{P},q),$$

$$\tilde{G}^1_{(0,1,0)}(\tilde{P},q) = -\alpha q + 2H^{[2]}_1(\tilde{P},q),$$

$$\tilde{G}^1_{(1,0,1,1)}(\tilde{P},q) = \tilde{G}^1_{(1,1,0,1)}(\tilde{P},q) = 0,$$

$$\tilde{G}^1_{(1,1,1,0)}(\tilde{P},q) = \tilde{G}^1_{(0,1,1,1)}(\tilde{P},q) = \tilde{G}^1_{(1,1,1,1,1)}(\tilde{P},q) = 0,$$

$$\tilde{G}^1_{(0,0,0)}(\tilde{P},q) = q^2 + \tilde{P}^2 + 6H^{[2]}_0(\tilde{P},q),$$

and the other coefficient $\tilde{G}^1_\beta(\tilde{P},q)$, such as $\tilde{G}^1_{(0,0,1,1)}(\tilde{P},q)$ and $\tilde{G}^1_{(1,1,1,1,0)}(\tilde{P},q)$,
needed for constructing the modified equation for the symplectic method of weak
second order equals zero. Comparing coefficients of h, h^2, h^3 within corresponding
pairs, we obtain the following equations

$$\frac{\partial H^{[1]}_1}{\partial q} = 0, \quad \frac{\partial H^{[1]}_1}{\partial p} = 0, \quad \frac{\partial H^{[1]}_0}{\partial q} = 0, \quad \frac{\partial H^{[1]}_0}{\partial p} = 0,$$

$$2\alpha - 6\frac{\partial H^{[2]}_1}{\partial q} = 0, \quad \frac{\partial H^{[2]}_1}{\partial p} = 0, \quad 2q + 6\frac{\partial H^{[2]}_0}{\partial q} = 0, \quad 2p + 6\frac{\partial H^{[2]}_0}{\partial p} = 0,$$

from which it follows that

$$\frac{\partial H_0^{[2]}}{\partial q} = -\frac{q}{3}, \quad \frac{\partial H_0^{[2]}}{\partial p} = -\frac{p}{3}, \quad \frac{\partial H_1^{[2]}}{\partial p} = 0, \quad \frac{\partial H_1^{[2]}}{\partial q} = \frac{\alpha}{3}.$$

Therefore the stochastic modified equation of (3.28) is

$$\begin{cases} dP(t) = \left(-Q(t) + \frac{h^2}{3} Q(t) \right) dt + \left(\alpha - \frac{h^2}{3}\alpha \right) dW(t), \quad P(0) = p_0, \\ dQ(t) = \left(P(t) - \frac{h^2}{3} P(t) \right) dt, \quad Q(0) = q_0. \end{cases}$$

Example 3.1.5 A model for synchrotron oscillations of particles in storage rings oscillator in [199] is

$$\begin{cases} dP(t) = -\alpha^2 \sin(Q(t))dt - \sigma_1 \cos(Q(t)) \circ dW_1(t) - \sigma_2 \sin(Q(t)) \circ dW_2(t), \\ dQ(t) = P(t)dt \end{cases}$$

with $\alpha \in \mathbb{R}$, $P(0) = p_0$, $Q(0) = q_0$. It is a stochastic Hamiltonian system with

$$H_0(P, Q) = -\alpha^2 \cos(Q) + \frac{1}{2}P^2, \quad H_1(P, Q) = \sigma_1 \sin(Q),$$

$$H_2(P, Q) = -\sigma_2 \cos(Q).$$

It can be verified that

$$G_{(0)}^1(P, q) = -\alpha^2 \cos(q) + \frac{1}{2}P^2, \quad G_{(1)}^1(P, q) = \sigma_1 \sin(q),$$

$$G_{(2)}^1(P, q) = -\sigma_2 \cos(q), \quad G_{(1,1)}^1(P, q) = G_{(2,2)}^1(P, q) = 0, \ldots$$

Thus the generating function S_1 for a symplectic method of weak order 1 is

$$S_1(P, q, h) = \left(G_{(0)}^1 + \frac{1}{2}G_{(1,1)}^1 + \frac{1}{2}G_{(2,2)}^1 \right)(P, q)h + G_{(1)}^1 \sqrt{h}\xi + G_{(2)}^1 \sqrt{h}\eta,$$

$$(3.29)$$

and the method generated by (3.29) via (3.14) is

$$P_{k+1} = P_k - (\alpha^2 \sin(Q_k)h + \sigma_1 \cos(Q_k)\sqrt{h}\xi_1^k + \sigma_2 \sin(Q_k)\sqrt{h}\xi_2^k),$$

$$Q_{k+1} = Q_k + P_{k+1}h, \quad k = 0, 1, \ldots, N-1. \tag{3.30}$$

Similar arguments lead to

$$\frac{\partial H_0^{[1]}}{\partial q} + \frac{1}{2}\left(-\sigma_1 \sin(q)\frac{\partial H_1^{[1]}}{\partial p} + \sigma_1 \cos(q)\frac{\partial^2 H_1^{[1]}}{\partial p \partial q}\right)$$

$$+ \frac{1}{2}\left(\sigma_2 \cos(q)\frac{\partial H_2^{[1]}}{\partial p} + \sigma_2 \sin(q)\frac{\partial^2 H_2^{[1]}}{\partial p \partial q}\right)$$

$$= -\frac{1}{2}\alpha^2 p \cos(q) + \frac{1}{2}\sigma_1^2 \sin(q)\cos(q) - \frac{1}{2}\sigma_2^2 \sin(q)\cos(q),$$

$$\frac{\partial H_0^{[1]}}{\partial p} + \frac{1}{2}\sigma_1 \cos(q)\frac{\partial^2 H_1^{[1]}}{\partial p^2} + \frac{1}{2}\sigma_2 \sin(q)\frac{\partial^2 H_2^{[1]}}{\partial p^2} = -\frac{1}{2}\alpha^2 \sin(q),$$

$$2\left(\sigma_1 \cos(q)\frac{\partial H_1^{[1]}}{\partial q} + \sigma_2 \sin(q)\frac{\partial H_2^{[1]}}{\partial q}\right) = (\sigma_1^2 - \sigma_2^2)p \sin(q)\cos(q),$$

$$2\left(\sigma_1 \cos(q)\frac{\partial H_1^{[1]}}{\partial p} + \sigma_2 \sin(q)\frac{\partial H_2^{[1]}}{\partial p}\right) = -\sigma_1^2 \cos^2(q) - \sigma_2^2 \sin^2(q).$$

As a result, we arrive at

$$\frac{\partial H_1^{[1]}}{\partial p} = -\frac{1}{2}\sigma_1 \cos(q), \qquad \frac{\partial H_1^{[1]}}{\partial q} = \frac{1}{2}\sigma_1 p \sin(q),$$

$$\frac{\partial H_2^{[1]}}{\partial p} = -\frac{1}{2}\sigma_2 \sin(q), \qquad \frac{\partial H_2^{[1]}}{\partial q} = -\frac{1}{2}\sigma_2 p \cos(q),$$

which yields

$$\frac{\partial H_0^{[1]}}{\partial q} = -\frac{1}{2}\alpha^2 p \cos(q), \qquad \frac{\partial H_0^{[1]}}{\partial p} = -\frac{1}{2}\alpha^2 \sin(q).$$

Thus, the stochastic modified equation of the numerical method (3.30) is

$$
\begin{cases}
dP(t) = (-\alpha^2 \sin(Q(t)) + \dfrac{h}{2}\alpha^2 P(t)\cos(Q(t)))dt \\
\qquad - (\sigma_1 \cos(Q(t)) + \dfrac{h}{2}\sigma_1 P(t)\sin(Q(t))) \circ dW_1(t) \\
\qquad - (\sigma_2 \sin(Q(t)) - \dfrac{h}{2}\sigma_2 P(t)\cos(Q(t))) \circ dW_2(t), \quad P(0) = p_0, \\
dQ(t) = (P(t) - \dfrac{h}{2}\alpha^2 \sin(Q(t)))dt - \dfrac{h}{2}\sigma_1 \cos(Q(t)) \circ dW_1(t) \\
\qquad - \dfrac{h}{2}\sigma_2 \sin(Q(t)) \circ dW_2(t), \quad Q(0) = q_0.
\end{cases}
$$

3.1.3 Modified Equation for Rough Symplectic Methods

In this part, we investigate the stochastic modified equation for rough symplectic methods applied to stochastic Hamiltonian systems driven by Gaussian rough paths

$$
\begin{cases}
dP_t = -\dfrac{\partial H_0(P_t, Q_t)}{\partial Q_t}dt - \displaystyle\sum_{r=1}^{m} \dfrac{\partial H_r(P_t, Q_t)}{\partial Q_t}d\mathbb{X}_t^r, \quad P_0 = p_0 \in \mathbb{R}^n, \\
dQ_t = \dfrac{\partial H_0(P_t, Q_t)}{\partial P_t}dt + \displaystyle\sum_{r=1}^{m} \dfrac{\partial H_r(P_t, Q_t)}{\partial P_t}d\mathbb{X}_t^r, \quad Q_0 = q_0 \in \mathbb{R}^n,
\end{cases} \tag{3.31}
$$

where $t \in (0, T]$ and $\mathbb{X}_t^r$, $r = 1, \ldots, m$, satisfy Assumption 1.4.1. Letting $Y := (P^\top, Q^\top)^\top$, $y := (p_0^\top, q_0^\top)^\top$ and $V_r := J_{2n}^{-1}\nabla H_r$, $r = 0, 1, \ldots, m$, we obtain a compact form

$$
\begin{cases}
dY_t = \displaystyle\sum_{r=0}^{m} V_r(Y_t)d\mathbb{X}_t^r = V(Y_t)d\mathbb{X}_t, \quad t \in (0, T], \\
Y_0 = y.
\end{cases} \tag{3.32}
$$

Let Y_k^h be given by a certain symplectic method of the rough Hamiltonian system, which is an approximation of Y_{t_k}, where $t_k = kh$, $k = 0, 1, \ldots, N$. Our main assumption on the numerical method is as follows.

Assumption 3.1.1 (See [60]) *The numerical approximation Y_{k+1}^h with $k = 0, 1, \ldots, N - 1$, can be expanded as an infinite series of functions of Y_k^h:*

$$Y_{k+1}^h = Y_k^h + \sum_{|\alpha|=1}^{\infty} d_\alpha(Y_k^h) h^{\alpha_0} (\mathbb{X}_{t_k, t_{k+1}}^1)^{\alpha_1} \cdots (\mathbb{X}_{t_k, t_{k+1}}^m)^{\alpha_m}, \qquad (3.33)$$

where $\alpha = (\alpha_0, \alpha_1, \ldots, \alpha_m) \in \mathbb{N}^{m+1}$, $|\alpha| := \alpha_0 + \alpha_1 + \cdots + \alpha_m \geq 1$ and $Y_0^h = y$. In addition, for $|\alpha| = 1$ with $\alpha_r = 1$, $r \in \{0, 1, \ldots, m\}$, define $V_\alpha(y) := V_r(y)$. It holds that

$$d_\alpha(y) = V_\alpha(y) \quad \forall \, |\alpha| = 1. \qquad (3.34)$$

To search for the stochastic modified equation of the symplectic method $\{Y_k^h\}_{0 \leq k \leq N}$ and overcome difficulties caused by both the non-differentiability and low regularity of $\mathbb{X}$, we begin from the Wong–Zakai approximation (see e.g., [23, 104] and references therein) of (3.32), i.e.,

$$\begin{cases} dy_t^h = V(y_t^h) dx_t^h = \displaystyle\sum_{r=0}^{m} V_r(y_t^h) dx_t^{h,r}, & t \in (0, T], \\ y_0^h = y, \end{cases} \qquad (3.35)$$

where $x^h = (x^{h,0}, x^{h,1}, \ldots, x^{h,m})$ is the piecewise linear approximation to $\mathbb{X}$ with

$$x_t^{h,r} := \mathbb{X}_{t_k}^r + \frac{t - t_k}{h} \mathbb{X}_{t_k, t_{k+1}}^r, \quad t \in (t_k, t_{k+1}], \quad k = 0, 1, \ldots, N - 1, \qquad (3.36)$$

for $r \in \{0, 1, \ldots, m\}$. Note that because the driving signal x^h is of bounded variation, it can be lifted to an η-rough path with $\eta \geq 1$. Thus, the Wong–Zakai approximation (3.35) is also a random differential equation

$$\begin{cases} \dot{y}_t^h = \displaystyle\sum_{r=0}^{m} V_r(y_t^h) \frac{\mathbb{X}_{t_k, t_{k+1}}^r}{h}, & t \in (t_k, t_{k+1}], \\ y_0^h = y, \end{cases} \qquad (3.37)$$

where $k = 0, 1, \ldots, N - 1$. By modifying the vector field of (3.37), we define the *stochastic modified equation* for a general rough numerical method satisfying (3.33) by the form

$$\begin{cases} \dot{\tilde{y}}_t = \displaystyle\sum_{|\alpha|=1}^{\infty} f_\alpha(\tilde{y}_t) h^{\alpha_0 - 1} (\mathbb{X}_{t_k, t_{k+1}}^1)^{\alpha_1} \cdots (\mathbb{X}_{t_k, t_{k+1}}^m)^{\alpha_m}, & t \in (t_k, t_{k+1}], \\ \tilde{y}_0 = y, \end{cases} \qquad (3.38)$$

where $k = 0, 1, \ldots, N - 1$, and $\tilde{y}$ is continuous on $[0, T]$. Since $|\alpha| \geq 1$, we have

$$0 \leq i(\alpha) := \min\{l : \alpha_l \geq 1, \, l = 0, 1, \ldots, m\} \leq m. \tag{3.39}$$

Consider $t \in (t_k, t_{k+1}]$ for $k \in \{0, 1, \ldots, N - 1\}$. The stochastic modified equation (3.38) can be rewritten in terms of an equation driven by x^h with the modified vector field $\bar{V}$

$$
\begin{aligned}
d\tilde{y}_t &= \sum_{|\alpha|=1}^{\infty} f_\alpha(\tilde{y}_t) h^{\alpha_0} (\mathbb{X}^1_{t_k,t_{k+1}})^{\alpha_1} \cdots (\mathbb{X}^{i(\alpha)}_{t_k,t_{k+1}})^{\alpha_{i(\alpha)}-1} \cdots (\mathbb{X}^m_{t_k,t_{k+1}})^{\alpha_m} \left(\frac{\mathbb{X}^{i(\alpha)}_{t_k,t_{k+1}}}{h} \right) dt \\
&= \sum_{|\alpha|=1}^{\infty} f_\alpha(\tilde{y}_t) h^{\alpha_0} (\mathbb{X}^1_{t_k,t_{k+1}})^{\alpha_1} \cdots (\mathbb{X}^{i(\alpha)}_{t_k,t_{k+1}})^{\alpha_{i(\alpha)}-1} \cdots (\mathbb{X}^m_{t_k,t_{k+1}})^{\alpha_m} dx_t^{h,i(\alpha)} \\
&= \sum_{l=0}^{m} \bar{V}_l(\tilde{y}_t) dx_t^{h,l} =: \bar{V}(\tilde{y}_t) dx_t^h. \tag{3.40}
\end{aligned}
$$

Due to the fact that the driving signal x^h can be lifted to an η-rough path with $\eta \geq 1$, the above equation can be interpreted in the rough path sense. It remains to determine the coefficient f_α. Taking the Taylor expansion and using the chain rule, we have

$$
\begin{aligned}
\tilde{y}_{t_{k+1}} = \tilde{y}_{t_k} &+ \sum_{|\alpha|=1}^{\infty} f_\alpha(\tilde{y}_{t_k}) h^{\alpha_0} (\mathbb{X}^1_{t_k,t_{k+1}})^{\alpha_1} \cdots (\mathbb{X}^m_{t_k,t_{k+1}})^{\alpha_m} \\
&+ \frac{1}{2!} \left(\frac{\partial}{\partial y} \left(\sum_{|\alpha|=1}^{\infty} f_\alpha(y) h^{\alpha_0} (\mathbb{X}^1_{t_k,t_{k+1}})^{\alpha_1} \cdots (\mathbb{X}^m_{t_k,t_{k+1}})^{\alpha_m} \right) \bigg|_{y=\tilde{y}_{t_k}} \right) \\
&\quad \times \left(\sum_{|\alpha|=1}^{\infty} f_\alpha(\tilde{y}_{t_k}) h^{\alpha_0} (\mathbb{X}^1_{t_k,t_{k+1}})^{\alpha_1} \cdots (\mathbb{X}^m_{t_k,t_{k+1}})^{\alpha_m} \right) \\
&+ \frac{1}{3!} \left(\frac{\partial}{\partial y} \left(\left(\frac{\partial}{\partial y} \left(\sum_{|\alpha|=1}^{\infty} f_\alpha(y) h^{\alpha_0} (\mathbb{X}^1_{t_k,t_{k+1}})^{\alpha_1} \cdots (\mathbb{X}^m_{t_k,t_{k+1}})^{\alpha_m} \right) \right) \right. \right. \\
&\quad \times \left. \left. \left(\sum_{|\alpha|=1}^{\infty} f_\alpha(y) h^{\alpha_0} (\mathbb{X}^1_{t_k,t_{k+1}})^{\alpha_1} \cdots (\mathbb{X}^m_{t_k,t_{k+1}})^{\alpha_m} \right) \right) \bigg|_{y=\tilde{y}_{t_k}} \right)
\end{aligned}
$$

$$\times \left(\sum_{|\alpha|=1}^{\infty} f_\alpha(\tilde{y}_{t_k}) h^{\alpha_0} (\mathbb{X}^1_{t_k,t_{k+1}})^{\alpha_1} \cdots (\mathbb{X}^m_{t_k,t_{k+1}})^{\alpha_m} \right) + \cdots$$

$$=: \tilde{y}_{t_k} + \sum_{|\alpha|=1}^{\infty} \tilde{f}_\alpha(\tilde{y}_{t_k}) h^{\alpha_0} (\mathbb{X}^1_{t_k,t_{k+1}})^{\alpha_1} \cdots (\mathbb{X}^m_{t_k,t_{k+1}})^{\alpha_m}. \tag{3.41}$$

Let $(D_{k^{i_1,i_2}} g)(y) := g'(y) f_{k^{i_1,i_2}}(y)$ with $k^{i_1,i_2} = \left(k_0^{i_1,i_2}, k_1^{i_1,i_2}, \ldots, k_m^{i_1,i_2} \right) \in \mathbb{N}^{m+1}$ and $|k^{i_1,i_2}| \geq 1$. Defining the set

$$O_i^\alpha := \Big\{ (k^{i,1}, \ldots, k^{i,i}) \,|\, k^{i,1}, \ldots, k^{i,i} \in \mathbb{N}^{m+1}, \ |k^{i,1}|, \ldots, |k^{i,i}| \geq 1,$$
$$k_l^{i,1} + \cdots + k_l^{i,i} = \alpha_l, \ l = 0, 1, \ldots, m \Big\},$$

we obtain

$$\tilde{f}_\alpha(y) = \begin{cases} f_\alpha(y), & \text{if } |\alpha| = 1; \\ f_\alpha(y) + \displaystyle\sum_{i=2}^{|\alpha|} \frac{1}{i!} \sum_{(k^{i,1},\ldots,k^{i,i}) \in O_i^\alpha} (D_{k^{i,1}} \cdots D_{k^{i,i-1}} f_{k^{i,i}})(y), & \text{if } |\alpha| \geq 2. \end{cases}$$

To ensure $Y_{k+1}^h = \tilde{y}_{t_{k+1}}$, we compare (3.33) and (3.41), and then need

$$\tilde{f}_\alpha(y) = d_\alpha(y) \quad \forall \alpha \in \mathbb{N}^{m+1}, \ |\alpha| \geq 1.$$

Therefore, the stochastic modified equation (3.38) is given by the recursion

$$f_\alpha(y) = \begin{cases} d_\alpha(y), & \text{if } |\alpha| = 1; \\ d_\alpha(y) - \displaystyle\sum_{i=2}^{|\alpha|} \frac{1}{i!} \sum_{(k^{i,1},\ldots,k^{i,i}) \in O_i^\alpha} (D_{k^{i,1}} \cdots D_{k^{i,i-1}} f_{k^{i,i}})(y), & \text{if } |\alpha| \geq 2. \end{cases}$$
$$\tag{3.42}$$

Note that f_α is determined by coefficients d_α and $f_{\alpha'}$ with $|\alpha'| < |\alpha|$.

Remark 3.1.2 The framework about the construction of the stochastic modified equation is also applicable for rough numerical methods with adaptive time-step sizes.

Based on orthogonal polynomials with respect to the measure induced by increments of noises, together with the integrability lemma in [116], [60] proves that the stochastic modified equation associated to a rough symplectic method is still a Hamiltonian system, which is stated in the following theorem.

Theorem 3.1.3 (See [60]) *Assume that V is bounded and continuously differentiable, and that all its derivatives are bounded. If $Y_1^h(y)$, the one-step numerical approximation, is given by applying a symplectic method satisfying Assumption 3.1.1 to (3.31), then the associated stochastic modified equation (3.38) is a perturbed Hamiltonian system. More precisely, for any multi-index α, there exists a Hamiltonian $\mathscr{H}_\alpha : \mathbb{R}^{2n} \to \mathbb{R}$ such that*

$$f_\alpha = J_{2n}^{-1} \nabla \mathscr{H}_\alpha, \tag{3.43}$$

and then the stochastic modified equation takes the following form

$$
\begin{cases}
\dot{\tilde{y}}_t = \displaystyle\sum_{|\alpha|=1}^{\infty} J_{2n}^{-1} \nabla \mathscr{H}_\alpha(\tilde{y}_t) h^{\alpha_0 - 1} (\mathbb{X}_{t_k,t_{k+1}}^1)^{\alpha_1} \cdots (\mathbb{X}_{t_k,t_{k+1}}^m)^{\alpha_m}, & t \in (t_k, t_{k+1}], \\
\tilde{y}_0 = y,
\end{cases}
$$

where $k = 0, 1, \ldots, N - 1$.

In general, the series given in (3.38) may not converge. For $\tilde{N} \geq 1$, we consider the $\tilde{N}$-truncated modified equation

$$
\begin{cases}
\dot{\tilde{y}}_t^{\tilde{N}} = \displaystyle\sum_{|\alpha|=1}^{\tilde{N}} f_\alpha(\tilde{y}_t^{\tilde{N}}) h^{\alpha_0 - 1} (\mathbb{X}_{t_k,t_{k+1}}^1)^{\alpha_1} \cdots (\mathbb{X}_{t_k,t_{k+1}}^m)^{\alpha_m}, & t \in [t_k, t_{k+1}], \\
\tilde{y}_0^{\tilde{N}} = y,
\end{cases}
\tag{3.44}
$$

where f_α is given by (3.42) with $|\alpha| = 1, \ldots, \tilde{N}$. Similar to (3.40), we rewrite (3.44) as

$$
\begin{aligned}
d\tilde{y}_t^{\tilde{N}} &= \sum_{|\alpha|=1}^{\tilde{N}} f_\alpha(\tilde{y}_t^{\tilde{N}}) h^{\alpha_0 - 1} (\mathbb{X}_{t_k,t_{k+1}}^1)^{\alpha_1} \cdots (\mathbb{X}_{t_k,t_{k+1}}^m)^{\alpha_m} dt \\
&= \sum_{|\alpha|=1}^{\tilde{N}} f_\alpha(\tilde{y}_t^{\tilde{N}}) h^{\alpha_0} (\mathbb{X}_{t_k,t_{k+1}}^1)^{\alpha_1} \cdots (\mathbb{X}_{t_k,t_{k+1}}^{i(\alpha)})^{\alpha_{i(\alpha)} - 1} \cdots (\mathbb{X}_{t_k,t_{k+1}}^m)^{\alpha_m} \left(\frac{\mathbb{X}_{t_k,t_{k+1}}^{i(\alpha)}}{h} \right) dt \\
&= \sum_{|\alpha|=1}^{\tilde{N}} f_\alpha(\tilde{y}_t^{\tilde{N}}) h^{\alpha_0} (\mathbb{X}_{t_k,t_{k+1}}^1)^{\alpha_1} \cdots (\mathbb{X}_{t_k,t_{k+1}}^{i(\alpha)})^{\alpha_{i(\alpha)} - 1} \cdots (\mathbb{X}_{t_k,t_{k+1}}^m)^{\alpha_m} dx_t^{h,i(\alpha)} \\
&=: \bar{V}^{\tilde{N}}(\tilde{y}_t^{\tilde{N}}) dx_t^h, \quad t \in (t_k, t_{k+1}],
\end{aligned}
$$

where x^h and $i(\alpha)$ are given by (3.36) and (3.39), respectively. It yields that (3.44) is equivalent to a rough differential equation with $\bar{V}^{\tilde{N}}$ and x_t^h. Below we shall give the convergence analysis on the error between Y_k^h and $\tilde{y}_{t_k}^{\tilde{N}}$, where $k = 1, \ldots, N$, in the case that $\mathbb{X}$ is a general Gaussian rough path satisfying Assumption 1.4.1.

Theorem 3.1.4 (See [60]) *Suppose that Assumption 1.4.1 holds and $V \in Lip^{\tilde{N}+1}$. Then for any $p > 2\rho$, there exists a random variable $C := C(\omega, p, \|V\|_{Lip^{\tilde{N}+1}}, \tilde{N})$ such that*

$$\|\tilde{y}_{t_1}^{\tilde{N}} - Y_1^h\| \leq Ch^{\frac{\tilde{N}+1}{p}}, \quad \text{a.s.,}$$

where $\tilde{y}^{\tilde{N}}$ is the solution of (3.44) and Y_1^h is defined by a numerical method satisfying (3.33).

Proof Consider the expansion

$$\tilde{y}_{t_1}^{\tilde{N}} = y + \sum_{|\alpha|=1}^{\infty} f_\alpha^{\tilde{N}}(y) h^{\alpha_0} (\mathbb{X}_{t_0,t_1}^1)^{\alpha_1} \cdots (\mathbb{X}_{t_0,t_1}^m)^{\alpha_m}.$$

Fix $p > 2\rho \geq 2$. Because the recursion (3.42) means $f_\alpha^{\tilde{N}} = \tilde{f}_\alpha = d_\alpha$ with $1 \leq |\alpha| \leq \tilde{N}$, and Assumption 1.4.1 implies $\|\mathbb{X}\|_{\frac{1}{p}\text{-Höl};[t_0,t_1]} < \infty$, we derive from the Taylor expansion that the leading term of the error between $\tilde{y}_{t_1}^{\tilde{N}}$ and Y_1^h is involved with $h^{\alpha_0} (\mathbb{X}_{t_0,t_1}^1)^{\alpha_1} \cdots (\mathbb{X}_{t_0,t_1}^m)^{\alpha_m}$, where $\alpha_0 = 0$ and $\alpha_1 + \cdots + \alpha_m = \tilde{N}+1$. Consequently,

$$\|\tilde{y}_{t_1}^{\tilde{N}} - Y_1^h\| \leq C(\omega, p, \|V\|_{Lip^{\tilde{N}+1}}, \tilde{N}) h^{\frac{\tilde{N}+1}{p}},$$

which finishes the proof. $\qquad\qquad\qquad\qquad\qquad\qquad\qquad\qquad\qquad\qquad$ □

Theorem 3.1.5 (See [60]) *Under Assumption 1.4.1, if $V \in Lip^{\tilde{N}+\gamma}$ with $\gamma > 2\rho$ and $\tilde{N} > 2\rho - 1$, then for any $p \in (2\rho, \gamma)$, there exists a random variable $C := C(\omega, p, \gamma, \|V\|_{Lip^{\tilde{N}+\gamma}}, \tilde{N}, T)$ such that*

$$\sup_{1 \leq k \leq N} \|\tilde{y}_{t_k}^{\tilde{N}} - Y_k^h\| \leq Ch^{\frac{\tilde{N}+1}{p}-1}, \quad \text{a.s.,}$$

where $\tilde{y}^{\tilde{N}}$ is the solution of (3.44) and Y_k^h is defined by a numerical method satisfying (3.33).

Proof Denote by $\pi(t_0, y_0, x^h)_t$, $t \geq t_0$, the flow with the initial value y_0 at time t_0. We arrive at

$$\|Y_k^h - \tilde{y}_{t_k}^{\tilde{N}}\| = \|\pi(t_k, Y_k^h, x^h)_{t_k} - \pi(t_0, Y_0^h, x^h)_{t_k}\|$$

$$\leq \sum_{s=1}^{k} \|\pi(t_s, Y_s^h, x^h)_{t_k} - \pi(t_{s-1}, Y_{s-1}^h, x^h)_{t_k}\|, \quad 1 \leq k \leq N.$$

Due to the Lipschitz continuity of the Itô–Lyons map (see [105, Theorem 10.26]), we have that for $1 \leq s < k$,

$$\|\pi(t_s, Y_s^h, x^h)_{t_k} - \pi(t_{s-1}, Y_{s-1}^h, x^h)_{t_k}\|$$

$$= \|\pi(t_{k-1}, \pi(t_s, Y_s^h, x^h)_{t_{k-1}}, x^h)_{t_k} - \pi(t_{k-1}, \pi(t_{s-1}, Y_{s-1}^h, x^h)_{t_{k-1}}, x^h)_{t_k}\|$$

$$\leq C \exp\left(C\bar{v}^p \|S_{[p]}(x^h)(\omega)\|_{p\text{-var};[t_{k-1},t_k]}^p\right) \|\pi(t_s, Y_s^h, x^h)_{t_{k-1}} - \pi(t_{s-1}, Y_{s-1}^h, x^h)_{t_{k-1}}\|,$$

where $C := C(p, \gamma)$ and $\bar{v} := \bar{v}(\|X\|_{\frac{1}{p}\text{-Höl};[0,T]}(\omega), \|V\|_{Lip^{\tilde{N}+\gamma}}, \tilde{N}) \geq \|\bar{V}\|_{Lip^\gamma}$. From

$$\|S_{[p]}(x^h)(\omega)\|_{p\text{-var};[u_1,u_2]}^p + \|S_{[p]}(x^h)(\omega)\|_{p\text{-var};[u_2,u_3]}^p \leq \|S_{[p]}(x^h)(\omega)\|_{p\text{-var};[u_1,u_3]}^p$$

with $0 \leq u_1 < u_2 < u_3 \leq T$, it follows that

$$\|\pi(t_s, Y_s^h, x^h)_{t_k} - \pi(t_{s-1}, Y_{s-1}^h, x^h)_{t_k}\|$$

$$\leq C \exp\left(C\bar{v}^p \|S_{[p]}(x^h)(\omega)\|_{p\text{-var};[t_s,t_k]}^p\right) \|\pi(t_s, Y_s^h, x^h)_{t_s} - \pi(t_{s-1}, Y_{s-1}^h, x^h)_{t_s}\|$$

$$\leq C \exp\left(C\bar{v}^p \|S_{[p]}(x^h)(\omega)\|_{p\text{-var};[0,T]}^p\right) \|Y_s^h - \pi(t_{s-1}, Y_{s-1}^h, x^h)_{t_s}\|, \quad 1 \leq s \leq k.$$

Since $\|S_{[p]}(x^h)(\omega)\|_{p\text{-var};[0,T]}^p$ is uniformly bounded with respect to h for almost all $\omega \in \Omega$ (see [105, Theorem 15.28]), by Theorem 3.1.4 and the fact that $\gamma > 2\rho \geq 2$, we deduce

$$\|Y_k^h - \tilde{y}_{t_k}^{\tilde{N}}\| \leq \sum_{s=1}^{k} C \exp\left(C\bar{v}^p \|S_{[p]}(x^h)(\omega)\|_{p\text{-var};[0,T]}^p\right) \|Y_s^h - \pi(t_{s-1}, Y_{s-1}^h, x^h)_{t_s}\|$$

$$\leq C(\omega, p, \gamma, \|V\|_{Lip^{\tilde{N}+\gamma}}, \tilde{N}, T) h^{\frac{\tilde{N}+1}{p}-1},$$

which completes the proof. $\qquad\qquad\qquad\qquad\qquad\qquad\qquad\qquad\qquad\qquad\qquad\qquad\quad\square$

In the case of additive noise, the diffusion part can be simulated exactly. Thus, Assumption 3.1.1 on the numerical method degenerates to

$$
Y_{k+1}^h = Y_k^h + \sum_{|\alpha|=1} V_\alpha(Y_k^h) h^{\alpha_0} (\mathbb{X}_{t_k,t_{k+1}}^1)^{\alpha_1} \cdots (\mathbb{X}_{t_k,t_{k+1}}^m)^{\alpha_m}
$$

$$
+ \sum_{|\alpha|=2,\alpha_0 \geq 1}^{\infty} d_\alpha(Y_k^h) h^{\alpha_0} (\mathbb{X}_{t_k,t_{k+1}}^1)^{\alpha_1} \cdots (\mathbb{X}_{t_k,t_{k+1}}^m)^{\alpha_m}. \tag{3.45}
$$

As a consequence, the convergence rate of the error between Y_k^h and $\tilde{y}_{t_k}^{\tilde{N}}$ could be improved, which is stated in the following corollary.

Corollary 3.1.1 (See [60]) *Let Assumption 1.4.1 hold and $V_i(y) = \sigma_i \in \mathbb{R}^{2n}$, $i = 1, \ldots, m$. If $V_0 \in Lip^{\tilde{N}+\gamma}$ with $\gamma > 2\rho$, then for any $p \in (2\rho, \gamma)$, there exists a random variable $C := C(\omega, p, \gamma, \|V_0\|_{Lip^{\tilde{N}+\gamma}}, \sigma_i, \tilde{N}, T)$ such that*

$$
\sup_{1 \leq k \leq N} \|\tilde{y}_{t_k}^{\tilde{N}} - Y_k^h\| \leq C h^{\frac{\tilde{N}}{p}}, \quad a.s.,
$$

where $\tilde{y}^{\tilde{N}}$ is the solution of (3.44) and Y_k^h is defined by a numerical method satisfying (3.45).

Proof Combining (3.45) with (3.42), we obtain that the leading term of the local error between $\tilde{y}_{t_1}^{\tilde{N}}$ and Y_1^h is involved with $h^{\alpha_0}(X_{t_0,t_1}^1)^{\alpha_1} \cdots (X_{t_0,t_1}^m)^{\alpha_m}$, where $\alpha_0 = 1$ and $\alpha_1 + \cdots + \alpha_m = \tilde{N}$. Then it follows that

$$
\|\tilde{y}_{t_1}^{\tilde{N}} - Y_1^h\| \leq C(\omega) h^{\frac{\tilde{N}}{p}+1}, \quad a.s.,
$$

from which we conclude the result by applying the same procedures as in the proof of Theorem 3.1.5. $\qquad\square$

3.2 Stochastic Numerical Methods Based on Modified Equation

Inspired by the theory of the stochastic modified equation, we introduce the methodology for constructing numerical methods with high weak order and less multiple stochastic integrals to be simulated for the stochastic differential equation in this section. The basic idea of the approach can be summarized as follows. Instead of utilizing the numerical method to the stochastic differential equation (3.1) directly, we apply it to a suitably modified stochastic differential equation, which is a perturbation of (3.1), so that the resulting numerical method turns to be a higher weak order approximation of the original stochastic differential equation. First, we introduce the general stochastic numerical method produced by using the backward Kolmogorov equation (see [1]). Then the high weak order symplectic method is proposed by means of the stochastic θ-generating function S_θ, $\theta \in [0, 1]$.

3.2.1 High Order Methods via Backward Kolmogorov Equation

Denote $\sigma = [\sigma_1, \ldots, \sigma_m]$ and $W = (W_1, \ldots, W_m)^\top$. Then (3.1) becomes

$$\begin{cases} dX(t) = a(X(t))dt + \sigma(X(t))dW(t), \\ X(0) = x. \end{cases} \tag{3.46}$$

Now we consider the stochastic differential equation (3.46) with suitably modified drift and diffusion functions

$$\begin{cases} d\widetilde{X}(t) = a_h(\widetilde{X}(t))dt + \sigma_h(\widetilde{X}(t))dW(t), \\ \widetilde{X}(0) = x, \end{cases} \tag{3.47}$$

where

$$a_h(\widetilde{X}) = a(\widetilde{X}) + \widetilde{a}_1(\widetilde{X})h + \widetilde{a}_2(\widetilde{X})h^2 + \cdots,$$
$$\sigma_h(\widetilde{X}) = \sigma(\widetilde{X}) + \widetilde{\sigma}_1(\widetilde{X})h + \widetilde{\sigma}_2(\widetilde{X})h^2 + \cdots$$

with $\widetilde{a}_i$, $\widetilde{\sigma}_i$, $i \in \mathbb{N}_+$ to be determined. Applying a weakly convergent numerical method of order $\gamma \in \mathbb{N}_+$ as follows

$$X_{k+1} = \Psi(a, \sigma, X_k, h, \triangle_k W), \quad k = 0, 1, \ldots, N - 1, \tag{3.48}$$

of (3.46), where $\Psi(a, \sigma, \cdot, h, \triangle_k W) : \mathbb{R}^d \to \mathbb{R}^d$ and $\triangle_k W = W(t_{k+1}) - W(t_k)$, to (3.47), we obtain the numerical method

$$\widetilde{X}_{k+1} = \Psi(a_h, \sigma_h, \widetilde{X}_k, h, \triangle_k W).$$

Our goal is to choose a_h, σ_h such that $\{\widetilde{X}_k\}_{k \geq 0}$ is a better weak approximation to the solution of (3.46), i.e.,

$$|\mathbb{E}[\phi(\widetilde{X}_N)] - \mathbb{E}[\phi(X(T))]| = O(h^{\gamma+\gamma'})$$

with $\phi \in \mathbf{C}_P^{2\gamma+2\gamma'+2}(\mathbb{R}^d, \mathbb{R})$ and $\gamma' \geq 1$.

Assumption 3.2.1 (See [1]) *The numerical solution of* (3.48) *satisfies*

$$u_{num}^{a,\sigma}(x, h) := \mathbb{E}[\phi(X_1)| \; X_0 = x]$$

$$= \phi(x) + h A_0(a, \sigma)\phi(x) + h^2 A_1(a, \sigma)\phi(x) + \cdots,$$

where $A_i(a, \sigma)$, $i = 0, 1, \ldots$, are linear differential operators depending on the drift and diffusion coefficients of the stochastic differential equation to which the numerical method is applied. We further assume that these differential operators $A_i(a, \sigma)$, $i = 0, 1, \ldots$, satisfy

$$A_i(a + \varepsilon \hat{a}, \sigma + \varepsilon \hat{\sigma}) = A_i(a, \sigma) + \varepsilon \hat{A}_i(a, \hat{a}, \sigma, \hat{\sigma}) + O\left(\varepsilon^2\right)$$

for any $a, \hat{a}, \sigma, \hat{\sigma}$ and sufficiently small $\varepsilon > 0$, where $\hat{A}_i(a, \hat{a}, \sigma, \hat{\sigma})$, $i = 0, 1, \ldots$, are differential operators.

The task now is to find a modified stochastic differential equation (3.47) such that

$$u_{num}^{a_h, \sigma_h}(x, h) := \mathbb{E}[\phi(\widetilde{X}_1)| \; \widetilde{X}_0 = x] = u(x, h) + O(h^{\gamma + \gamma' + 1}),$$

i.e., a numerical approximation $\widetilde{X}_k$ of weak order $\gamma + \gamma'$ with $\gamma' > 0$ for (3.46). In addition, assume that the numerical method is of weak order at least one. This assumption implies $A_0(a, \sigma)\phi = \mathbb{L}\phi$ and $A_0(a_h, \sigma_h)\phi = \widetilde{\mathbb{L}}\phi$, where

$$\widetilde{\mathbb{L}}\phi = a_h^\top \frac{\partial \phi}{\partial x} + \frac{1}{2}\text{Tr}\left(\sigma_h \sigma_h^\top \nabla^2 \phi\right)$$

with ∇^2 being the Hessian matrix operator. Substituting the expressions of a_h and σ_h yields

$$\widetilde{\mathbb{L}} = \mathbb{L} + h\widetilde{\mathbb{L}}_1 + h^2 \widetilde{\mathbb{L}}_2 + \cdots$$

with $\widetilde{\mathbb{L}}_j$, $j \in \mathbb{N}_+$, satisfying

$$\widetilde{\mathbb{L}}_j \phi = \tilde{a}_j^\top \frac{\partial \phi}{\partial x} + \frac{1}{2}\sum_{i=0}^{j}\text{Tr}\left(\tilde{\sigma}_{j-i}\tilde{\sigma}_i^\top \nabla^2 \phi\right), \tag{3.49}$$

where $\tilde{a}_0 := a$ and $\tilde{\sigma}_0 := \sigma$. We now show that under suitable assumptions, the weak order γ of the numerical method (3.48) can be improved to $\gamma + \gamma'$ with $\gamma' \geq 1$ by applying it to a suitably modified stochastic differential equation with modified drift and diffusion coefficients of the form

$$a_{h,\gamma+\gamma'-1}(\widetilde{X}) = a(\widetilde{X}) + \widetilde{a}_1(\widetilde{X})h + \cdots + \widetilde{a}_{\gamma+\gamma'-1}(\widetilde{X})h^{\gamma+\gamma'-1},$$

$$\sigma_{h,\gamma+\gamma'-1}(\widetilde{X}) = \sigma(\widetilde{X}) + \widetilde{\sigma}_1(\widetilde{X})h + \cdots + \widetilde{\sigma}_{\gamma+\gamma'-1}(\widetilde{X})h^{\gamma+\gamma'-1}.$$

The numerical method with improved weak order γ' can be rewritten as

$$\widetilde{X}_{k+1} = \Psi(a_{h,\gamma+\gamma'-1}, \sigma_{h,\gamma+\gamma'-1}, \widetilde{X}_k, h, \Delta_k W), \quad k = 0, 1, \ldots, N-1. \tag{3.50}$$

Theorem 3.2.1 (See [1]) *Assume that the numerical method (3.48) has weak order $\gamma \geq 1$ and that Assumption 3.2.1 holds. Let $\gamma' \geq 1$ and assume that functions $\widetilde{a}_j$ and $\widetilde{\sigma}_j$, where $j = 1, \ldots, \gamma + \gamma' - 2$, have been constructed such that*

$$\widetilde{X}_{k+1} = \Psi(a_{h,\gamma+\gamma'-2}, \sigma_{h,\gamma+\gamma'-2}, \widetilde{X}_k, h, \Delta_k W), \quad k = 0, 1, \ldots, N-1,$$

is of weak order $\gamma + \gamma' - 1$. Consider the differential operator defined as

$$\overset{\approx}{\mathbb{L}}_{\gamma+\gamma'-1}\phi = \lim_{h \to 0} \frac{u(x,h) - u_{num}^{a_{h,\gamma+\gamma'-2},\sigma_{h,\gamma+\gamma'-2}}(x,h)}{h^{\gamma+\gamma'}}. \tag{3.51}$$

If there exist functions $\widetilde{a}_{\gamma+\gamma'-1}: \mathbb{R}^d \to \mathbb{R}^d$ and $\widetilde{\sigma}_{\gamma+\gamma'-1}: \mathbb{R}^d \to \mathbb{R}^{d \times m}$ such that $\overset{\approx}{\mathbb{L}}_{\gamma+\gamma'-1}\phi$ can be rewritten in the form

$$\overset{\approx}{\mathbb{L}}_{\gamma+\gamma'-1}\phi = \widetilde{a}_{\gamma+\gamma'-1}^{\top} \frac{\partial\phi}{\partial x} + \frac{1}{2}\sum_{j=0}^{\gamma+\gamma'-1} \mathrm{Tr}(\widetilde{\sigma}_{\gamma+\gamma'-1-j}\widetilde{\sigma}_j^{\top}\nabla_x^2\phi),$$

then (3.50) applied to the stochastic differential equation with the modified drift and diffusion has weak order of accuracy $\gamma + \gamma'$ for (3.46) provided $\widetilde{a}_{h,\gamma+\gamma'-1} \in \mathbf{C}_P^{2(\gamma+\gamma'+1)}(\mathbb{R}^d, \mathbb{R}^d)$ and $\widetilde{\sigma}_{h,\gamma+\gamma'-1} \in \mathbf{C}_P^{2(\gamma+\gamma'+1)}(\mathbb{R}^d, \mathbb{R}^{d \times m})$. Furthermore, the weak error satisfies

$$|\mathbb{E}[\phi(\widetilde{X}_N)] - \mathbb{E}[\phi(X(T))]| = O(h^{\gamma+\gamma'}) \tag{3.52}$$

for all functions $\phi \in \mathbf{C}_P^{2(\gamma+\gamma'+1)}(\mathbb{R}^d, \mathbb{R})$.

Proof By means of the induction hypothesis and the fact that $\widetilde{X}_k, k = 0, 1, \ldots, N$, is of weak order $\gamma + \gamma' - 1$, we obtain

$$u_{num}^{a_{h,\gamma+\gamma'-2},\sigma_{h,\gamma+\gamma'-2}}(x,h) = \phi(x) + hA_0(a_{h,\gamma+\gamma'-2}, \sigma_{h,\gamma+\gamma'-2})\phi(x) + \cdots + O(h^{\gamma+\gamma'+1})$$

$$= \phi(x) + h\mathbb{L}\phi(x) + \ldots + \frac{h^{\gamma+\gamma'-1}}{(\gamma+\gamma'-1)!}\mathbb{L}^{\gamma+\gamma'-1}\phi(x) + h^{\gamma+\gamma'}B_{\gamma+\gamma'}\phi(x) + O(h^{\gamma+\gamma'+1}),$$

where $B_{\gamma+\gamma'}$ is a certain differential operator. According to Assumption 3.2.1, we derive

$$A_i(a_{h,\gamma+\gamma'-1}, \sigma_{h,\gamma+\gamma'-1}) = A_i(a_{h,\gamma+\gamma'-2}, \sigma_{h,\gamma+\gamma'-2}) + O(h^{\gamma+\gamma'-1}),$$

where $i \geq 1$. Then the weak Taylor expansion of the numerical method (3.50) can be rewritten as

$$u_{num}^{a_{h,\gamma+\gamma'-1}, \sigma_{h,\gamma+\gamma'-1}}(x, h)$$

$$= \phi(x) + hA_0(a_{h,\gamma+\gamma'-1}, \sigma_{h,\gamma+\gamma'-1})\phi(x) + \cdots + O(h^{\gamma+\gamma'+1})$$

$$= \phi(x) + hA_0(a_{h,\gamma+\gamma'-2}, \sigma_{h,\gamma+\gamma'-2})\phi(x) + h^{\gamma+\gamma'}\widetilde{\mathbb{L}}_{\gamma+\gamma'-1}\phi(x) + \cdots + O(h^{\gamma+\gamma'+1})$$

$$= \phi(x) + h\mathbb{L}\phi(x) + \ldots + \frac{h^{\gamma+\gamma'-1}}{(\gamma+\gamma'-1)!}\mathbb{L}^{\gamma+\gamma'-1}\phi(x)$$

$$+ h^{\gamma+\gamma'}(\widetilde{\mathbb{L}}_{\gamma+\gamma'-1} + B_{\gamma+\gamma'})\phi(x) + O(h^{\gamma+\gamma'+1}),$$

where $\widetilde{\mathbb{L}}_{\gamma+\gamma'-1}$ is defined in (3.49). If $\widetilde{a}_{\gamma+\gamma'-1}$ and $\widetilde{\sigma}_{\gamma+\gamma'-1}$ are such that

$$\widetilde{\mathbb{L}}_{\gamma+\gamma'-1} = \frac{\mathbb{L}^{\gamma+\gamma'}}{(\gamma+\gamma')!} - B_{\gamma+\gamma'},$$

then the local weak order of $\widetilde{X}_k$, $k = 0, 1, \ldots, N$, is $\gamma+\gamma'+1$. Now observing that the right-hand side of the above equality is equal to the right-hand side of (3.51), together with Remark 2.1.2, proves the theorem.   □

Remark 3.2.1 We would like to emphasize that the aim and theory for high weak order numerical methods based on the modified stochastic differential equations and those for the stochastic modified equations of stochastic numerical methods are different. In the former approach, the modified stochastic differential equation constitutes only a surrogate to obtain a better numerical approximation of the solution of the original stochastic differential equation. In the latter approach, the stochastic modified equation is a tool to better understand the behavior of the numerical method applied to the original stochastic differential equation.

Now we illustrate the methodology by constructing high weak order numerical methods for the following 1-dimensional stochastic differential equation which exactly preserves all quadratic invariants

$$\begin{cases} dX(t) = \sigma_0(X(t))dt + \sigma_1(X(t)) \circ dW(t), \\ X(0) = x, \end{cases} \tag{3.53}$$

where $\sigma_0, \sigma_1 \in \mathbf{C}^\infty(\mathbb{R}, \mathbb{R})$. Let us consider the stochastic midpoint method

$$X_{k+1} = X_k + \sigma_0 \left(\frac{X_k + X_{k+1}}{2} \right) h + \sigma_1 \left(\frac{X_k + X_{k+1}}{2} \right) \Delta_k \widehat{W}$$

as a basic numerical scheme. Using the framework of numerical methods constructed by the modified equation (see [1]), we introduce the following new numerical method

$$X_{k+1} = X_k + \sigma_{h,1}^0 \left(\frac{X_k + X_{k+1}}{2} \right) h + \sigma_{h,1}^1 \left(\frac{X_k + X_{k+1}}{2} \right) \Delta_k \widehat{W}, \qquad (3.54)$$

where $\sigma_{h,1}^0 = \sigma_0 + \widetilde{\sigma}_1^0 h$ and $\sigma_{h,1}^1 = \sigma_1 + \widetilde{\sigma}_1^1 h$ with

$$\widetilde{\sigma}_1^0 = \frac{1}{4} \left(\frac{1}{2} \sigma_0'' \sigma_1^2 - \sigma_1' \sigma_0' \sigma_1 \right), \quad \widetilde{\sigma}_1^1 = \frac{1}{4} \left(\frac{1}{2} \sigma_1'' \sigma_1^2 - (\sigma_1')^2 \sigma_1 \right).$$

Theorem 3.2.2 (See [1]) *The numerical method (3.54) for (3.53) has weak order 2. Moreover, it exactly conserves all quadratic invariants of (3.53).*

A natural extension would be to search for modified stochastic differential equations to design numerical methods of higher weak order or numerical methods preserving some stability and geometric properties, and we refer to [1] for more details.

3.2.2 High Order Symplectic Methods via Generating Function

In this subsection, we construct high weak order stochastic symplectic methods based on the modified stochastic Hamiltonian system and the stochastic θ-generating function with $\theta \in [0, 1]$. Consider the perturbed stochastic Hamiltonian system

$$\begin{cases} d\tilde{P}(t) = -\dfrac{\partial \tilde{H}_0(\tilde{P}(t), \tilde{Q}(t))}{\partial \tilde{Q}} dt - \displaystyle\sum_{r=1}^{m} \dfrac{\partial \tilde{H}_r(\tilde{P}(t), \tilde{Q}(t))}{\partial \tilde{Q}} \circ dW_r(t), \quad \tilde{P}(0) = p, \\[4mm] d\tilde{Q}(t) = \dfrac{\partial \tilde{H}_0(\tilde{P}(t), \tilde{Q}(t))}{\partial \tilde{P}} dt + \displaystyle\sum_{r=1}^{m} \dfrac{\partial \tilde{H}_r(\tilde{P}(t), \tilde{Q}(t))}{\partial \tilde{P}} \circ dW_r(t), \quad \tilde{Q}(0) = q, \end{cases}$$

$$(3.55)$$

where $t \in (0, T]$, and

$$\tilde{H}_r(\tilde{P}, \tilde{Q}) = H_r(\tilde{P}, \tilde{Q}) + H_r^{[1]}(\tilde{P}, \tilde{Q})h + H_r^{[2]}(\tilde{P}, \tilde{Q})h^2 + \cdots, \quad r = 0, 1, \ldots, m,$$

$$(3.56)$$

with undetermined functions $H_j^{[i]}$ for $j = 0, 1, \ldots, m$ and $i \in \mathbb{N}_+$. Let $\check{P}(t) = (1 - \theta)p + \theta\tilde{P}(t)$, $\check{Q}(t) = (1 - \theta)\tilde{Q}(t) + \theta q$ with $\theta \in [0, 1]$. As shown in Chap. 1, the associated stochastic θ-generating function $\tilde{S}_\theta(\check{P}(t), \check{Q}(t), t)$, $\theta \in [0, 1]$, could be exploited to express the exact solution as follows

$$\tilde{P}(t) = p - \frac{\partial \tilde{S}_\theta(\check{P}(t), \check{Q}(t), t)}{\partial \check{Q}}, \quad \tilde{Q}(t) = q + \frac{\partial \tilde{S}_\theta(\check{P}(t), \check{Q}(t), t)}{\partial \check{P}}, \quad t \in (0, T],$$

with

$$\tilde{S}_\theta(\check{P}, \check{Q}, t) = \sum_\alpha \tilde{G}_\alpha^\theta(\check{P}, \check{Q}, h) J_\alpha(t).$$

Here, $\tilde{G}_\alpha^\theta(\check{P}, \check{Q}, h)$ is defined by using similar arguments as in (1.34). Taking advantage of the truncated generating function

$$\grave{S}_\theta(\grave{P}, \grave{Q}, h) = \sum_\alpha \tilde{G}_\alpha^\theta(\grave{P}, \grave{Q}, h) \sum_{l(\beta) \leq \gamma} C_\alpha^\beta I_\alpha(h)$$

with $\grave{P} = (1 - \theta)p + \theta\acute{P}$ and $\grave{Q} = (1 - \theta)\acute{Q} + \theta q$, we obtain the one-step approximation

$$\acute{P} = p - \frac{\partial \grave{S}_\theta(\grave{P}, \grave{Q}, h)}{\partial \grave{Q}}, \quad \acute{Q} = q + \frac{\partial \grave{S}_\theta(\grave{P}, \grave{Q}, h)}{\partial \grave{P}}. \tag{3.57}$$

The numerical method based on the above one-step approximation (3.57) is of weak order $\gamma \in \mathbb{N}_+$ for the stochastic Hamiltonian system (3.55). It suffices to find these unknown functions $H_j^{[i]}$, $j = 0, 1, \ldots, m$, $i = 1, 2, \ldots$, such that the applied numerical method based on (3.57) is a better week approximation for the original stochastic system (3.13), i.e.,

$$|\mathbb{E}[\phi(\acute{P}_N, \acute{Q}_N)] - \mathbb{E}[\phi(P(T), Q(T))]| = O(h^{\gamma + \gamma'}), \tag{3.58}$$

where $\phi \in \mathbf{C}_P^{2(\gamma + \gamma' + 1)}$ with $\gamma' \geq 1$. According to the relationship between the global weak error and local weak error, we could get the function $H_j^{[i]}$ for $j = 0, 1, \ldots, m$, and $i \in \mathbb{N}_+$ by comparing the local error between $\mathbb{E}[\phi(\acute{P}, \acute{Q})]$ and $\mathbb{E}[\phi(P(h), Q(h))]$. For simplicity, we denote $P(h) = P$, and $Q(h) = Q$, and then have

$$P = p - \frac{\partial S_\theta(\hat{P}, \hat{Q}, h)}{\partial \hat{Q}}, \quad Q = q + \frac{\partial S_\theta(\hat{P}, \hat{Q}, h)}{\partial \hat{P}} \tag{3.59}$$

with $\hat{P} = (1 - \theta)p + \theta P$ and $\hat{Q} = \theta q + (1 - \theta)Q$. Taking expectation leads to

$$\mathbb{E}[\phi(P, Q)] = \phi(p, q) + \sum_{i=1}^{n} \frac{\partial \phi}{\partial p_i}(p, q)\mathbb{E}\left[-\frac{\partial S_\theta}{\partial \hat{Q}_i} \right] + \sum_{i=1}^{n} \frac{\partial \phi}{\partial q_i}(p, q)\mathbb{E}\left[\frac{\partial S_\theta}{\partial \hat{P}_i} \right]$$

$$+ \frac{1}{2}\sum_{i,j=1}^{n} \frac{\partial^2 \phi(p, q)}{\partial p_i \partial p_j}\mathbb{E}\left[\frac{\partial S_\theta}{\partial \hat{Q}_i}\frac{\partial S_\theta}{\partial \hat{Q}_j} \right] - \sum_{i,j=1}^{n} \frac{\partial^2 \phi(p, q)}{\partial q_i \partial p_j}\mathbb{E}\left[\frac{\partial S_\theta}{\partial \hat{P}_i}\frac{\partial S_\theta}{\partial \hat{Q}_j} \right]$$

$$+ \frac{1}{2}\sum_{i,j=1}^{n} \frac{\partial^2 \phi(p, q)}{\partial q_i \partial q_j}\mathbb{E}\left[\frac{\partial S_\theta}{\partial \hat{P}_i}\frac{\partial S_\theta}{\partial \hat{P}_j} \right] + \cdots,$$

where the derivatives of S_θ take values at $(\hat{P}, \hat{Q}, h)$. Similarly,

$$\mathbb{E}[\phi(\grave{P}, \grave{Q})] = \phi(p, q) + \sum_{i=1}^{n} \frac{\partial \phi}{\partial p_i}(p, q)\mathbb{E}\left[-\frac{\partial \grave{S}_\theta}{\partial \grave{Q}_i} \right] + \sum_{i=1}^{n} \frac{\partial \phi}{\partial q_i}(p, q)\mathbb{E}\left[\frac{\partial \grave{S}_\theta}{\partial \grave{P}_i} \right]$$

$$+ \frac{1}{2}\sum_{i,j=1}^{n} \frac{\partial^2 \phi(p, q)}{\partial p_i \partial p_j}\mathbb{E}\left[\frac{\partial \grave{S}_\theta}{\partial \grave{Q}_i}\frac{\partial \grave{S}_\theta}{\partial \grave{Q}_j} \right] - \sum_{i,j=1}^{n} \frac{\partial^2 \phi(p, q)}{\partial q_i \partial p_j}\mathbb{E}\left[\frac{\partial \grave{S}_\theta}{\partial \grave{P}_i}\frac{\partial \grave{S}_\theta}{\partial \grave{Q}_j} \right]$$

$$+ \frac{1}{2}\sum_{i,j=1}^{n} \frac{\partial^2 \phi(p, q)}{\partial q_i \partial q_j}\mathbb{E}\left[\frac{\partial \grave{S}_\theta}{\partial \grave{P}_i}\frac{\partial \grave{S}_\theta}{\partial \grave{P}_j} \right] + \cdots,$$

where the derivatives of $\grave{S}_\theta$ take values at $(\grave{P}, \grave{Q}, h)$. To satisfy the condition

$$\left| \mathbb{E}[\phi(P, Q)] - \mathbb{E}[\phi(\grave{P}, \grave{Q})] \right| = O(h^{\gamma + \gamma' + 1}), \tag{3.60}$$

we need to compute the left hand side of the above equation up to the terms of mean-square order $\gamma + \gamma'$.

Now we illustrate this approach with the construction of a numerical method of weak order 2 for the stochastic Hamiltonian system

$$\begin{cases} dP(t) = -\dfrac{\partial H_0(P(t), Q(t))}{\partial Q}dt - \displaystyle\sum_{r=1}^{m} \dfrac{\partial H_r(Q(t))}{\partial Q} \circ dW_r(t), \quad P(0) = p, \\[3mm] dQ(t) = \dfrac{\partial H_0(P(t), Q(t))}{\partial P}dt, \quad Q(0) = q, \end{cases}$$
$$\tag{3.61}$$

where Hamiltonians H_r, $r = 0, 1, \ldots, m$, are smooth enough. It means that both γ and γ' are equal to 1, and we need to determine derivatives of $H_r^{[1]}$ for $r \in \{0, 1, \ldots, m\}$. Via (3.57), one can verify that for $i \in \{1, \ldots, n\}$,

$$\acute{Q}_i - q_i = \left(\frac{\partial H_0}{\partial \hat{P}_i} + h\frac{\partial H_0^{[1]}}{\partial \hat{P}_i} + hC_\theta \sum_{r=1}^{m}\sum_{k=1}^{n} \frac{\partial^2 H_r^{[1]}}{\partial \hat{P}_i \partial \hat{P}_k} \frac{\partial H_r}{\partial \hat{Q}_k}\right)h + \sum_{r=1}^{m}\left(h\frac{\partial H_r^{[1]}}{\partial \hat{P}_i}\right)I_{(r)}(h) + O(h^{\frac{5}{2}}),$$

$$p_i - \acute{P}_i = \left(\frac{\partial H_0}{\partial \hat{Q}_i} + h\frac{\partial H_0^{[1]}}{\partial \hat{Q}_i} + hC_\theta \sum_{r=1}^{m}\sum_{k=1}^{n}\left(\frac{\partial^2 H_r^{[1]}}{\partial \hat{Q}_i \partial \hat{P}_k} \frac{\partial H_r}{\partial \hat{Q}_k} + \frac{\partial H_r^{[1]}}{\partial \hat{P}_k} \frac{\partial^2 H_r}{\partial \hat{Q}_k \partial \hat{Q}_i}\right)\right)h$$

$$+ \sum_{r=1}^{m}\left(\frac{\partial H_r}{\partial \hat{Q}_i} + h\frac{\partial H_r^{[1]}}{\partial \hat{Q}_i}\right)I_{(r)}(h) + O(h^{\frac{5}{2}}),$$

where $C_\theta = \frac{2\theta - 1}{2}$ and the derivatives of H_r and $H_r^{[1]}$ take values at $(\hat{P}, \hat{Q}, h)$ with $r = 0, 1, \ldots, m$. By means of (1.34), we obtain the coefficient $G_\alpha^\theta(\hat{P}, \hat{Q})$ in the case of (3.61) as follows

$$G_{(r,s)}^\theta(\hat{P}, \hat{Q}) = 0, \qquad G_{(r,s,u)}^\theta(\hat{P}, \hat{Q}) = 0, \qquad G_{(r,s,u,v)}^\theta(\hat{P}, \hat{Q}) = 0,$$

$$G_{(r,0)}^\theta(\hat{P}, \hat{Q}) = (\theta - 1)\sum_{k=1}^{n} \frac{\partial H_0(\hat{P}, \hat{Q})}{\partial \hat{P}_k} \frac{\partial H_r(\hat{P}, \hat{Q})}{\partial \hat{Q}_k},$$

$$G_{(0,r)}^\theta(\hat{P}, \hat{Q}) = \theta \sum_{k=1}^{n} \frac{\partial H_0(\hat{P}, \hat{Q})}{\partial \hat{P}_k} \frac{\partial H_r(\hat{P}, \hat{Q})}{\partial \hat{Q}_k},$$

$$G_{(0,r,s)}^\theta(\hat{P}, \hat{Q}) = \theta^2 \sum_{k,j=1}^{n} \frac{\partial H_s(\hat{P}, \hat{Q})}{\partial \hat{Q}_k} \frac{\partial H_r(\hat{P}, \hat{Q})}{\partial \hat{Q}_j} \frac{\partial^2 H_0(\hat{P}, \hat{Q})}{\partial \hat{P}_k \partial \hat{P}_j},$$

$$G_{(r,0,s)}^\theta(\hat{P}, \hat{Q}) = \theta(\theta - 1) \sum_{k,j=1}^{n} \frac{\partial H_s(\hat{P}, \hat{Q})}{\partial \hat{Q}_k} \frac{\partial H_r(\hat{P}, \hat{Q})}{\partial \hat{Q}_j} \frac{\partial^2 H_0(\hat{P}, \hat{Q})}{\partial \hat{P}_k \partial \hat{P}_j},$$

$$G_{(r,s,0)}^\theta(\hat{P}, \hat{Q}) = (\theta - 1)^2 \sum_{k,j=1}^{n} \frac{\partial H_s(\hat{P}, \hat{Q})}{\partial \hat{Q}_k} \frac{\partial H_r(\hat{P}, \hat{Q})}{\partial \hat{Q}_j} \frac{\partial^2 H_0(\hat{P}, \hat{Q})}{\partial \hat{P}_k \partial \hat{P}_j}$$

for $r, s, u, v \in \{1, \ldots, m\}$. Differentiating the stochastic θ-generating function with $\theta \in [0, 1]$ yields that for $i \in \{1, \ldots, n\}$,

$$\frac{\partial S_\theta}{\partial \hat{P}_i} = \frac{\partial H_0}{\partial \hat{P}_i}h + \sum_{r=1}^{m}\sum_{k=1}^{n} \frac{\partial H_r}{\partial \hat{Q}_k} \frac{\partial^2 H_0}{\partial \hat{P}_k \partial \hat{P}_i}(\theta J_{(0,r)}(h) + (\theta - 1)J_{(r,0)}(h))$$

$$+ \sum_{r,s=1}^{m}\sum_{k,l=1}^{n} \frac{\partial H_r}{\partial \hat{Q}_k} \frac{\partial H_s}{\partial \hat{Q}_l} \frac{\partial^3 H_0}{\partial \hat{P}_l \partial \hat{P}_k \partial \hat{P}_i}(\theta^2 J_{(0,s,r)}(h) + (\theta - 1)^2 J_{(s,r,0)}(h))$$

$$+ \theta(\theta - 1)J_{(s,0,r)}(h)) + \frac{2\theta - 1}{2}\sum_{k=1}^{n}\Big(\frac{\partial H_0}{\partial \hat{Q}_k}\frac{\partial^2 H_0}{\partial \hat{P}_k \partial \hat{P}_i} + \frac{\partial^2 H_0}{\partial \hat{Q}_k \partial \hat{P}_i}\frac{\partial H_0}{\partial \hat{P}_k}\Big)h^2 + \cdots,$$

$$\frac{\partial S_\theta}{\partial \hat{Q}_i} = \frac{\partial H_0}{\partial \hat{Q}_i}h + \sum_{r=1}^{m}\frac{\partial H_r}{\partial \hat{Q}_i}J_{(r)}(h) + \sum_{r=1}^{m}\sum_{k=1}^{n}\frac{\partial H_r}{\partial \hat{Q}_k}\frac{\partial^2 H_0}{\partial \hat{P}_k \partial \hat{Q}_i}(\theta J_{(0,r)}(h) + (\theta - 1)J_{(r,0)}(h))$$

$$+ \sum_{r,s=1}^{m}\sum_{k,l=1}^{n}\Big(\frac{\partial H_r}{\partial \hat{Q}_k}\frac{\partial H_s}{\partial \hat{Q}_l}\frac{\partial^3 H_0}{\partial \hat{P}_l \partial \hat{P}_k \partial \hat{Q}_i} + \frac{\partial^2 H_r}{\partial \hat{Q}_k \partial \hat{Q}_i}\frac{\partial H_s}{\partial \hat{Q}_l}\frac{\partial^2 H_0}{\partial \hat{P}_l \partial \hat{P}_k} + \frac{\partial H_r}{\partial \hat{Q}_k}\frac{\partial^2 H_s}{\partial \hat{Q}_l \partial \hat{Q}_i}$$

$$\times \frac{\partial^2 H_0}{\partial \hat{P}_l \partial \hat{P}_k}\Big)(\theta^2 J_{(0,s,r)}(h) + (\theta - 1)^2 J_{(s,r,0)}(h) + \theta(\theta - 1)J_{(s,0,r)}(h))$$

$$+ \frac{2\theta - 1}{2}\sum_{k=1}^{n}\Big(\frac{\partial H_0}{\partial \hat{Q}_k}\frac{\partial^2 H_0}{\partial \hat{P}_k \partial \hat{Q}_i} + \frac{\partial^2 H_0}{\partial \hat{Q}_k \partial \hat{Q}_i}\frac{\partial H_0}{\partial \hat{P}_k}\Big)h^2 + \cdots$$

with the derivatives of H_r and $H_r^{[1]}$ taking values at $(\hat{P}, \hat{Q}, h)$ with $r = 0, 1, \ldots, m$. To obtain (3.60), we need to let coefficients of h and h^2 be equal within the following pairs, respectively, for $i, j = 1, \ldots, n$,

$$\mathbb{E}\Big[\frac{\partial S_\theta}{\partial \hat{Q}_i}\Big] \text{ and } \mathbb{E}\Big[\frac{\partial \dot{S}_\theta}{\partial \hat{Q}_i}\Big]; \quad \mathbb{E}\Big[\frac{\partial S_\theta}{\partial \hat{P}_i}\Big] \text{ and } \mathbb{E}\Big[\frac{\partial \dot{S}_\theta}{\partial \hat{P}_i}\Big]; \quad \mathbb{E}\Big[\frac{\partial S_\theta}{\partial \hat{Q}_i}\frac{\partial S_\theta}{\partial \hat{Q}_j}\Big] \text{ and } \mathbb{E}\Big[\frac{\partial \dot{S}_\theta}{\partial \hat{Q}_i}\frac{\partial \dot{S}_\theta}{\partial \hat{Q}_j}\Big];$$

$$\mathbb{E}\Big[\frac{\partial S_\theta}{\partial \hat{Q}_i}\frac{\partial S_\theta}{\partial \hat{P}_j}\Big] \text{ and } \mathbb{E}\Big[\frac{\partial \dot{S}_\theta}{\partial \hat{Q}_i}\frac{\partial \dot{S}_\theta}{\partial \hat{P}_j}\Big]; \quad \mathbb{E}\Big[\frac{\partial S_\theta}{\partial \hat{P}_i}\frac{\partial S_\theta}{\partial \hat{P}_j}\Big] \text{ and } \mathbb{E}\Big[\frac{\partial \dot{S}_\theta}{\partial \hat{P}_i}\frac{\partial \dot{S}_\theta}{\partial \hat{P}_j}\Big].$$

Comparing the above pairs and performing the Taylor expansion of the partial derivatives of S_θ and those of $\dot{S}_\theta$ at (p, q, h) recursively, where $\theta \in [0, 1]$, we get

$$\frac{\partial H_r^{[1]}}{\partial Q_i} = \frac{(2\theta - 1)}{2}\sum_{k=1}^{n}\frac{\partial^2 H_0}{\partial P_k \partial Q_i}\frac{\partial H_r}{\partial Q_k}, \quad \frac{\partial H_r^{[1]}}{\partial P_i} = \frac{(2\theta - 1)}{2}\sum_{k=1}^{n}\frac{\partial^2 H_0}{\partial P_k \partial P_i}\frac{\partial H_r}{\partial Q_k},$$

$$\frac{\partial H_0^{[1]}}{\partial Q_i} = \frac{(2\theta - 1)}{2}\sum_{k=1}^{n}\Big(\frac{\partial H_0}{\partial Q_k}\frac{\partial^2 H_0}{\partial P_k \partial Q_i} + \frac{\partial^2 H_0}{\partial Q_k \partial Q_i}\frac{\partial H_0}{\partial P_k}\Big)$$

$$+ \frac{1}{4}\sum_{r=1}^{m}\sum_{k,l=1}^{n}\frac{\partial H_r}{\partial Q_k}\frac{\partial H_r}{\partial Q_l}\frac{\partial^3 H_0}{\partial P_l \partial P_k \partial Q_i},$$

$$\frac{\partial H_0^{[1]}}{\partial P_i} = \frac{(2\theta - 1)}{2}\sum_{k=1}^{n}\Big(\frac{\partial H_0}{\partial Q_k}\frac{\partial^2 H_0}{\partial P_k \partial P_i} + \frac{\partial^2 H_0}{\partial Q_k \partial P_i}\frac{\partial H_0}{\partial P_k}\Big)$$

$$+ \frac{1}{4}\sum_{r=1}^{m}\sum_{k,l=1}^{n}\frac{\partial H_r}{\partial Q_k}\frac{\partial H_r}{\partial Q_l}\frac{\partial^3 H_0}{\partial P_l \partial P_k \partial P_i}$$

for $r = 1, \ldots, m$, and $i = 1, \ldots, n$. Based on the above relations and the one-step approximation (3.57), one can derive a stochastic symplectic method of weak order 2.

Remark 3.2.2 For the general stochastic Hamiltonian system, the midpoint method is the priority selection to approximate the modified stochastic Hamiltonian system. Moreover, we would like to mention that the above approach is available for larger γ and γ', following the same procedures as the case of $\gamma = \gamma' = 1$.

3.3   Conformal Symplectic and Ergodic Methods for Stochastic Langevin Equation

The discretization of the modified stochastic differential equation, which is constructed by modifying the drift and diffusion coefficients as polynomials with respect to the time-step size h, represents a powerful tool for obtaining high weak order methods. In this section, we will use the modified stochastic differential equation and the stochastic generating function to construct numerical methods, which are of high weak order and conformally symplectic, for the following stochastic Langevin equation with additive noise

$$
\begin{cases}
dP(t) = -f(Q(t))dt - vP(t)dt - \sum_{r=1}^{m} \sigma_r dW_r(t), & P(0) = p \in \mathbb{R}^n, \\
dQ(t) = MP(t)dt, & Q(0) = q \in \mathbb{R}^n,
\end{cases}
\tag{3.62}
$$

where $f \in \mathbf{C}^\infty(\mathbb{R}^n, \mathbb{R}^n)$, $M \in \mathbb{R}^{n \times n}$ is a positive definite symmetric matrix, $v > 0$ and $\sigma_r \in \mathbb{R}^n$ with $r \in \{1, \ldots, m\}$, $m \geq n$ and $\text{rank}\{\sigma_1, \ldots, \sigma_m\} = n$. In addition, assume that there exists a scalar function $F \in \mathbf{C}^\infty(\mathbb{R}^n, \mathbb{R})$ satisfying

$$
f_i(Q) = \frac{\partial F(Q)}{\partial Q_i}, \quad i = 1, \ldots, n.
$$

It has been shown in Chap. 1 that the considered stochastic Langevin equation is a dissipative Hamiltonian system, whose phase flow preserves the stochastic conformal symplectic structure

$$
dP(t) \wedge dQ(t) = \exp(-vt)dp \wedge dq \quad \forall\, t \geq 0,
$$

and is ergodic with a unique invariant measure, i.e., the Boltzmann–Gibbs measure, if Assumption 1.3.1 holds and $M = I_n$. This property implies that the temporal average of the solution will converge to its spatial average, which is also known as the ergodic limit, with respect to the invariant measure over long time.

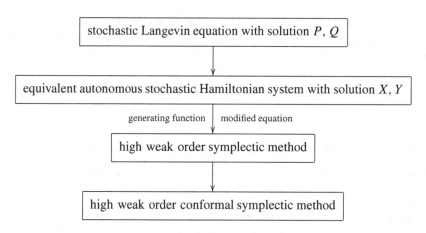

To achieve high weak order conformal symplectic methods that accurately approximate the exact solution, while preserving both the geometric structure and dynamical behavior of the system, we introduce a transformation from the stochastic Langevin equation to a stochastic Hamiltonian system. It then suffices to construct high order symplectic methods for the autonomous Hamiltonian system, which turn out to be conformal symplectic methods for the original system based on the inverse transformation of the phase spaces. To this end, denoting $X_i(t) = \exp(vt)P_i(t)$ and $Y_i(t) = Q_i(t)$ and applying the Itô formula to $X_i(t)$ and $Y_i(t)$ for $i = 1, \ldots, n$, where $t \geq 0$, one can rewrite (3.62) as

$$
\begin{cases}
dX_i(t) = -\exp(vt) f_i(Y_1(t), \ldots, Y_n(t))dt - \exp(vt) \sum_{r=1}^{m} \sigma_r dW_r(t), \quad X_i(0) = p_i, \\[2ex]
dY_i(t) = \exp(-vt) \sum_{j=1}^{n} M_{ij} X_j(t)dt, \quad Y_i(0) = q_i.
\end{cases}
$$

$$(3.63)$$

It can be seen that (3.63) is a non-autonomous stochastic Hamiltonian system with time-dependent Hamiltonian functions

$$
\bar{H}_0(X, Y) = \exp(vt) F(Y_1, \ldots, Y_n) + \frac{1}{2} \exp(-vt) \sum_{i,j=1}^{n} X_i M_{ij} X_j,
$$

$$
\bar{H}_r(X, Y) = \exp(vt) \sum_{i=1}^{n} \sigma_r^i Y_i, \quad r = 1, \ldots, m.
$$

To obtain an autonomous stochastic Hamiltonian system, we introduce two new variables $X_{n+1} \in \mathbb{R}$ and $Y_{n+1} \in \mathbb{R}$ as the $(n+1)$th component of X and Y, respectively, satisfying

$$dY_{n+1}(t) = dt, \quad dX_{n+1}(t) = -\frac{\partial \bar{H}_0(X(t), Y(t))}{\partial t} dt - \sum_{r=1}^{m} \frac{\partial \bar{H}_r(X(t), Y(t))}{\partial t} \circ dW_r(t)$$

with $Y_{n+1}(0) = 0$ and $X_{n+1}(0) = F(q_1, \ldots, q_n) + \frac{1}{2} \sum_{i,j=1}^{n} p_i M_{ij} p_j + \sum_{r=1}^{m} \sum_{i=1}^{n} \sigma_r^i q_i$.

Then we derive the $2n + 2$-dimensional autonomous stochastic Hamiltonian system

$$\begin{cases} dX(t) = -\dfrac{\partial H_0(X(t), Y(t))}{\partial Y} dt - \displaystyle\sum_{r=1}^{m} \dfrac{\partial H_r(X(t), Y(t))}{\partial Y} \circ dW_r(t), \\[2mm] dY(t) = \dfrac{\partial H_0(X(t), Y(t))}{\partial X} dt + \displaystyle\sum_{r=1}^{m} \dfrac{\partial H_r(X(t), Y(t))}{\partial X} \circ dW_r(t) \end{cases} \tag{3.64}$$

with $X(0) = (X_1(0), \ldots, X_{n+1}(0))^\top$, $Y(0) = (Y_1(0), \ldots, Y_{n+1}(0))^\top$, and the associated Hamiltonians are

$$H_0(X, Y) = \exp(v Y_{n+1}) F(Y_1, \ldots, Y_n) + \frac{1}{2} \exp(-v Y_{n+1}) \sum_{i,j=1}^{n} X_i M_{ij} X_j + X_{n+1},$$

$$H_r(X, Y) = \exp(v Y_{n+1}) \sum_{i=1}^{n} \sigma_r^i Y_i, \quad r = 1, \ldots, m.$$

Notice that the motion of the system can be described by the stochastic θ-generating functions with $\theta \in [0, 1]$, and here we concentrate on the type-I stochastic generating function, which has a series expansion

$$S_1(X, y, t) = \sum_{\alpha} G_\alpha^1(X, y) J_\alpha(t). \tag{3.65}$$

Denote $X(0) = x$ and $Y(0) = y$. Then the mapping $(x^\top, y^\top)^\top \mapsto (X(t)^\top, Y(t)^\top)^\top$, $t \geq 0$, defined by

$$X(t) = x - \frac{\partial S_1(X(t), y, t)}{\partial y}, \quad Y(t) = y + \frac{\partial S_1(X(t), y, t)}{\partial X} \tag{3.66}$$

is the stochastic flow of (3.64). For a fixed small time-step size h, making use of (3.65) and taking Taylor expansion to $\frac{\partial S_1}{\partial y_i} := \frac{\partial S_1}{\partial y_i}(X, y, h)$ and $\frac{\partial S_1}{\partial X_i} := \frac{\partial S_1}{\partial X_i}(X, y, h)$ at (x, y, h) for $i = 1, \ldots, n$, we obtain

$$
\frac{\partial S_1}{\partial y_i} = \exp(vy_{n+1})\left(\sum_{r=1}^{m} \sigma_r^i (J_{(r)}(h) + vJ_{(0,r)}(h)) + f_i(y)\left(h + \frac{vh^2}{2}\right)\right)
$$
$$
+ \frac{h^2}{2} \sum_{j,k=1}^{n} \frac{\partial^2 F(y)}{\partial y_i \partial y_j} M_{jk} x_k + R_1^i,
$$

$$
\frac{\partial S_1}{\partial X_i} = \exp(-vy_{n+1}) \sum_{j=1}^{n} M_{ij} x_j \left(h - \frac{vh^2}{2}\right) - \sum_{j=1}^{n}\sum_{r=1}^{m} M_{ij}\sigma_r^j J_{(r,0)}(h)
$$
$$
- \frac{h^2}{2} \sum_{j=1}^{n} M_{ij} f_j(y) + R_2^i,
$$

where every term in R_1^i and R_2^i contains the product of multiple stochastic integrals whose the lowest mean-square order is at least $\frac{5}{2}$ and so are the following remainder term R_3 in the sequel. Furthermore, $\frac{\partial S_1}{\partial X_{n+1}}(X, y, h) = h$ and

$$
\frac{\partial S_1}{\partial y_{n+1}} = vh\left(\exp(vy_{n+1})F(y) - \frac{\exp(-vy_{n+1})}{2}\sum_{i,j=1}^{n} x_i M_{ij} x_j\right)\left(1 + \frac{vh}{2}\right)
$$
$$
+ v \exp(vy_{n+1}) \sum_{r=1}^{m}\sum_{i=1}^{n} \sigma_r^i y_i (J_{(r)}(h) + vJ_{(0,r)}(h))
$$
$$
+ \sum_{i,j=1}^{n}\sum_{r=1}^{m} v\sigma_r^i M_{ij} x_j h J_{(r)}(h) + v \exp(vy_{n+1}) \sum_{r_1,r_2=1}^{m} \sigma_{r_1}^\top M \sigma_{r_2} J_{(0,r_1,r_2)}(h)
$$
$$
+ v \sum_{i,j=1}^{n} \left(\exp(-vy_{n+1})\frac{\partial F(y)}{\partial y_i} M_{ij} x_j h^2\right)
$$
$$
- \frac{1}{2} \exp(vy_{n+1}) \sum_{r_1,r_2=1}^{m} \sigma_{r_1}^i M_{ij} \sigma_{r_2}^j h J_{(r_1)}(h) J_{(r_2)}(h)\right) + R_3,
$$

where $\frac{\partial S_1}{\partial y_{n+1}}$ takes the value at (X, y, h).

To construct high weak order symplectic methods for (3.64), we modify the Hamiltonians first, and introduce the following $2n + 2$-dimensional stochastic Hamiltonian system

$$
\begin{cases}
d\tilde{X}(t) = -\dfrac{\partial \tilde{H}_0(\tilde{X}(t), \tilde{Y}(t))}{\partial \tilde{Y}} dt - \displaystyle\sum_{r=1}^{m} \dfrac{\partial \tilde{H}_r(\tilde{X}(t), \tilde{Y}(t))}{\partial \tilde{Y}} \circ dW_r(t), \\[4mm]
d\tilde{Y}(t) = \dfrac{\partial \tilde{H}_0(\tilde{X}(t), \tilde{Y}(t))}{\partial \tilde{X}} dt + \displaystyle\sum_{r=1}^{m} \dfrac{\partial \tilde{H}_r(\tilde{X}(t), \tilde{Y}(t))}{\partial \tilde{X}} \circ dW_r(t),
\end{cases}
\tag{3.67}
$$

where $\tilde{X}(0) = x$, $\tilde{Y}(0) = y$ and

$$
\tilde{H}_r(\tilde{X}, \tilde{Y}) = H_r(\tilde{X}, \tilde{Y}) + H_r^{[1]}(\tilde{X}, \tilde{Y})h + \cdots + H_r^{[\iota]}(\tilde{X}, \tilde{Y})h^{\iota}, \quad r = 0, 1, \ldots, m
\tag{3.68}
$$

with functions $H_j^{[i]}$, $j = 0, 1, \ldots, m$, $i = 1, \ldots, \iota$, $\iota \in \mathbb{N}_+$ to be determined. Now we exploit the methodology of constructing higher weak order symplectic methods in Sect. 3.2.2 to choose undetermined functions such that the proposed method is of weak order $\gamma + \gamma'$ when approximating (3.64), even though it is only a weak γth order method of (3.67) for some $\gamma, \gamma' \in \mathbb{N}_+$. According to the truncated modified generating function

$$
S^G(X^G, y, t) = \sum_{\alpha} G_\alpha^G(X^G, y) \sum_{l(\beta) \leq \gamma} C_\alpha^\beta I_\beta(t),
\tag{3.69}
$$

which contains undetermined functions $H_j^{[i]}$, $j = 0, 1, \ldots, m$, $i = 1, \ldots, \iota$, in (3.68), we deduce the following one-step approximation

$$
X^G = x - \frac{\partial S^G(X^G, y, h)}{\partial y}, \quad Y^G = y + \frac{\partial S^G(X^G, y, h)}{\partial X^G},
\tag{3.70}
$$

which preserves the stochastic symplectic structure and is of weak order γ for (3.67). To specify high weak order symplectic methods, we need to determine all the terms $H_j^{[i]}$ such that (3.70) satisfies

$$
\left| \mathbb{E}[\phi(X(h), Y(h))] - \mathbb{E}[\phi(X^G, Y^G)] \right| = O(h^{\gamma + \gamma' + 1})
\tag{3.71}
$$

for $\phi \in \mathbf{C}_P^{2\gamma + 2\gamma' + 1}(\mathbb{R}^{2n+2}, \mathbb{R})$, that is, the numerical method based on (3.70) is of weak order $\gamma + \gamma'$ for (3.64).

Now we illustrate this approach by a specific case $\gamma = \gamma' = 1$ and show that the proposed method for this particular case admits an invariant measure. In this case, the one-step approximation (3.70) becomes

$$X^G = x - \left(\frac{\partial \tilde{H}_0(X^G, y)}{\partial y} + \frac{1}{2}\sum_{r=1}^{m}\frac{\partial G_{(r,r)}^G(X^G, y)}{\partial y}\right)h - \sum_{r=1}^{m}\frac{\partial \tilde{H}_r(X^G, y)}{\partial y}J_{(r)}(h),$$

$$Y^G = y + \left(\frac{\partial \tilde{H}_0(X^G, y)}{\partial X^G} + \frac{1}{2}\sum_{r=1}^{m}\frac{\partial G_{(r,r)}^G(X^G, y)}{\partial X^G}\right)h + \sum_{r=1}^{m}\frac{\partial \tilde{H}_r(X^G, y)}{\partial X^G}J_{(r)}(h),$$

$$(3.72)$$

where

$$G_{(r,r)}^G = \exp(vy_{n+1})\sum_{i=1}^{n}\sigma_r^i\left(\frac{\partial H_r^{[1]}}{\partial X_i^G} + vy_i\frac{\partial H_r^{[1]}}{\partial X_{n+1}^G}\right)h + \sum_{i=1}^{n+1}\frac{\partial H_r^{[1]}}{\partial y_i}\frac{\partial H_r^{[1]}}{\partial X_i^G}h^2.$$

Denote $\frac{\partial S^G}{\partial y_j} := \frac{\partial S^G}{\partial y_j}(X^G, y, h)$, $\frac{\partial S^G}{\partial X_j^G} := \frac{\partial S^G}{\partial X_j^G}(X^G, y, h)$, $\frac{\partial H_r^{[1]}}{\partial y_j} := \frac{\partial H_r^{[1]}}{\partial y_j}(x, y)$ and $\frac{\partial H_r^{[1]}}{\partial x_j} := \frac{\partial H_r^{[1]}}{\partial x_j}(x, y)$ for $j = 1, \ldots, n+1$ and $r = 0, 1, \ldots, m$. Performing the Taylor expansion to $\phi(X(h), Y(h))$ and $\phi(X^G, Y^G)$ at (x, y) and taking the expectation, we have

$$\mathbb{E}[\phi(X(h), Y(h))] - \mathbb{E}[\phi(X^G, Y^G)]$$

$$= \sum_{i=1}^{n+1}\frac{\partial \phi(x, y)}{\partial x_i}\mathbb{E}\left[\frac{\partial S^G}{\partial y_i} - \frac{\partial S_1}{\partial y_i}\right] + \sum_{i=1}^{n+1}\frac{\partial \phi(x, y)}{\partial y_i}\mathbb{E}\left[\frac{\partial S_1}{\partial X_i} - \frac{\partial S^G}{\partial X_i^G}\right]$$

$$+ \frac{1}{2}\sum_{i,j=1}^{n+1}\frac{\partial^2\phi(x, y)}{\partial x_i\partial x_j}\mathbb{E}\left[\frac{\partial S_1}{\partial y_i}\frac{\partial S_1}{\partial y_j} - \frac{\partial S^G}{\partial y_i}\frac{\partial S^G}{\partial y_j}\right] \qquad (3.73)$$

$$+ \sum_{i,j=1}^{n+1}\frac{\partial^2\phi(x, y)}{\partial y_i\partial x_j}\mathbb{E}\left[\frac{\partial S^G}{\partial X_i^G}\frac{\partial S^G}{\partial y_j} - \frac{\partial S_1}{\partial X_i}\frac{\partial S_1}{\partial y_j}\right]$$

$$+ \frac{1}{2}\sum_{i,j=1}^{n+1}\frac{\partial^2\phi(x, y)}{\partial y_i\partial y_j}\mathbb{E}\left[\frac{\partial S_1}{\partial X_i}\frac{\partial S_1}{\partial X_j} - \frac{\partial S^G}{\partial X_i^G}\frac{\partial S^G}{\partial X_j^G}\right] + \cdots.$$

To make the stochastic symplectic method be of weak order 2, we choose $H_j^{[i]}$, $j = 0, 1, \ldots, m$, $i = 1, \ldots, \iota$, such that the terms containing h and h^2 in the right hand side of (3.73) vanish. Notice that the coefficients of $J_{(r)}(h)$ and h in $\frac{\partial S^G}{\partial X_i^G}$ and

$\frac{\partial S^G}{\partial y_i}$ are equal to those in $\frac{\partial S_1}{\partial X_i}$ and $\frac{\partial S_1}{\partial y_i}$ with $i = 1, \ldots, n+1$, and $r = 1, \ldots, m$, respectively. Then we get

$$\mathbb{E}\left[\frac{\partial S^G}{\partial X_{n+1}^G}\frac{\partial S^G}{\partial y_{n+1}} - \frac{\partial S_1}{\partial X_{n+1}}\frac{\partial S_1}{\partial y_{n+1}}\right] = \sum_{r=1}^{m}\sum_{i=1}^{n} v\exp(vy_{n+1})\sigma_r^i y_i \frac{\partial H_r^{[1]}}{\partial x_{n+1}}h^2 + e_1(x, y)h^3,$$

where $e_1(x, y)$ denotes the coefficient of the term containing h^3 and can be calculated by means of the expression of the partial derivatives of S^G and S_1, and so are the other remainder terms e_l, $l = 2, \ldots, 7$, in the sequel. Thus, we let $\frac{\partial H_r^{[1]}}{\partial x_{n+1}} = 0$ for $r = 1, \ldots, m$. Substituting $\frac{\partial H_r^{[1]}}{\partial x_{n+1}} = 0$ into $\frac{\partial S^G}{\partial X_{n+1}^G}$, we obtain

$$\mathbb{E}\left[\frac{\partial S^G}{\partial X_{n+1}^G} - \frac{\partial S_1}{\partial X_{n+1}}\right] = \frac{\partial H_0^{[1]}}{\partial x_{n+1}}h^2 + e_2(x, y)h^3,$$

which leads us to make $\frac{\partial H_0^{[1]}}{\partial x_{n+1}} = 0$. Analogously, by $\frac{\partial H_r^{[1]}}{\partial x_{n+1}} = 0$ for $r = 0, 1, \ldots, m$, we derive

$$\mathbb{E}\left[\frac{\partial S_1}{\partial y_i}\frac{\partial S_1}{\partial y_j} - \frac{\partial S^G}{\partial y_i}\frac{\partial S^G}{\partial y_j}\right]$$
$$= \exp(vy_{n+1})\sum_{r=1}^{m}\left(v\exp(vy_{n+1})\sigma_r^i\sigma_r^j - \sigma_r^i\frac{\partial H_r^{[1]}}{\partial y_j} - \sigma_r^j\frac{\partial H_r^{[1]}}{\partial y_i}\right)h^2 + e_3(x, y)h^3$$

and

$$\mathbb{E}\left[\frac{\partial S_1}{\partial y_i}\frac{\partial S_1}{\partial X_j} - \frac{\partial S^G}{\partial y_i}\frac{\partial S^G}{\partial X_j^G}\right] = \exp(vy_{n+1})\sum_{r=1}^{m}\sigma_r^i\left(\frac{1}{2}\sum_{k=1}^{n}M_{jk}\sigma_r^k - \frac{\partial H_r^{[1]}}{\partial x_j}\right)h^2 + e_4(x, y)h^3$$

with $i, j = 1, \ldots, n$, and hence choose

$$\frac{\partial H_r^{[1]}}{\partial y_i} = \frac{1}{2}v\exp(vy_{n+1})\sigma_r^i, \qquad \frac{\partial H_r^{[1]}}{\partial x_i} = \frac{1}{2}\sum_{j=1}^{n}M_{ij}\sigma_r^j, \qquad r = 1, \ldots, m.$$

Moreover, because

$$\mathbb{E}\left[\frac{\partial S_1}{\partial X_i}\frac{\partial S_1}{\partial X_j} - \frac{\partial S^G}{\partial X_i^G}\frac{\partial S^G}{\partial X_j^G}\right] = e_5(x, y)h^3, \qquad i, j = 1, \ldots, n+1,$$

it has no influence on determining the unknown functions. Since $\frac{\partial H_r^{[1]}}{\partial y_i}$ and $\frac{\partial H_r^{[1]}}{\partial x_i}$ with $r = 0, 1, \ldots, m$, are independent of x_i and y_i, we deduce

$$\mathbb{E}\left[\frac{\partial S_1}{\partial y_i} - \frac{\partial S^G}{\partial y_i}\right] = \left(\frac{1}{2}\sum_{j,k=1}^n \frac{\partial^2 F(y)}{\partial y_i \partial y_j} M_{jk}x_k + \frac{1}{2}v\exp(vy_{n+1})f_i(y) - \frac{\partial H_0^{[1]}}{\partial y_i}\right)h^2 + e_6(x,y)h^3,$$

$$\mathbb{E}\left[\frac{\partial S}{\partial X_i} - \frac{\partial S^G}{\partial X_i}\right] = \left(\frac{1}{2}\sum_{j=1}^n M_{ij}f_j(y) - \frac{1}{2}\sum_{j=1}^n v\exp(-vy_{n+1})M_{ij}x_j - \frac{\partial H_0^{[1]}}{\partial x_i}\right)h^2 + e_7(x,y)h^3$$

for $i = 1, \ldots, n$. We choose $H_0^{[1]}$ such that the above terms containing h^2 vanish, i.e.,

$$\frac{\partial H_0^{[1]}}{\partial y_i} = \frac{1}{2}\sum_{j,k=1}^n \frac{\partial^2 F(y)}{\partial y_i \partial y_j} M_{jk}x_k + \frac{1}{2}v\exp(vy_{n+1})f_i(y),$$

$$\frac{\partial H_0^{[1]}}{\partial x_i} = \frac{1}{2}\sum_{j=1}^n M_{ij}\left(f_j(y) - v\exp(-vy_{n+1})x_j\right)$$

for $i = 1, \ldots, n$. Substituting the above results on the partial derivatives of $H_r^{[1]}$, $r = 0, 1, \ldots, m$, into (3.72), we arrive at the numerical method of (3.67) as follows

$$X_i^G = x_i - \sum_{r=1}^m \exp(vt_k)\sigma_r^i I_{(r)}(h) - \exp(vt_k)f_i(y)h - \frac{1}{2}\sum_{r=1}^m v\exp(vt_k)\sigma_r^i h I_{(r)}(h)$$

$$- \frac{1}{2}\sum_{j,k=1}^n \frac{\partial^2 F(y)}{\partial y_i \partial y_j} M_{jk}X_k^G h^2 - \frac{1}{2}v\exp(vt_k)f_i(y)h^2,$$

$$Y_i^G = y_i + \sum_{j=1}^n \exp(-vt_k)M_{ij}X_j^G h + \frac{1}{2}\sum_{r=1}^m\sum_{j=1}^n M_{ij}\sigma_r^j I_{(r)}(h)h$$

$$+ \frac{1}{2}\sum_{j=1}^n M_{ij}\left(f_j(y) - v\exp(-vt_k)X_j^G\right)h^2, \tag{3.74}$$

where $x_i = X_i(t_k)$, $y_i = Y_i(t_k)$ for $i = 1, \ldots, n$, and $y_{n+1} = t_k$ for $k \in \{0, 1, \cdots, N-1\}$. To transform (3.74) into an equivalent method of (3.62), we denote by x_-, X_-^G, y_- and Y_-^G the first n components of x, X^G, y and Y^G, respectively, and denote $P_k := \exp(-vt_k)x_-$, $P_{k+1} := \exp(-vt_{k+1})X_-^G$, $Q_k := y_-$

and $Q_{k+1} = Y_-^G$. Based on the transformation between two phase spaces of (3.62) and (3.64), we obtain

$$
\begin{aligned}
P_{k+1} =& \exp(-vh)P_k - \frac{h^2}{2}\nabla^2 F(Q_k)MP_{k+1} - h\left(1 + \frac{vh}{2}\right)\exp(-vh)f(Q_k) \\
& - \left(1 + \frac{vh}{2}\right)\exp(-vh)\sigma\triangle_k W, \\
Q_{k+1} =& Q_k + h\left(1 - \frac{vh}{2}\right)\exp(vh)MP_{k+1} + \frac{h^2}{2}Mf(Q_k) + \frac{h}{2}M\sigma\triangle_k W,
\end{aligned}
$$
(3.75)

where $\sigma = (\sigma_1, \ldots, \sigma_m)$ and $\triangle_k W = W(t_{k+1}) - W(t_k)$ with $W = (W_1, \ldots, W_m)^\top$.

Theorem 3.3.1 (See [128]) *Assume that the coefficient f of (3.62) satisfies the global Lipschitz condition and the linear growth condition, i.e.,*

$$
\|f(u) - f(w)\| \le L\|u - w\|, \quad \|f(u)\| \le C_f(1 + \|u\|)
$$

for some constants $L > 0$ and $C_f \ge 0$, and any $u, w \in \mathbb{R}^n$. Then there exists a positive constant h_0 such that for any $h \le h_0$ and $s \ge 1$, it holds that

$$
\sup_{k \in \{1, \ldots, N\}} \mathbb{E}\left[\|P_k\|^s + \|Q_k\|^s\right] < \infty. \tag{3.76}
$$

Moreover, the proposed method (3.75) is of weak order 2. More precisely,

$$
|\mathbb{E}[\phi(P(T), Q(T))] - \mathbb{E}[\phi(P_N, Q_N)]| = O(h^2)
$$

for all $\phi \in \mathbf{C}_P^6(\mathbb{R}^{2n}, \mathbb{R})$.

Proof For any fixed initial value $z = (p^\top, q^\top)^\top$, random variable $\xi := \xi^1$ and h, we arrive at

$$
\begin{aligned}
\|P_1 - p\| \le& |\exp(-vh) - 1|\|p\| + h\left(1 + \frac{vh}{2}\right)\|f(q)\| + \sqrt{h}\left(1 + \frac{vh}{2}\right)\|\sigma\xi\| \\
& + \frac{h^2}{2}\|\nabla^2 F(q)\|\|M\|\|p\| + \frac{h^2}{2}\|\nabla^2 F(q)\|\|M\|\|P_1 - p\|,
\end{aligned}
$$

based on (3.75). Let $C_v = 1 + \frac{vh}{2}$. Applying the global Lipschitz condition and the mean value theorem, we obtain

$$\|P_1 - p\| \leq |-vh\exp(-v\theta h)|\|p\| + hC_f(1 + \|z\|) + \sqrt{h}C_v\|\sigma\xi\|$$
$$+ \frac{h^2}{2}L\|M\|\|z\| + \frac{h^2}{2}L\|M\|\|P_1 - p\|$$
$$\leq C(1 + \|z\|)(\|\xi\|\sqrt{h} + h) + L\|M\|\|P_1 - p\|\frac{h^2}{2}$$

for some $\theta \in (0, 1)$. It is obvious that there exists $h_0 > 0$ such that for any $h \in (0, h_0)$, $L\|M\|\frac{h^2}{2} \leq \frac{1}{2}$, which implies $\|P_1 - p\| \leq 2C(1 + \|z\|)(\|\xi\|\sqrt{h} + h)$. On the other hand, for $h \leq h_0$, we get

$$\|\mathbb{E}[P_1 - p]\| \leq vh\|p\| + hL\|M\|\|p\| + hC_fC_v(1 + \|z\|) + \frac{h^2}{2}L\|M\|\|\mathbb{E}[P_1 - p]\|,$$

which yields

$$\|\mathbb{E}[P_1 - p]\| \leq C(1 + \|z\|)h.$$

Similar to the estimation of $P_1 - p$, we derive

$$\|Q_1 - q\| \leq C(1 + \|z\|)(\|\xi\|\sqrt{h} + h), \quad \|\mathbb{E}[Q_1 - q]\| \leq C(1 + \|z\|)h.$$

We can conclude that, for $Z_1 = (P_1^\top, Q_1^\top)^\top$,

$$\|Z_1 - z\| \leq C(\|\xi\| + \sqrt{h})(1 + \|z\|)\sqrt{h} \leq C(\|\xi\| + 1)(1 + \|z\|)\sqrt{h}. \qquad (3.77)$$

Thus, (3.76) can be obtained with the help of [193, Lemma 9.1]. For the weak error estimate, one can verify that

$$|\mathbb{E}[\phi(P(h), Q(h))] - \mathbb{E}[\phi(P_1, Q_1)]| = O(h^3), \qquad (3.78)$$

which, together with [193, Theorem 9.1], yields the global weak order 2 for the proposed method (3.75). $\qquad\square$

Now we prove the conformal symplecticity of the proposed method (3.75) as well as its ergodicity.

Theorem 3.3.2 *The numerical method (3.75) preserves the stochastic conformal symplectic structure, i.e,*

$$dP_{k+1} \wedge dQ_{k+1} = \exp(-vh)dP_k \wedge dQ_k, \quad a.s.,$$

where $k = 0, 1, \ldots, N - 1$.

Proof Based on (3.75), we obtain

$$\begin{aligned}
\mathrm{d}P_{k+1} \wedge \mathrm{d}Q_{k+1} &= \mathrm{d}P_{k+1} \wedge \mathrm{d}Q_k + \frac{1}{2}h^2 \mathrm{d}P_{k+1} \wedge M\nabla^2 F(Q_k)\mathrm{d}Q_k \\
&= \exp(-vh)\mathrm{d}P_k \wedge \mathrm{d}Q_k - \frac{h^2}{2}\mathrm{d}\left[\nabla^2 F(Q_k)M P_{k+1}\right] \wedge \mathrm{d}Q_k \\
&\quad + \frac{h^2}{2}\mathrm{d}P_{k+1} \wedge M\nabla^2 F(Q_k)\mathrm{d}Q_k.
\end{aligned}$$

Denoting $\tilde{M}(Q_k, P_{k+1}) := \nabla F(Q_k)^\top M P_{k+1}$, we have

$$\mathrm{d}\left[\nabla^2 F(Q_k)M P_{k+1}\right] \wedge \mathrm{d}Q_k = D_{QQ}\tilde{M}\mathrm{d}Q_k \wedge \mathrm{d}Q_k + \nabla^2 F(Q_k)M \mathrm{d}P_{k+1} \wedge \mathrm{d}Q_k,$$

which means

$$\mathrm{d}P_{k+1} \wedge \mathrm{d}Q_{k+1} = \exp(-vh)\mathrm{d}P_k \wedge \mathrm{d}Q_k.$$

Thus, the proof is finished.   □

Remark 3.3.1 The numerical method (3.75) also has the exponentially dissipative phase volume. Namely, denoting $D(q) = \left(I_n + \frac{h^2}{2}\nabla^2 F(q)M\right)^{-1}$, then we obtain

$$\begin{aligned}
\det\begin{pmatrix} \frac{\partial P_1}{\partial p} & \frac{\partial P_1}{\partial q} \\ \frac{\partial Q_1}{\partial p} & \frac{\partial Q_1}{\partial q} \end{pmatrix} &= \det\begin{pmatrix} \exp(-vh)D(q) & \frac{\partial P_1}{\partial q} \\ h(1-\frac{vh}{2})MD(q) & D(q)^{-\top} + h(1-\frac{vh}{2})\exp(vh)M\frac{\partial P_1}{\partial q} \end{pmatrix} \\
&= \det(\exp(-vh)I_n)\det(D(q))\det(D(q)^{-\top}) = \exp(-vnh).
\end{aligned}$$

Furthermore, $\det\begin{pmatrix} \frac{\partial P_N}{\partial p} & \frac{\partial P_N}{\partial q} \\ \frac{\partial Q_N}{\partial p} & \frac{\partial Q_N}{\partial q} \end{pmatrix} = \exp(-vnT)$.

Now our objective is to study the ergodicity of (3.75). To this end, we first introduce some sufficient conditions of deducing the existence of a minorization condition on a compact set $\mathscr{G}$, together with a Lyapunov function inducing repeated returns into $\mathscr{G}$, which are key points to proving the ergodicity of the numerical method (see e.g., [131, 188] and references therein).

Condition 3.3.1 *For some fixed compact set $\mathscr{G}$ belonging to the Borel σ-algebra $\mathscr{B}(\mathbb{R}^{2n})$, the Markov chain $Z_k := (P_k^\top, Q_k^\top)^\top$, $k \in \mathbb{N}$, with transition kernel $\mathscr{P}_k(z, A) := \mathbb{P}(Z_k \in A \mid Z_0 = z)$ satisfies*

(1) for some $z^* \in \mathscr{G}^\circ$ and for any $\delta > 0$, there exists a positive integer $k_1 := k_1(\delta)$ such that

$$\mathscr{P}_{k_1}(z, B_\delta(z^*)) > 0 \quad \forall z \in \mathscr{G},$$

where $\mathscr{G}°$ is the interior of $\mathscr{G}$, and $B_\delta(z^*)$ denotes the open ball of radius δ centered at z^*;

(2) for any $k \in \mathbb{N}$, the transition kernel $\mathscr{P}_k(z, A)$ possesses a density $\rho_k(z, w)$, i.e.,

$$\mathscr{P}_k(z, A) = \int_A \rho_k(z, y)dy \quad \forall z \in \mathscr{G}, \ A \in \mathscr{B}\left(\mathbb{R}^{2n}\right) \cap \mathscr{B}(\mathscr{G})$$

with $\rho_k(z, w)$ jointly continuous in $(z, w) \in \mathscr{G} \times \mathscr{G}$.

Consider the Markov chain given by sampling at the rate K with the kernel $\mathscr{P}(z, A) := \mathscr{P}_K(z, A)$, where $K \in \mathbb{N}_+$. The basic conclusion of the next lemma is known as *the minorization condition* (see e.g., [188, 190] and references therein).

Lemma 3.3.1 (See [188]) *Suppose that Condition 3.3.1 is satisfied. There are a choice of $K \in \mathbb{N}_+$, a positive constant η, and a probability measure v, with $v(\mathscr{G}^c) = 0$ and $v(\mathscr{G}) = 1$, such that*

$$\mathscr{P}(z, A) \geqslant \eta v(A) \quad \forall A \in \mathscr{B}\left(\mathbb{R}^{2n}\right), \ z \in \mathscr{G}.$$

Denote the general family of numerical methods for (3.62) by

$$Z_{k+1} = \Psi\left(Z_k, \Delta_k W\right), \quad Z_0 = z$$

with $Z_k := (P_k^\top, Q_k^\top)^\top$ and $k = 0, 1, \ldots, N - 1$.

Condition 3.3.2 *The function $\Psi \in \mathbf{C}^\infty(\mathbb{R}^{2n} \times \mathbb{R}^m, \mathbb{R}^{2n})$ and satisfies*

(1) there exist $K_1 > 0$ and $\epsilon > 0$ which are independent of h, such that $\mathbb{E}[\|Z(h) - Z_1\|^2] \leq K_1(1 + \|z\|^2)h^{\epsilon+2}$ for all $z \in \mathbb{R}^{2n}$, where $Z(h) = (P(h)^\top, Q(h)^\top)^\top$;
(2) there exists $K_2 = K(\gamma) > 0$ which is independent of h, such that $\mathbb{E}[\|Z_1\|^\gamma] \leq k_2(1 + \|z\|^\gamma)$ for all $z \in \mathbb{R}^{2n}$ and $\gamma \geq 1$.

The next proposition provides conditions under which the numerical method preserves a Lyapunov function. We say that $V : \mathbb{R}^{2n} \to \mathbb{R}$ is an *essentially quadratic function* if there are some constants $C_1, C_2, C_3 > 0$, such that for any $z \in \mathbb{R}^{2n}$,

$$C_1\left(1 + \|z\|^2\right) \leq V(z) \leq C_2\left(1 + \|z\|^2\right), \quad \|\nabla V(z)\| \leq C_3(1 + \|z\|).$$

Proposition 3.3.1 *Let $V : \mathbb{R}^{2n} \to [1, \infty)$ be essentially quadratic and satisfy*

$$\lim_{z \to \infty} V(z) = \infty, \quad \mathbb{L}V(z) \leqslant -C_4 V(z) + C_5$$

for some constants C_4, $C_5 > 0$, where $\mathbb{L}$ is the generator for (3.3). If Condition 3.3.2 holds, then the numerical approximation Z_k satisfies

$$\mathbb{E}[V(Z_{k+1})] \leq \alpha \mathbb{E}[V(Z_k)] + \beta$$

for $k \in \{0, 1, \ldots, N-1\}$ and some real numbers $\alpha \in (0, 1)$ and $\beta \geq 0$.

This proposition implies the Lyapunov condition, that is, there exists a positive constant C such that

$$\sup_{k \in \{1, \ldots, N\}} \mathbb{E}[V(Z_k)] \leq C.$$

This condition is always applied to a discrete Markov chain to gain the existence of the invariant measure. In conclusion, the minorization condition, together with the Lyapunov condition, yields the existence and uniqueness of the invariant measure based on [188, Theorem 2.5].

Theorem 3.3.3 (See [128]) *Assume that the vector field f is globally Lipschitz continuous, $M = I_n$, and Assumption 1.3.1 holds. Then $\{(P_k, Q_k)\}_{k=0}^{N}$ given by (3.75), which is an $\mathscr{F}_{t_k}$-adapted Markov chain, satisfies Condition 3.3.2 and hence admits an invariant measure μ_h on $(\mathbb{R}^{2n}, \mathscr{B}(\mathbb{R}^{2n}))$. In addition, if f is a linear function, then Condition 3.3.1 is satisfied and the invariant measure is unique, which imply that (3.75) is ergodic.*

Proof Based on local error estimates, it can be checked that Condition 3.3.2(2) holds. Rewrite (3.62) into

$$P(h) = p - \int_0^h \exp(-v(h-s))f(Q(s))ds - \int_0^h \exp(-v(h-s))\sigma dW(s),$$

$$Q(h) = q + \int_0^h P(s)ds$$

with $P(0) = p$ and $Q(0) = q$. By means of (3.75), we get

$$P(h) - P_1$$

$$= \left(h\left(1 + \frac{vh}{2}\right)\exp(-vh)f(q) + \frac{h^2}{2}\nabla^2 F(q)P_1 - \int_0^h \exp(-v(h-s))f(Q(s))ds \right)$$

$$+ \left(\left(1 + \frac{vh}{2}\right)\exp(-vh)\sigma \triangle_0 W - \int_0^h \exp(-v(h-s))\sigma dW(s) \right) =: \mathrm{I} + \mathrm{II},$$

$$Q(h) - Q_1$$

$$= \left(\int_0^h P(s)ds - h(1 - \frac{vh}{2})\exp(vh)P_1 \right) - \left(\frac{h}{2}M\sigma \triangle_0 W + \frac{h^2}{2}f(q) \right) =: \mathrm{III} + \mathrm{IV}.$$

Now we show the estimations of above terms respectively. Taking the expectation leads to

$$
\mathbb{E}[\|\mathrm{I}\|^2] \leq C\mathbb{E}\left[\left\|\frac{h^2}{2}\nabla^2 F(q)P_1\right\|^2\right] + C\mathbb{E}\left[\left\|\int_0^h \exp(-v(h-s))\left(f(Q(s)) - f(q)\right)ds\right\|^2\right]
$$

$$
+ C\left\|\int_0^h \exp(-v(h-s))ds f(q) - h\left(1 + \frac{vh}{2}\right)\exp(-vh)f(q)\right\|^2
$$

$$
\leq Ch^3(1 + \|z\|^2) + C\mathbb{E}\left[\int_0^h \|Q(s) - Q_1\|^2 ds\right].
$$

For the term II, according to the Itô isometry,

$$
\mathbb{E}[\|\mathrm{II}\|^2] \leq \int_0^h \left(\left(1 + \frac{vh}{2}\right)\exp(-vh) - \exp(-v(h-s))\right)^2 ds \mathrm{Tr}\left(\sigma\sigma^\top\right) \leq Ch^3.
$$

Similarly, we obtain

$$
\mathbb{E}[\|\mathrm{III}\|^2] \leq C\mathbb{E}\left[\left\|\int_0^h P(s) - P_1 ds\right\|^2\right] + C\mathbb{E}\left[\left\|h\left(1 - \left(1 - \frac{vh}{2}\right)\exp(vh)\right)P_1\right\|^2\right]
$$

$$
\leq C\mathbb{E}\left[\int_0^h \|P(s) - P_1\|^2 ds\right] + Ch^4(1 + \|z\|^2),
$$

and $\mathbb{E}[\|\mathrm{IV}\|^2] \leq Ch^3(1 + \|q\|^2)$. Therefore, we conclude

$$
\mathbb{E}[\|Z(h) - Z_1\|^2] \leq C\int_0^h \mathbb{E}[\|Z(s) - Z_1\|^2]ds + Ch^3(1 + \|z\|^2),
$$

which, together with the Grönwall inequality, yields Condition 3.3.2(1) with $\epsilon = 1$. It can be verified that there exist $\tilde{\alpha} \in (0, 1)$ and $\tilde{\beta} \in [0, \infty)$ such that

$$
\mathbb{E}(V(Z_{k+1})|\mathscr{F}_{t_k}) \leq \tilde{\alpha}V(Z_k) + \tilde{\beta}
$$

for the essential quadratic function $V: \mathbb{R}^{2n} \to \mathbb{R}$, which is defined by $V(z) = \frac{1}{2}\|p\|^2 + F(q) + \frac{v}{2}p^\top q + \frac{v^2}{4}\|q\|^2 + 1$ with $z = (p^\top, q^\top)^\top$ (see [188, Theorem 7.2]). Hence,

$$
\mathbb{E}[V(Z_{k+1})] \leq \tilde{\alpha}\mathbb{E}[V(Z_k)] + \tilde{\beta} \leq \tilde{\alpha}^{k+1}\mathbb{E}[V(Z_0)] + \tilde{\beta}\frac{1 - \tilde{\alpha}^k}{1 - \tilde{\alpha}} \leq C(Z_0),
$$

which induces the existence of the invariant measure based on [83, Proposition 7.10].

We now consider the chain $\{Z_{2k}\}_{k\in\mathbb{N}_+}$ sampled at rate $K = 2$ and verify Condition 3.3.1 when f is linear with a constant $C_f := \nabla f = \nabla^2 F$. Let

$$\mathscr{G} := \left\{ (P^\top, Q^\top)^\top \in \mathbb{R}^{2n} : Q = 0, \|P\| \le 1 \right\},$$

which is a compact set. For any $z = (p^\top, 0)^\top \in \mathscr{G}$ and $w = (w_1^\top, w_2^\top)^\top \in B$ with $B \in \mathscr{B}(\mathbb{R}^{2n})$, both $\triangle_0 W$ and $\triangle_1 W$ can be properly chosen to ensure that $P_2 = w_1$ and $Q_2 = w_2$ starting from $(P_0^\top, Q_0^\top)^\top = z$. In detail, denoting $L_h = h\left(1 - \frac{vh}{2}\right)\exp(vh)$, from (3.75), we obtain

$$w_1 = \exp(-vh)P_1 - \frac{h^2}{2}C_f w_1 - h\left(1 + \frac{vh}{2}\right)\exp(-vh)f(Q_1)$$

$$- \left(1 + \frac{vh}{2}\right)\exp(-vh)\sigma\triangle_1 W, \tag{3.79}$$

$$w_2 = Q_1 + L_h w_1 + \frac{h^2}{2}f(Q_1) + \frac{h}{2}\sigma\triangle_1 W$$

$$= Q_1 + L_h w_1 + \frac{h}{2}\left(1 + \frac{vh}{2}\right)^{-1}\exp(vh)\left(\exp(-vh)P_1 - w_1 - \frac{h^2}{2}C_f w_1\right), \tag{3.80}$$

$$P_1 = \exp(-vh)p - \frac{h^2}{2}C_f P_1 - h\left(1 + \frac{vh}{2}\right)\exp(-vh)f(0)$$

$$- \left(1 + \frac{vh}{2}\right)\exp(-vh)\sigma\triangle_0 W, \tag{3.81}$$

$$Q_1 = L_h P_1 + \frac{h^2}{2}f(0) + \frac{h}{2}\sigma\triangle_0 W$$

$$= L_h P_1 + \frac{h}{2}\left(1 + \frac{vh}{2}\right)^{-1}\exp(vh)\left(\exp(-vh)p - P_1 - \frac{h^2}{2}C_f P_1\right). \tag{3.82}$$

Noticing that (3.80) and (3.82) form a linear system, from which we can solve P_1 and Q_1. Then $\triangle_1 W$ and $\triangle_0 W$ can be uniquely determined by (3.79) and (3.81), respectively. Condition 3.3.1(1) is then ensured via the property that Brownian motions hit a cylinder set with positive probability. For Condition 3.3.1(2), from the equation

$$P_1 = D(q)\exp(-vh)\left(p - \left(1 + \frac{vh}{2}\right)\sigma\triangle_0 W - h\left(1 + \frac{vh}{2}\right)f(q)\right),$$

$$\tag{3.83}$$

$$Q_1 = q + h\left(1 - \frac{vh}{2}\right)\exp(vh)P_1 + \frac{h^2}{2}f(q) + \frac{h}{2}\sigma\triangle_0 W$$

with $D(q) = \left(I_n + \frac{h^2}{2} \nabla^2 F(q) \right)^{-1}$, it can be found out that P_1 has a $\mathbf{C}^\infty$ density based on the facts that $\triangle_0 W$ has a $\mathbf{C}^\infty$ density, σ is full rank, and $D(q)$ is positive definite for any $q \in \mathbb{R}^n$. Thus, Q_1 also has a $\mathbf{C}^\infty$ density. Applying Lemma 3.3.1 and [188, Theorem 7.3], we complete the proof. □

Remark 3.3.2 For larger γ and γ', choosing undetermined functions such that the error in (3.71) is of higher order, we can also deduce higher weak order symplectic methods for (3.64), which turn to be high weak order conformal symplectic methods for (3.62) via the inverse transformation $(X, Y) \mapsto (P, Q)$. However, the ergodicity of higher weak order methods is unknown as far as we have known.

Now we verify the above theoretical results by performing numerical tests to the following 2-dimensional stochastic linear Langevin equation

$$\begin{cases} dP(t) = -Q(t)dt - vP(t)dt - \sigma dW(t), & P(0) = p, \\ dQ(t) = P(t)dt, & Q(0) = q, \end{cases} \tag{3.84}$$

where $v > 0$ and $\sigma \neq 0$ are constants and $W(\cdot)$ is a 1-dimensional standard Wiener process. Let $H(P, Q) := \frac{1}{2}P^2 + \frac{1}{2}Q^2$. Then the above system can be rewritten as

$$\begin{cases} dP(t) = -\left(\dfrac{\partial H(P(t), Q(t))}{\partial Q} + v \dfrac{\partial H(P(t), Q(t))}{\partial P} \right) dt - \sigma dW(t), \\ dQ(t) = \dfrac{\partial H(P(t), Q(t))}{\partial P} dt. \end{cases} \tag{3.85}$$

By using the Itô formula, we obtain

$$H(P(t), Q(t)) = H(P(0), Q(0)) + \int_0^t \mathbb{L}H(P(s), Q(s))ds + \frac{\sigma^2}{2} \int_0^t P(s)dW(s),$$

where

$$\mathbb{L} = P\frac{\partial}{\partial Q} - (vP + Q)\frac{\partial}{\partial P} + \frac{\sigma^2}{2}\frac{\partial^2}{\partial^2 P}.$$

Taking the expectation yields

$$\mathbb{E}[H(P(t), Q(t))] = \mathbb{E}[H(P(0), Q(0))] + \mathbb{E}\left[\int_0^t \mathbb{L}H(P(s), Q(s))ds \right],$$

and

$$\mathbb{E}\left[\int_0^t \mathbb{L}H(P, Q)ds \right] = \int_0^t \left(\int_{\mathbb{R}^2} \mathbb{L}H(P, Q)\rho(P, Q, s)dPdQ \right) ds,$$

where ρ satisfies the Fokker–Planck equation with initial data $\rho(P(0), Q(0), 0) = \rho_0(P(0), Q(0))$. Due to the Fokker–Planck equation, we obtain

$$\int_{\mathbb{R}^2} \mathbb{L}H(P, Q)\rho(P, Q, s)dPdQ = \int_{\mathbb{R}^2} H(P, Q)\mathbb{L}^*\rho(P, Q, s)dPdQ = 0.$$

It can be verified that if $\rho(P, Q, s) = \exp\left(-\frac{2v}{\sigma^2}H(P(s), Q(s))\right)$, then

$$\mathbb{L}^*\rho := -\frac{\partial}{\partial Q}(P\rho) + \frac{\partial}{\partial P}((vP + Q)\rho) + \frac{\sigma^2}{2}\frac{\partial^2}{\partial^2 P}\rho = 0.$$

Assume that the initial value obeys the Boltzmann–Gibbs distribution

$$\rho_0(p, q) = \Theta \exp\left(-\frac{2v}{\sigma^2}H(p, q)\right) \tag{3.86}$$

with $\Theta = \left(\int_{\mathbb{R}^2} \exp\left(-\frac{v(p^2+q^2)}{\sigma^2}\right)dpdq\right)^{-1}$ being a renormalization constant. It follows that $\mathbb{L}^*\rho_0 = 0$ and the solution to (3.84) possesses a unique invariant measure μ_1 satisfying

$$d\mu_1 = \rho_0(p, q)dpdq.$$

The proposed method applied to (3.84) yields

$$P_{k+1} = \exp(-vh)P_k - \frac{h^2}{2}P_{k+1} - h\left(1 + \frac{vh}{2}\right)\exp(-vh)Q_k$$

$$- \left(1 + \frac{vh}{2}\right)\exp(-vh)\sigma\Delta_k W, \tag{3.87}$$

$$Q_{k+1} = Q_k + h\left(1 - \frac{vh}{2}\right)\exp(vh)P_{k+1} + \frac{h^2}{2}Q_k + \frac{h}{2}\sigma\Delta_k W.$$

Based on Theorems 3.3.2 and 3.3.3, (3.87) inherits both the conformal symplecticity and ergodicity of the original system. To verify these properties numerically, we choose $p = 3$ and $q = 1$. In all the experiments, the expectation is approximated by taking the average over 5000 realizations.

Figure 3.1 shows the value $\frac{S_k \exp(vt_k)}{S_0}$ of a weak order 2 method generated by truncating the Taylor expansion and the proposed method (3.87) with v being different dissipative scales and S_k being the triangle square at step k. We choose the original triangle which is produced by three points $(-1, 5)$, $(20, 2)$, $(0, 30)$. It can be observed that the discrete phase area of the proposed method is exponential decay, i.e., $S_k = \exp(-vt_k)S_0$ with the same dissipative coefficient v as the continuous case, while the weak Taylor 2 method does not.

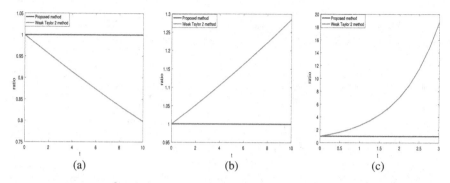

Fig. 3.1 Evolution of $\frac{S_k \exp(\upsilon t_k)}{S_0}$ of two numerical methods ($\sigma = 1$). (a) $\upsilon = 1$. (b) $\upsilon = 2$. (c) $\upsilon = 4$

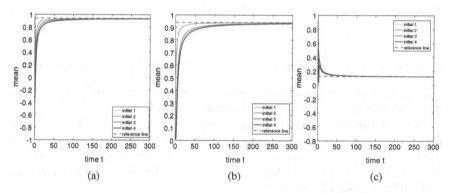

Fig. 3.2 Temporal average $\frac{1}{N} \sum_{k=1}^{N} \mathbb{E}[\phi(P_k, Q_k)]$ starting from different initial values ($\upsilon = 2$, $\sigma = 0.5$ and $T = 300$). (a) $\phi(p,q) = \cos(p+q)$. (b) $\phi(p,q) = \exp(-\frac{p^2}{2} - \frac{q^2}{2})$. (c) $\phi(p,q) = \sin(p^2+q^2)$

For the ergodicity and weak convergence of the proposed method, we have taken three different kinds of test functions (a) $\phi(p,q) = \cos(p+q)$, (b) $\phi(p,q) = \exp\left(-\frac{p^2}{2} - \frac{q^2}{2}\right)$ and (c) $\phi(p,q) = \sin(p^2 + q^2)$. To verify whether the temporal averages starting from different initial values converge to the spatial average, i.e., the ergodic limit

$$\int_{\mathbb{R}^2} \phi(p,q)d\mu_1 = \int_{\mathbb{R}^2} \phi(p,q)\rho_1(p,q)dpdq,$$

we introduce the reference value for a specific test function ϕ to represent the ergodic limit: since the function ϕ is uniformly bounded and the density function ρ_1 dissipates exponentially, the integrator is almost zero when $p^2 + q^2$ is sufficiently large. Thus, we choose $\int_{-10}^{10} \int_{-10}^{10} \phi(p,q)\rho_1(p,q)dpdq$ as the reference value, which appears as the dashed line in Fig. 3.2. We can tell from Fig. 3.2 that the

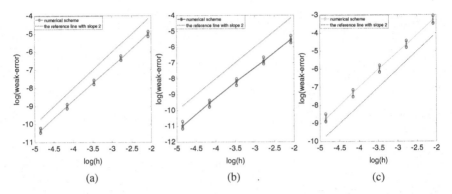

Fig. 3.3 Rate of convergence in weak sense ($v = 2$ and $\sigma = 0.5$). (**a**) $\phi(p, q) = \cos(p + q)$. (**b**) $\phi(p, q) = \exp(-\frac{p^2}{2} - \frac{q^2}{2})$. (**c**) $\phi(p, q) = \sin(p^2 + q^2)$

temporal averages $\frac{1}{N} \sum\limits_{k=1}^{N} \mathbb{E}[\phi(P_k, Q_k)]$ of the proposed method starting from four different initial values initial(1) $= (-10, 1)^\top$, initial(2) $= (2, 0)^\top$, initial(3) $= (0, 3)^\top$ and initial(4) $= (4, 2)^\top$ converge to the reference line.

Figure 3.3 plots the value $\log |\mathbb{E}[\phi(P(T), Q(T))] - \mathbb{E}[\phi(P_N, Q_N)]|$ against $\log h$ for five different time-step sizes $h = 2^{-i}$, $i = 3, \ldots, 7$, at $T = 1$, where $(P(T), Q(T))$ and (P_N, Q_N) represent the exact and numerical solutions at time T, respectively. It can be seen that the weak order of (3.87) is 2, which is indicated by the reference line of slope 2.

3.4 End Notes

The stochastic modified differential equation can be viewed as a small perturbation of the original equation that the numerical method under consideration approximates more accurately. This chapter is concerned with the stochastic modified equation in both the Wiener process and rough path settings, and its applications in constructing high weak order numerical methods. Especially, we present stochastic modified equations of stochastic (rough) symplectic methods.

There have been a lot of works concentrating on the stochastic modified equation and its combination with invariant measures (see e.g., [2, 3, 9, 90, 92] and references therein). For instance, [9] shows an expansion of the weak error and expansions of ergodic averages associated with the numerical method. Debussche and Faou in [90] show that the generator associated with the Euler–Maruyama method for the elliptic or hypoelliptic stochastic differential equation coincides with the solution of a modified Kolmogorov equation up to high order terms concerning the time-step size, and presents that every invariant measure of the Euler–Maruyama method is close to a modified invariant measure obtained by asymptotic expansion. Abdulle et al. in [2] provide a systematic procedure based on the framework of modified stochastic differential equations for the construction of high order stochastic numerical methods that capture the invariant measure of a wide class of ergodic stochastic differential equations with accuracy independent of the weak order of the underlying method.

Generally speaking, stochastic modified equation is commonly understood in the weak convergent sense, and thus in practice often encounters one inevitable problem, the efficiency of simulating some physical quantities. Recently, the multilevel Monte Carlo (MLMC) method has become a state-of-the-art approach that reduces the sampling cost and improves the efficiency (see e.g., [6, 23, 108, 109] and references therein). In [205], both the weak approximation theory and stochastic modified equation analysis are utilized to develop an alternative method of analysis for MLMC method. This gives greater freedom in the application of MLMC methods. Inspired by the above result, [147] presents the simplified version of the complexity theorem for the MLMC method via the stochastic modified differential equation.

Beyond that the stochastic modified differential equation also plays an important role in the study of stochastic gradient descent, which is widely used to solve optimization problems. It has been pointed out in [173] there generally lacked a systematic approach to study precise dynamical properties of stochastic gradient descent and its variants. To address this issue, the stochastic modified equation approach is introduced in [172], where a weak approximation result for the stochastic gradient descent is given in a finite-sum-objective setting. Further, the general mathematical foundation for analyzing stochastic gradient descent is developed in [173] via the stochastic modified equation. The advantage of this technique is that the fluctuation in the stochastic gradient descent iteration can be well captured by the fluctuation in the stochastic modified equation (see [273]).

However, due to the non-differentiable property of the solution to the stochastic differential equation, the existence of stochastic modified equation of a given stochastic numerical method in strong or weak convergent sense is still unclear and needs also to be further investigated.

Chapter 4
Infinite-Dimensional Stochastic Hamiltonian Systems

The infinite-dimensional stochastic Hamiltonian systems arise in various fields such as stochastic electromagnetics, quantum physics and stochastic fluid mechanics, etc. As in the finite-dimensional case, one of inherent canonical properties of infinite-dimensional stochastic Hamiltonian systems is that the phase flow preserves the infinite-dimensional stochastic symplectic structure. In this chapter, we will investigate this geometric structure, provide some specific stochastic structure-preserving methods, and analyze their strong convergence rates. Moreover, we will prove that some symplectic discretizations of the stochastic linear Schrödinger equation can approximate the large deviations rate function of the observable, which provides a method of approximating large deviations rate function from the viewpoint of numerical discretization.

4.1 Infinite-Dimensional Stochastic Hamiltonian Systems

In this section, we concentrate on both the stochastic variational principle and stochastic symplecticity of the infinite-dimensional stochastic Hamiltonian system as follows

$$
\begin{cases}
dP(t) = -\dfrac{\delta H_0(P(t), Q(t))}{\delta Q}dt - \dfrac{\delta H_1(P(t), Q(t))}{\delta Q} \circ dW(t), & P(0) = p, \\[4mm]
dQ(t) = \dfrac{\delta H_0(P(t), Q(t))}{\delta P}dt + \dfrac{\delta H_1(P(t), Q(t))}{\delta P} \circ dW(t), & Q(0) = q,
\end{cases}
$$

$$(4.1)$$

where $P, Q \colon \mathscr{O} \times [0, T] \to \mathbb{R}$ with $\mathscr{O} \subset \mathbb{R}^d$, $d \in \mathbb{N}_+$, and H_0 and H_1 are Hamiltonian functionals. Moreover, $W(\cdot)$ is a U-valued $\mathbf{Q}$-Wiener process with U being a separable Hilbert space, i.e., there exist an orthonormal basis $\{e_k\}_{k \in \mathbb{N}_+}$ of

© The Author(s), under exclusive license to Springer Nature Singapore Pte Ltd. 2022
J. Hong, L. Sun, *Symplectic Integration of Stochastic Hamiltonian Systems*,
Lecture Notes in Mathematics 2314, https://doi.org/10.1007/978-981-19-7670-4_4

U and a sequence of mutually independent real-valued Brownian motions $\{\beta_k\}_{k\in\mathbb{N}_+}$ such that $W(t) = \sum_{k\in\mathbb{N}_+} \mathbf{Q}^{\frac{1}{2}} e_k \beta_k(t)$ with $\mathbf{Q}$ being symmetric, positive definite and of finite trace. Now we recall some properties of the $\mathbf{Q}$-Wiener process.

Proposition 4.1.1 (See [85]) *Given two separable Hilbert spaces $\mathcal{H}_1$ and $\mathcal{H}_2$, let $W(\cdot)$ be an $\mathcal{H}_1$-valued $\mathbf{Q}$-Wiener process and define*

$$\mathcal{N}_W^2(0, T; \mathscr{L}_2(\mathbf{Q}^{\frac{1}{2}}(\mathcal{H}_1), \mathcal{H}_2))$$

$$:= \left\{ \Phi : [0, T] \times \Omega \to \mathscr{L}_2(\mathbf{Q}^{\frac{1}{2}}(\mathcal{H}_1), \mathcal{H}_2)) \mid \Phi \text{ is predicable and} \right.$$

$$\left. \mathbb{E}\left[\int_0^T \|\Phi(s) \circ \mathbf{Q}^{\frac{1}{2}}\|_{\mathscr{L}_2(\mathcal{H}_1, \mathcal{H}_2)}^2 ds \right] < \infty \right\},$$

where $\mathscr{L}_2(\mathcal{H}_1, \mathcal{H}_2)$ denotes the space of Hilbert–Schmidt operators from $\mathcal{H}_1$ to $\mathcal{H}_2$. Assume that $\Phi_1, \Phi_2 \in \mathcal{N}_W^2(0, T; \mathscr{L}_2(\mathbf{Q}^{\frac{1}{2}}(\mathcal{H}_1), \mathcal{H}_2))$, then the correlation operators

$$V(t, s) = \mathbb{COR}(\Phi_1 \cdot W(t), \Phi_2 \cdot W(s)), \quad t, s \in [0, T]$$

are given by the formula

$$V(t, s) = \mathbb{E}\left[\int_0^{t \wedge s} \Phi_2(r) \mathbf{Q}(\Phi_1(r))^* dr \right].$$

Here, the operator $V(t, s)$ is defined by

$$\langle V(t, s)a, b \rangle_{\mathcal{H}_2} = \mathbb{E}\left[\langle \Phi_1 \cdot W(t), a \rangle_{\mathcal{H}_2} \langle \Phi_2 \cdot W(s), b \rangle_{\mathcal{H}_2} \right] \quad \forall\, a, b \in \mathcal{H}_2.$$

Proposition 4.1.2 (See [83]) *Assume that $\mathbf{Q}$ is a nonnegative symmetric operator on a real separable Hilbert space $\mathbb{H}$ with finite trace. Let $\eta_1 \geq \eta_2 \geq \cdots \geq \eta_n \geq \cdots$ be the eigenvalues of $\mathbf{Q}$. Define the determinant of $I - 2\epsilon\mathbf{Q}$ by setting*

$$\det(I - 2\epsilon\mathbf{Q}) := \lim_{n\to\infty} \prod_{k=1}^n (1 - 2\epsilon\eta_k) := \prod_{k=1}^\infty (1 - 2\epsilon\eta_k).$$

Let $\mu = \mathcal{N}(0, \mathbf{Q})$ be the symmetric Gaussian measure on $\mathbb{H}$. Then for every $\epsilon \in \mathbb{R}$,

$$\int_{\mathbb{H}} \exp(\epsilon \|x\|_{\mathbb{H}}^2) \mu(dx) = \begin{cases} (\det(I - 2\epsilon\mathbf{Q}))^{-\frac{1}{2}}, & \text{if } \epsilon < \frac{1}{2\eta_1}; \\ +\infty, & \text{otherwise.} \end{cases} \qquad (4.2)$$

Unlike the finite-dimensional case, there have not been general existence and uniqueness theorems for solutions of infinite-dimensional stochastic Hamiltonian systems. Below we shall illustrate (4.1) through several physical examples.

Example 4.1.1 (Stochastic Wave Equation with Additive Noise) As a kind of commonly observed physical phenomenon, the wave motions are usually described by the stochastic partial differential equation of hyperbolic type. Consider the following stochastic wave equation with cubic nonlinearity driven by an additive noise

$$
\begin{cases}
du(t) = v(t)dt, & \text{in } \mathscr{O} \times (0, T], \\
dv(t) = \Delta u(t)dt - f(u(t))dt + dW(t), & \text{in } \mathscr{O} \times (0, T], \quad (4.3) \\
u(0) = u_0, \quad v(0) = v_0, & \text{in } \mathscr{O},
\end{cases}
$$

where $\mathscr{O} = [0, 1]^d$ with $d = 1, 2$, Δ is the Laplace operator with homogeneous Dirichlet boundary condition, and the nonlinear term $f(u) = c_\rho u^\rho + \cdots + c_1 u + c_0$ is assumed to be a polynomial with odd degree $\rho \leq 3$ and $c_\rho > 0$. Furthermore, $W(\cdot)$ is an $\mathbf{L}^2(\mathscr{O})$-valued $\mathbf{Q}$-Wiener process, where $\mathbf{L}^2(\mathscr{O}) := \mathbf{L}^2(\mathscr{O}, \mathbb{R})$. It can be verified that (4.3) is rewritten in the form (4.1) with Hamiltonian functionals

$$
H_0(u, v) = \int_{\mathscr{O}} \left(\frac{1}{2}|\nabla u|^2 + \frac{1}{2}|v|^2 + \sum_{i=0}^{\rho} \frac{c_i}{i+1} u^{i+1} \right) dx, \quad H_1(u) = -\int_{\mathscr{O}} u dx.
$$

Example 4.1.2 (Stochastic Schrödinger Equation with Multiplicative Noise) The stochastic cubic Schrödinger equation perturbed by the linear multiplicative noise

$$
\begin{cases}
du(t) = \mathbf{i}\left(\Delta u(t) + \lambda |u(t)|^2 u(t) \right) dt + \mathbf{i} u(t) \circ dW(t), & \text{in } \mathscr{O} \times (0, T], \\
u(0) = u_0, & \text{in } \mathscr{O}
\end{cases}
$$

$$(4.4)$$

with $\mathscr{O} = [0, 1]$, $\lambda = \pm 1$ and $\mathbf{Q}$-Wiener process $W(t) = \sum_{k \in \mathbb{N}_+} \mathbf{Q}^{\frac{1}{2}} e_k \beta_k(t)$, where $\{e_k\}_{k \in \mathbb{N}_+}$ is an orthonormal basis in $\mathbf{L}^2(\mathscr{O})$, has been studied to motivate the possible role of noise to prevent or delay collapse formation. Denoting by P and Q the real and imaginary parts of u, respectively, we have that (4.4) is equivalent to

$$
\begin{cases}
dP(t) = -\left(\Delta Q(t) + \lambda \left(P^2(t) + Q^2(t) \right) Q(t) \right) dt - Q(t) \circ dW(t), & P(0) = p, \\
dQ(t) = \left(\Delta P(t) + \lambda \left(P^2(t) + Q^2(t) \right) P(t) \right) dt + P(t) \circ dW(t), & Q(0) = q.
\end{cases}
$$

$$(4.5)$$

It is not hard to check that the associated Hamiltonian functionals are

$$H_0(P, Q) = -\frac{1}{2}\int_{\mathscr{O}}(|\nabla P|^2 + |\nabla Q|^2)dx + \frac{\lambda}{4}\int_{\mathscr{O}}\left(P^2 + Q^2\right)^2 dx,$$

and

$$H_1(P, Q) = \frac{1}{2}\int_{\mathscr{O}}\left(P^2 + Q^2\right)dx.$$

Now we turn to the stochastic variational principle for (4.1). Define the generalized stochastic action functional as

$$\mathbb{S}(P, Q) := \int_{\mathscr{O}}\int_0^T P \circ dQ dx - \int_0^T H_0(P, Q)dt - \int_0^T H_1(P, Q) \circ dW(t),$$

and denote the variations of P and Q by δP and δQ, respectively. Then the variation $\delta S(P, Q)$ of $\mathbb{S}(P, Q)$ satisfies

$$\delta\mathbb{S}(P, Q) = \frac{d}{d\epsilon}\Big|_{\epsilon=0}\mathbb{S}(P + \epsilon\delta P, Q + \epsilon\delta Q)$$

$$= \int_{\mathscr{O}}\left(\int_0^T \delta P(t)\left(\circ dQ(t) - \frac{\delta H_0}{\delta P}(P(t), Q(t))dt - \frac{\delta H_1}{\delta P}(P(t), Q(t)) \circ dW(t)\right)\right.$$

$$\left. - \int_0^T \delta Q(t)\left(\circ dP(t) + \frac{\delta H_0}{\delta Q}(P(t), Q(t))dt + \frac{\delta H_1}{\delta Q}(P(t), Q(t)) \circ dW(t)\right)\right)dx$$

under the condition of fixed endpoints, i.e., $\delta Q(0) = \delta Q(T) = 0$.

Theorem 4.1.1 *If $(P(t), Q(t))$ satisfies the infinite-dimensional stochastic Hamiltonian system (4.1) for $t \in [0, T]$, then the pair (P, Q) is a critical point of the stochastic action functional $\mathbb{S}$, i.e., $\delta\mathbb{S}(P, Q) = 0$ for all $(\delta P, \delta Q)$ such that $\delta Q(0) = \delta Q(T) = 0$.*

Remark 4.1.1 Similar to the finite-dimensional case, if $H_0 = H_1 = H$, then H is an invariant of system (4.1). In fact,

$$dH = \int_{\mathscr{O}}\left(\frac{\delta H}{\delta P}dP + \frac{\delta H}{\delta Q}dQ\right)dx$$

$$= \int_{\mathscr{O}}\left(-\frac{\delta H}{\delta P}\frac{\delta H}{\delta Q}dt - \frac{\delta H}{\delta P}\frac{\delta H}{\delta Q} \circ dW + \frac{\delta H}{\delta Q}\frac{\delta H}{\delta P}dt + \frac{\delta H}{\delta Q}\frac{\delta H}{\delta P} \circ dW\right)dx = 0.$$

The symplectic form for the infinite-dimensional stochastic Hamiltonian system (4.1) is proposed in [47] and reads

$$\varpi_2(t) = \int_{\mathcal{O}} dP(t) \wedge dQ(t)dx \quad \forall\, t \in [0, T], \tag{4.6}$$

where the overbar on ω_2 is a reminder that the differential 2-form $dP \wedge dQ$ is integrated over the space. Preservation of the symplectic form (4.6) means that the spatial integral of the oriented areas of projections onto the coordinate planes (p, q) is an integral invariant (see [47]). One can also verify the preservation of the symplecticity by exploiting the multi-symplectic conservation law (see [75]). Next, we investigate the symplecticity in the case of the stochastic Schrödinger equation with multiplicative noise (4.4) by an alternative approach.

Theorem 4.1.2 (See [47]) *Under the zero boundary condition, the phase flow of the stochastic Schödinger equation (4.4) preserves the stochastic symplectic structure almost surely, that is,*

$$\varpi_2(t) = \int_{\mathcal{O}} dP(t) \wedge dQ(t)dx = \int_{\mathcal{O}} dp \wedge dqdx = \varpi_2(0), \quad a.s.,$$

for $t \in [0, T]$.

Proof Using the formula of change of variables in differential forms, we obtain that for any $t \in [0, T]$,

$$\varpi_2(t) = \int_{\mathcal{O}} dP(t) \wedge dQ(t)dx$$
$$= \int_{\mathcal{O}} \left(\frac{\partial P(t)}{\partial p} \frac{\partial Q(t)}{\partial q} - \frac{\partial P(t)}{\partial q} \frac{\partial Q(t)}{\partial p} \right) dp \wedge dqdx.$$

Differentiating the above equation leads to

$$\frac{d\varpi_2(t)}{dt} = \int_{\mathcal{O}} \frac{d}{dt} \left(\frac{\partial P}{\partial p} \frac{\partial Q}{\partial q} - \frac{\partial P}{\partial q} \frac{\partial Q}{\partial p} \right) dp \wedge dqdx. \tag{4.7}$$

Denote $P_p := \frac{\partial P}{\partial p}$, $P_q := \frac{\partial P}{\partial q}$, $Q_p := \frac{\partial Q}{\partial p}$, $Q_q := \frac{\partial Q}{\partial q}$ and let $\Psi(P, Q) = \frac{\lambda}{4}(P^2 + Q^2)^2$. From the differentiability with respect to the initial data of stochastic infinite-dimensional equations (see [85, Chapter 9]) it follows that P_p, Q_p, P_q and Q_q obey the following formulae

$$dP_p = -\left(\Delta Q_p + \frac{\partial^2 \Psi}{\partial P \partial Q} P_p + \frac{\partial^2 \Psi}{\partial Q^2} Q_p \right) dt - Q_p \circ dW(t), \quad P_p(0) = I,$$

$$dQ_p = \left(\Delta P_p + \frac{\partial^2 \Psi}{\partial P^2} P_p + \frac{\partial^2 \Psi}{\partial P \partial Q} Q_p \right) dt + P_p \circ dW(t), \quad Q_p(0) = 0,$$

$$dP_q = -\left(\Delta Q_q + \frac{\partial^2 \Psi}{\partial P \partial Q}P_q + \frac{\partial^2 \Psi}{\partial Q^2}Q_q\right)dt - Q_q \circ dW(t), \quad P_q(0) = 0,$$

$$dQ_q = \left(\Delta P_q + \frac{\partial^2 \Psi}{\partial P^2}P_q + \frac{\partial^2 \Psi}{\partial P \partial Q}Q_q\right)dt + P_q \circ dW(t), \qquad Q_q(0) = I,$$

which yields

$$d\left(\frac{\partial P}{\partial p}\frac{\partial Q}{\partial q} - \frac{\partial P}{\partial q}\frac{\partial Q}{\partial p}\right)$$

$$= \left(-(\Delta Q_p + \frac{\partial^2 \Psi}{\partial P \partial Q}P_p + \frac{\partial^2 \Psi}{\partial Q^2}Q_p)Q_q + (\Delta Q_q + \frac{\partial^2 \Psi}{\partial P \partial Q}P_q + \frac{\partial^2 \Psi}{\partial Q^2}Q_q)Q_p\right.$$

$$\left. + (\Delta P_q + \frac{\partial^2 \Psi}{\partial P^2}P_q + \frac{\partial^2 \Psi}{\partial P \partial Q}Q_q)P_p - (\Delta P_p + \frac{\partial^2 \Psi}{\partial P^2}P_p + \frac{\partial^2 \Psi}{\partial P \partial Q}Q_p)P_q\right)dt$$

$$+ \left(Q_pQ_q - Q_qQ_p - P_pP_q + P_qP_p\right) \circ dW(t)$$

$$= \left(-(\Delta Q_p)Q_q + (\Delta Q_q)Q_p + (\Delta P_q)P_p - (\Delta P_p)P_q\right)dt.$$

A direct calculations leads to

$$\frac{d\varpi_2(t)}{dt} = \int_{\mathscr{O}}\left(-\Delta Q_pQ_q + \Delta Q_qQ_p + \Delta P_qP_p - \Delta P_pP_q\right)dp \wedge dqdx$$

$$= -\int_{\mathscr{O}}[d(\Delta Q) \wedge dQ + d(\Delta P) \wedge dP]dx$$

$$= -\int_{\mathscr{O}}\frac{\partial}{\partial x}[d(Q_x) \wedge dQ + d(P_x) \wedge dP]dx.$$

Due to the zero boundary condition, we arrive at $\frac{d\varpi_2(t)}{dt} = 0$, which completes the proof. $\square$

The stochastic wave equation (4.3), the stochastic Schrödinger equation (4.4) and the forthcoming stochastic Maxwell equation (4.9) all describe the evolution in time of processes with values in function spaces or, in other words, random fields in which one coordinate–'time'–is distinguished. Moreover, they can be rewritten as the semilinear stochastic evolution equation

$$\begin{cases} dX(t) = -AX(t)dt + f(X(t))dt + g(X(t))dW(t), \quad t \in (0, T], \\ X(0) = X_0, \end{cases} \tag{4.8}$$

where A is a linear operator, f and g are drift and diffusion nonlinearities, respectively (see e.g., [85, 161, 176] and references therein).

A commonly used approach to discretizing (4.8) is combining temporal discrete methods for stochastic ordinary differential equations with spatial discrete methods for deterministic partial differential equations. Given an interval $[0, T]$, we take a partition $0 = t_0 \leq t_1 \leq \cdots \leq t_N = T$ with a uniform time-step size $h = t_{n+1} - t_n$, $n = 0, 1, \ldots, N - 1$. For the sake of simplicity, we consider the case that $\mathscr{O} \subset \mathbb{R}^1$, let $M \in \mathbb{N}_+$ be the number of orthogonal modes in spectral methods or discretization steps in space ($M\tau = |\mathscr{O}|$, $|\mathscr{O}|$ is the length of the interval $\mathscr{O}$) for finite difference methods or spectral Galerkin methods, and denote a full discretization of (4.8) by X_n^M with $n \in \{0, 1, \ldots, N\}$. Now we introduce the strong convergence of X_n^M for stochastic partial differential equations (see e.g., [161, 270] and references therein).

Definition 4.1.1 (Strong Convergence in $\mathbf{L}^p(\Omega, \mathbb{H})$) Assume that X_N^M is a numerical approximation of the solution $X(T)$ to (4.8) at time T. If there exists a constant C independent of h and τ such that

$$\left(\mathbb{E}\left[\left\| X_N^M - X(T) \right\|_{\mathbb{H}}^p \right] \right)^{1/p} \leq C \left(\tau^{p_1} + h^{p_2} \right),$$

where $p_1, p_2 > 0$ and $p \geq 2$, then the numerical method is strongly convergent to the solution to (4.8). The strong convergence order in space is p_1 and the convergence order in time is p_2.

To meet the needs of applications, it is natural and important to design numerical methods to inherit properties of the original system as much as possible. Below we shall provide some structure-preserving stochastic numerical discretizations for stochastic Maxwell equations, stochastic Schrödinger equations, and stochastic cubic wave equations successively.

4.2 Stochastic Maxwell Equation

To physically model the interaction of electromagnetic waves with complex media in applications, such as radar imaging and remote sensing, optical imaging, laser beam propagation through the atmosphere, and communications, Maxwell equations have been always taken into consideration (see e.g., [7, 33, 229] and references therein). However, in practical circumstances, stochasticity originating due to factors such as random perturbations of the electric current density or the magnetic current density is common in electromagnetic signal transmissions. Such kinds of stochasticity are ubiquitous in engineering applications and are often modeled by some stochastic processes. The resulting stochastic model is the stochastic Maxwell equation (see e.g., [165, 210] and references therein).

In this section, we consider the stochastic Maxwell equation driven by multiplicative noise

$$\begin{cases} d\mathbb{U}(t) = A\mathbb{U}(t)dt + \mathbb{F}(\mathbb{U}(t))dt + \mathbb{G}(\mathbb{U}(t))dW(t), & t \in (0, T], \\ \mathbb{U}(0) = (\mathbf{e}^\top, \mathbf{h}^\top)^\top \end{cases} \tag{4.9}$$

supplemented with the boundary condition of a perfect conductor $\mathbf{n} \times \mathbf{E} = 0$ on $(0, T] \times \partial \mathcal{O}$, where $\mathbb{U} = (\mathbf{E}^{\top}, \mathbf{H}^{\top})^{\top}$ is a $U := (\mathbf{L}^2(\mathcal{O}))^6$-valued function with $\mathbf{E}$ and $\mathbf{H}$ representing the electric field and magnetic field, respectively, $\mathbf{n}$ is the unit outward normal of $\partial \mathcal{O}$, and $\mathcal{O}$ is a bounded and simply connected domain of $\mathbb{R}^3$ with smooth boundary $\partial \mathcal{O}$. Moreover,

$$W(t) = \sum_{k \in \mathbb{N}_+} \mathbf{Q}^{\frac{1}{2}} e_k \beta_k(t),$$

and $\{e_k\}_{k \in \mathbb{N}_+}$ is an orthonormal basis of U consisting of eigenfunctions of $\mathbf{Q}$ which is symmetric, nonnegative and of finite trace. The Maxwell operator A is defined by

$$A \begin{bmatrix} \mathbf{E} \\ \mathbf{H} \end{bmatrix} := \begin{bmatrix} 0 & \epsilon^{-1} \nabla \times \\ -\mu^{-1} \nabla \times & 0 \end{bmatrix} \begin{bmatrix} \mathbf{E} \\ \mathbf{H} \end{bmatrix} = \begin{bmatrix} \epsilon^{-1} \nabla \times \mathbf{H} \\ -\mu^{-1} \nabla \times \mathbf{E} \end{bmatrix}, \qquad (4.10)$$

where $\epsilon, \mu \in \mathbf{L}^{\infty}(\mathcal{O})$ satisfy $\epsilon, \mu \geq \kappa > 0$ with κ being a positive constant. It has the domain $\mathrm{Dom}(A) := H_0(\mathrm{curl}, \mathcal{O}) \times H(\mathrm{curl}, \mathcal{O})$, where

$$H(\mathrm{curl}, \mathcal{O}) := \{\mathbf{U} \in (\mathbf{L}^2(\mathcal{O}))^3 : \nabla \times \mathbf{U} \in (\mathbf{L}^2(\mathcal{O}))^3\}$$

is termed by the curl-space and

$$H_0(\mathrm{curl}, \mathcal{O}) := \{\mathbf{U} \in H(\mathrm{curl}, \mathcal{O}) : \mathbf{n} \times \mathbf{U}|_{\partial \mathcal{O}} = 0\}$$

is the subspace of $H(\mathrm{curl}, \mathcal{O})$ with zero tangential trace. With the inner product

$$\langle u, v \rangle_{H(\mathrm{curl}, \mathcal{O})} := \langle u, v \rangle_{(\mathbf{L}^2(\mathcal{O}))^3} + \langle \nabla \times u, \nabla \times v \rangle_{(\mathbf{L}^2(\mathcal{O}))^3}, \quad u, v \in H(\mathrm{curl}, \mathcal{O}),$$

$H(\mathrm{curl}, \mathcal{O})$ is a Hilbert space, and it is also the closure of $(\mathbf{C}^{\infty}(\bar{\mathcal{O}}))^3$ with respect to the graph norm

$$\| \cdot \|_{H(\mathrm{curl}, \mathcal{O})} := \left(\| \cdot \|_{(\mathbf{L}^2(\mathcal{O}))^3}^2 + \| \nabla \times \cdot \|_{(\mathbf{L}^2(\mathcal{O}))^3}^2 \right)^{1/2}.$$

For the space $H_0(\mathrm{curl}, \mathcal{O})$, it is the closure of $(\mathbf{C}_0^{\infty}(\bar{\mathcal{O}}))^3$ with respect to $\| \cdot \|_{H(\mathrm{curl}, \mathcal{O})}$. In addition, assumptions on ϵ, μ ensure that the Hilbert space $V := (\mathbf{L}^2(\mathcal{O}))^3 \times (\mathbf{L}^2(\mathcal{O}))^3$ is equipped with the weighted scalar product

$$\left\langle \begin{bmatrix} \mathbf{E}_1 \\ \mathbf{H}_1 \end{bmatrix}, \begin{bmatrix} \mathbf{E}_2 \\ \mathbf{H}_2 \end{bmatrix} \right\rangle_V = \int_{\mathcal{O}} (\mu \langle \mathbf{H}_1, \mathbf{H}_2 \rangle + \epsilon \langle \mathbf{E}_1, \mathbf{E}_2 \rangle) \, dx,$$

where $\langle \cdot, \cdot \rangle$ stands for the standard Euclidean inner product. This weighted scalar product is equivalent to the standard inner product on U. Moreover, the corresponding norm, which represents the electromagnetic energy of the physical system, induced by the inner product is

$$\left\| \begin{bmatrix} \mathbf{E} \\ \mathbf{H} \end{bmatrix} \right\|_V^2 = \int_{\mathscr{O}} \left(\mu \|\mathbf{H}\|^2 + \epsilon \|\mathbf{E}\|^2 \right) dx$$

with $\| \cdot \|$ being the Euclidean norm. According to the norm $\| \cdot \|_V$, the associated graph norm of A is defined by

$$\|\mathbb{V}\|_{\mathrm{Dom}(A)}^2 := \|\mathbb{V}\|_V^2 + \|A\mathbb{V}\|_V^2.$$

The Maxwell operator A is closed and $\mathrm{Dom}(A)$ equipped with the graph norm is a Banach space (see [204]). Moreover, A is skew-adjoint, that is, for any $\mathbb{V}_1, \mathbb{V}_2 \in \mathrm{Dom}(A)$,

$$\langle A\mathbb{V}_1, \mathbb{V}_2 \rangle_V = -\langle \mathbb{V}_1, A\mathbb{V}_2 \rangle_V,$$

and generates a unitary C_0-group $\{E(t)\}_{t \in \mathbb{R}}$ with $E(t) := \exp(tA)$ via Stone's theorem (see [121]). Based on the property of the unitary group, it can be verified that

$$\|E(t)\mathbb{V}\|_V = \|\mathbb{V}\|_V \quad \forall \, \mathbb{V} \in V, \, t \in [0, T], \tag{4.11}$$

which means that the electromagnetic energy is preserved (see [120]). Besides, the unitary group $\{E(t)\}_{t \in \mathbb{R}}$ satisfies the following properties.

Lemma 4.2.1 (See [37]) *Let I be the identity operator on V. For $t \geq 0$, it holds that*

$$\|E(t) - I\|_{\mathscr{L}(\mathrm{Dom}(A), V)} \leq Ct, \tag{4.12}$$

where the constant C does not depend on t. Here, $\mathscr{L}(\mathrm{Dom}(A), V)$ denotes the space of bounded linear operators from $\mathrm{Dom}(A)$ to V.

Denote by $\mathscr{L}_2(U_0, U)$ the space of Hilbert–Schmidt operators from U_0 to U with $U_0 = \mathbf{Q}^{\frac{1}{2}}(U)$. As a consequence of Lemma 4.2.1, we obtain the following lemma.

Lemma 4.2.2 (See [68]) *For any $\Phi \in \mathscr{L}_2(U_0, \mathrm{Dom}(A))$ and any $t \geq 0$, we have*

$$\|(E(t) - I)\Phi\|_{\mathscr{L}_2(U_0, V)} \leq Ct \|\Phi\|_{\mathscr{L}_2(U_0, \mathrm{Dom}(A))}. \tag{4.13}$$

Proof From Lemma 4.2.1 and the definition of the Hilbert–Schmidt norm it follows that

$$\| (E(t) - I) \, \Phi \|^2_{\mathscr{L}_2(U_0, V)} = \sum_{k \in \mathbb{N}_+} \| (E(t) - I) \, \Phi \mathbf{Q}^{\frac{1}{2}} e_k \|^2_V$$

$$\leq C t^2 \sum_{k \in \mathbb{N}_+} \| \Phi \mathbf{Q}^{\frac{1}{2}} e_k \|^2_{\text{Dom}(A)} \leq C t^2 \| \Phi \|^2_{\mathscr{L}_2(U_0, \text{Dom}(A))},$$

which finishes the proof.   □

To guarantee the existence and uniqueness of the strong solution to (4.9), we make the following assumptions.

Assumption 4.2.1 (Initial Value) *The initial value* $\mathbb{U}(0)$ *of (4.9) is a* $\text{Dom}(A)$-*valued random variable with* $\mathbb{E}\big[\|\mathbb{U}(0)\|^p_{\text{Dom}(A)} \big] < \infty$ *for any* $p \geq 1$.

Assumption 4.2.2 (Drift Coefficient) *There exist constants* $C_{\mathbb{F}}, C^1_{\mathbb{F}} > 0$ *such that the operator* $\mathbb{F} \colon V \to V$ *satisfies*

$$\|\mathbb{F}(\mathbb{V}_1) - \mathbb{F}(\mathbb{V}_2)\|_V \leq C_{\mathbb{F}} \|\mathbb{V}_1 - \mathbb{V}_2\|_V \quad \forall\, \mathbb{V}_1, \mathbb{V}_2 \in V,$$

$$\|\mathbb{F}(\mathbb{V}_1) - \mathbb{F}(\mathbb{V}_2)\|_{\text{Dom}(A)} \leq C^1_{\mathbb{F}} \|\mathbb{V}_1 - \mathbb{V}_2\|_{\text{Dom}(A)} \quad \forall\, \mathbb{V}_1, \mathbb{V}_2 \in \text{Dom}(A),$$

$$\|\mathbb{F}(\mathbb{V})\|_V \leq C_{\mathbb{F}}(1 + \|\mathbb{V}\|_V) \quad \forall\, \mathbb{V} \in V,$$

$$\|\mathbb{F}(\mathbb{V})\|_{\text{Dom}(A)} \leq C^1_{\mathbb{F}} \big(1 + \|\mathbb{V}\|_{\text{Dom}(A)} \big) \quad \forall\, \mathbb{V} \in \text{Dom}(A).$$

The widespread linear drift term

$$\mathbb{F}(\mathbb{U}) = B(x)\mathbb{U},$$

where $B(x) = diag(\sigma_1(x), \sigma_1(x), \sigma_1(x), \sigma_2(x), \sigma_2(x), \sigma_2(x))$ is a 6×6 matrix with smooth functions σ_1 and σ_2, satisfies Assumption 4.2.2.

Assumption 4.2.3 (Diffusion Coefficient) *The operator* $\mathbb{G} \colon V \to \mathscr{L}_2(U_0, V)$ *satisfies*

$$\|\mathbb{G}(\mathbb{V}_1) - \mathbb{G}(\mathbb{V}_2)\|_{\mathscr{L}_2(U_0, V)} \leq C_{\mathbb{G}} \|\mathbb{V}_1 - \mathbb{V}_2\|_V \quad \forall\, \mathbb{V}_1, \mathbb{V}_2 \in V,$$

$$\|\mathbb{G}(\mathbb{V}_1) - \mathbb{G}(\mathbb{V}_2)\|_{\mathscr{L}_2(U_0, \text{Dom}(A))} \leq C^1_{\mathbb{G}} \|\mathbb{V}_1 - \mathbb{V}_2\|_{\text{Dom}(A)} \quad \forall\, \mathbb{V}_1, \mathbb{V}_2 \in \text{Dom}(A),$$

$$\|\mathbb{G}(\mathbb{V})\|_{\mathscr{L}_2(U_0, V)} \leq C_{\mathbb{G}}(1 + \|\mathbb{V}\|_V) \quad \forall\, \mathbb{V} \in V, \tag{4.14}$$

$$\|\mathbb{G}(\mathbb{V})\|_{\mathscr{L}_2(U_0, \text{Dom}(A))} \leq C^1_{\mathbb{G}}(1 + \|\mathbb{V}\|_{\text{Dom}(A)}) \quad \forall\, \mathbb{V} \in \text{Dom}(A),$$

where $C_{\mathbb{G}}, C^1_{\mathbb{G}} > 0$ *depend on* $\mathbf{Q}$.

Example 4.2.1 We first take the stochastic Maxwell equation (4.9) driven by additive noise into account. In detail, we let $\mathcal{O} = [0, 1]^3$, $\epsilon = \mu = 1$ and consider

$\mathbb{G}$ as the Nemytskij operator of diag $(\lambda, \lambda, \lambda, \lambda, \lambda, \lambda)$ for a real number $\lambda \neq 0$. In this case, one can choose the orthonormal basis of U as

$$
\begin{aligned}
e_{(i_1, j_1, k_1, \dots, i_6, j_6, k_6)}&(x_1, x_2, x_3) \\
&= (e_{(i_1, j_1, k_1)}(x_1, x_2, x_3), \dots, e_{(i_6, j_6, k_6)}(x_1, x_2, x_3)) \\
&= (\sin(i_l \pi x_1) \sin(j_l \pi x_2) \sin(k_l \pi x_3))_{l=1}^6
\end{aligned}
$$

for $i_l, j_l, k_l \in \mathbb{N}_+$ and $x_1, x_2, x_3 \in [0, 1]$, and let the operator $\mathbf{Q}$ meet

$$
\mathbf{Q}^{\frac{1}{2}} e_{(i_1, j_1, k_1, \dots, i_6, j_6, k_6)} = \mu_{i_1, j_1, k_1} \delta_{i_1, \dots, i_6} \delta_{j_1, \dots, j_6} \delta_{k_1, \dots, k_6} e_{(i_1, j_1, k_1, \dots, i_6, j_6, k_6)},
$$

where $\mu_{i_1, j_1, k_1} \in \mathbb{R}$, $\delta_{i_1, \dots, i_6} = 1$ if $i_1 = \cdots = i_6$, otherwise $\delta_{i_1, \dots, i_6} = 0$. Furthermore, suppose that $\|\mathbf{Q}^{\frac{1}{2}}\|_{\mathscr{L}_2(U, (\mathbf{H}_0^1(\mathscr{O}))^6)} < \infty$, where $\mathbf{H}_0^1(\mathscr{O}) = \{u \in \mathbf{H}^1(\mathscr{O}) : u = 0 \text{ on } \partial\mathscr{O}\}$. Then the operator $\mathbb{G}$ satisfies inequalities in (4.14).

Example 4.2.2 Now we consider the stochastic Maxwell equation (4.9) with multiplicative noise, where $\mathbb{G}$ is the Nemytskij operator of diag$((\mathbf{E}^\top, \mathbf{H}^\top))$ and $\mathscr{O} = [0, 1]^3$. Assume that the orthonormal basis of U is the same as the one in Example 4.2.1, and in addition that

$$
\mathbf{Q}^{\frac{1}{2}} e_{(i_1, j_1, k_1, \dots, i_6, j_6, k_6)} = \lambda_{i_1, j_1, k_1} \delta_{i_1, \dots, i_6} \delta_{j_1, \dots, j_6} \delta_{k_1, \dots, k_6} e_{(i_1, j_1, k_1, \dots, i_6, j_6, k_6)}
$$

for some λ_{i_1, j_1, k_1} and $\mathbf{Q}^{\frac{1}{2}} \in \mathscr{L}_2(U, (\mathbf{H}^{1+\gamma}(\mathscr{O}))^6)$ with $\gamma > \frac{3}{2}$. In this case, we only present that the last inequality in (4.14) holds, and the others can be obtained by similar arguments.

To show the last inequality in (4.14), it suffices to prove that

$$
\|\mathbb{G}(\mathbb{V})\|_{\mathscr{L}_2(U_0, \text{Dom}(A))} \leq C \|\mathbf{Q}^{\frac{1}{2}}\|_{\mathscr{L}_2(U, (\mathbf{H}^{1+\gamma}(\mathscr{O}))^6)} (1 + \|\mathbb{V}\|_{\text{Dom}(A)}). \tag{4.15}
$$

Rearranging the order of the basis $e_k = (e_{(i_1, j_1, k_1)}, \dots, e_{(i_6, j_6, k_6)})$ and taking advantage of the definition of the graph norm yield

$$
\|\mathbb{G}(\mathbb{V})\|_{\mathscr{L}_2(U_0, \text{Dom}(A))}^2 = \sum_{k \in \mathbb{N}_+} \|\mathbb{G}(\mathbb{V})\mathbf{Q}^{\frac{1}{2}} e_k\|_V^2 + \sum_{k \in \mathbb{N}_+} \|A(\mathbb{G}(\mathbb{V})\mathbf{Q}^{\frac{1}{2}} e_k)\|_V^2.
$$

Furthermore, we arrive at

$$
\begin{aligned}
&\|\mathbb{G}(\mathbb{V})\|_{\mathscr{L}_2(U_0, \text{Dom}(A))}^2 \\
&\leq C \sum_{k \in \mathbb{N}_+} \left(\|\mathbf{Q}^{\frac{1}{2}} e_k\|_{(\mathbf{L}^\infty(\mathscr{O}))^6}^2 \|\mathbb{V}\|_V^2 + \|\nabla \times (\mathbf{E}\mathbf{Q}^{\frac{1}{2}} e_k)\|_{(\mathbf{L}^2(\mathscr{O}))^3}^2 + \|\nabla \times (\mathbf{H}\mathbf{Q}^{\frac{1}{2}} e_k)\|_{(\mathbf{L}^2(\mathscr{O}))^3}^2 \right),
\end{aligned}
$$

where

$$\mathbf{E}\mathbf{Q}^{\frac{1}{2}}e_k = (\tilde{\mathbf{E}}_1(\mathbf{Q}^{\frac{1}{2}}e_k)_1, \tilde{\mathbf{E}}_2(\mathbf{Q}^{\frac{1}{2}}e_k)_2, \tilde{\mathbf{E}}_3(\mathbf{Q}^{\frac{1}{2}}e_k)_3),$$

$$\mathbf{H}\mathbf{Q}^{\frac{1}{2}}e_k = (\tilde{\mathbf{H}}_1(\mathbf{Q}^{\frac{1}{2}}e_k)_4, \tilde{\mathbf{H}}_2(\mathbf{Q}^{\frac{1}{2}}e_k)_5, \tilde{\mathbf{H}}_3(\mathbf{Q}^{\frac{1}{2}}e_k)_6)$$

with $(\mathbf{Q}^{\frac{1}{2}}e_k)_i$ denoting the ith component of the vector function $\mathbf{Q}^{\frac{1}{2}}e_k$ for $i \in \{1, \ldots, 6\}$. Based on the definition of the curl operator, we obtain

$$\left\| \nabla \times (\mathbf{E}\mathbf{Q}^{\frac{1}{2}}e_k) \right\|_{(\mathbf{L}^2(\mathscr{O}))^3}^2$$

$$\leq C \left\| \mathbf{Q}^{\frac{1}{2}}e_k \right\|_{(\mathbf{L}^\infty(\mathscr{O}))^6}^2 \left(\left\| \frac{\partial}{\partial x_2}\tilde{\mathbf{E}}_3 - \frac{\partial}{\partial x_3}\tilde{\mathbf{E}}_2 \right\|_{\mathbf{L}^2(\mathscr{O})}^2 \right.$$

$$+ \left\| \frac{\partial}{\partial x_1}\tilde{\mathbf{E}}_3 - \frac{\partial}{\partial x_3}\tilde{\mathbf{E}}_1 \right\|_{\mathbf{L}^2(\mathscr{O})}^2 + \left\| \frac{\partial}{\partial x_1}\tilde{\mathbf{E}}_2 - \frac{\partial}{\partial x_2}\tilde{\mathbf{E}}_1 \right\|_{\mathbf{L}^2(\mathscr{O})}^2 \right)$$

$$+ C \left(\left\| \frac{\partial}{\partial x_1}(\mathbf{Q}^{\frac{1}{2}}e_k)_2 \right\|_{\mathbf{L}^\infty(\mathscr{O})}^2 + \left\| \frac{\partial}{\partial x_1}(\mathbf{Q}^{\frac{1}{2}}e_k)_3 \right\|_{\mathbf{L}^\infty(\mathscr{O})}^2 \right.$$

$$+ \left\| \frac{\partial}{\partial x_2}(\mathbf{Q}^{\frac{1}{2}}e_k)_1 \right\|_{\mathbf{L}^\infty(\mathscr{O})}^2 + \left\| \frac{\partial}{\partial x_2}(\mathbf{Q}^{\frac{1}{2}}e_k)_3 \right\|_{\mathbf{L}^\infty(\mathscr{O})}^2$$

$$+ \left\| \frac{\partial}{\partial x_3}(\mathbf{Q}^{\frac{1}{2}}e_k)_1 \right\|_{\mathbf{L}^\infty(\mathscr{O})}^2 + \left\| \frac{\partial}{\partial x_3}(\mathbf{Q}^{\frac{1}{2}}e_k)_2 \right\|_{\mathbf{L}^\infty(\mathscr{O})}^2 \right) \left\| \mathbf{E} \right\|_{(\mathbf{L}^2(\mathscr{O}))^3}^2$$

$$\leq C \left\| \mathbf{Q}^{\frac{1}{2}}e_k \right\|_{(\mathbf{L}^\infty(\mathscr{O}))^6}^2 \left\| \nabla \times \mathbf{E} \right\|_{(\mathbf{L}^2(\mathscr{O}))^3}^2 + C \left\| \mathbf{Q}^{\frac{1}{2}}e_k \right\|_{(\mathbf{W}^{1,\infty}(\mathscr{O}))^6}^2 \left\| \mathbf{E} \right\|_{(\mathbf{L}^2(\mathscr{O}))^3}^2,$$

which leads to

$$\|\mathbb{G}(\mathbb{V})\|_{\mathscr{L}_2(U_0,\mathrm{Dom}(A))}^2 \leq C \sum_{k \in \mathbb{N}_+} \left\| \mathbf{Q}^{\frac{1}{2}}e_k \right\|_{(\mathbf{L}^\infty(\mathscr{O}))^6}^2 \left(\|\mathbb{V}\|_V^2 + \|A\mathbb{V}\|_V^2 \right)$$

$$+ C \sum_{k \in \mathbb{N}_+} \left\| \mathbf{Q}^{\frac{1}{2}}e_k \right\|_{(\mathbf{W}^{1,\infty}(\mathscr{O}))^6}^2 \|\mathbb{V}\|_V^2.$$

Making use of the Sobolev embedding $\mathbf{H}^{\gamma+1}(\mathscr{O}) \hookrightarrow \mathbf{L}^\infty(\mathscr{O})$ for any $\gamma > \frac{3}{2}$, we deduce (4.15) and the linear growth property of $\mathbb{G}$. By employing the Weyl's law, after rearranging the order of the indices, if the growth of $\lambda_k = \lambda_{i_1,j_1,k_1}$ satisfies $\sum_{k=1}^\infty \lambda_k^2 k^{\frac{2+2\gamma}{3}} < \infty$, then the operator $\mathbb{G}$ satisfies all the conditions in Assumption 4.2.3.

Based on the above assumptions, we now establish well-posedness and regularity results of (4.9).

Lemma 4.2.3 (See [68]) *Let $T > 0$. Under Assumptions 4.2.1–4.2.3, the solution of (4.9) is*

$$\mathbb{U}(t) = E(t)\mathbb{U}(0) + \int_0^t E(t-s)\mathbb{F}(\mathbb{U}(s))ds + \int_0^t E(t-s)\mathbb{G}(\mathbb{U}(s))dW(s)$$

and satisfies

$$\mathbb{E}\left[\sup_{0 \le t \le T} \|\mathbb{U}(t)\|^p_{\mathrm{Dom}(A)}\right] < C\left(1 + \mathbb{E}\left[\|\mathbb{U}(0)\|^p_{\mathrm{Dom}(A)}\right]\right)$$

for any $p \ge 2$, where the constant $C := C(p, T, \mathbf{Q}, \mathbb{U}(0), \mathbb{F}, \mathbb{G}) > 0$.

The well-posedness of stochastic Maxwell equations can be given by the similar arguments as in [51, Corollary 3.1] and [174, Theorem 9], by a refined Faedo–Galerkin method and spectral multiplier theorem in [135], and by the stochastically perturbed partial differential equations approach in [247]. Subsequently we show a lemma on the Hölder regularity in time of the solution to (4.9).

Lemma 4.2.4 (See [68]) *Let $T > 0$. Under Assumptions 4.2.1-4.2.3, the solution $\mathbb{U}$ of the stochastic Maxwell equation (4.9) satisfies*

$$\mathbb{E}\left[\|\mathbb{U}(t) - \mathbb{U}(s)\|^{2p}_V\right] \le C|t - s|^p$$

for any $0 \le s, t \le T$, and $p \ge 1$, where the constant $C := C(p, T, \mathbf{Q}, \mathbb{U}(0), \mathbb{F}, \mathbb{G}) > 0$.

When $\epsilon = \mu = 1$ and take $\mathbb{F} = 0$ and $\mathbb{G}$ as in Example 4.2.1, (4.9) becomes the linear stochastic Maxwell eauation with additive noise

$$\begin{cases} d\begin{bmatrix} \mathbf{E}(t) \\ \mathbf{H}(t) \end{bmatrix} = \begin{bmatrix} 0 & \nabla\times \\ -\nabla\times & 0 \end{bmatrix}\begin{bmatrix} \mathbf{E}(t) \\ \mathbf{H}(t) \end{bmatrix} dt + \lambda dW(t), & \text{in } \mathscr{O} \times (0, T], \\ \mathbf{E}(0) = \mathbf{e}, \quad \mathbf{H}(0) = \mathbf{h}, & \text{in } \mathscr{O}, \end{cases} \tag{4.16}$$

where $\lambda > 0$, and $W(\cdot)$ is an $(L^2(\mathscr{O}))^6$-valued $\mathbf{Q}$-Wiener process. The linear stochastic Maxwell equation (4.16) is an important example of applications in statistical radiophysics (see [229, Chapter 3] and references therein). It can be verified that (4.16) is an infinite-dimensional stochastic Hamiltonian system (4.1) with

$$H_0(\mathbf{E}, \mathbf{H}) = -\frac{1}{2}\int_{\mathscr{O}} (\mathbf{E} \cdot \nabla \times \mathbf{E} + \mathbf{H} \cdot \nabla \times \mathbf{H})\, dx,$$

$$H_1(\mathbf{E}, \mathbf{H}) = \int_{\mathscr{O}} \left((\lambda, \lambda, \lambda)^\top \cdot \mathbf{E} - (\lambda, \lambda, \lambda)^\top \cdot \mathbf{H}\right) dx.$$

Lemma 4.2.5 (See [52]) *Assume that $\mathcal{O} = [0,1]^3$, and $\mathbf{Q}^{\frac{1}{2}} \in \mathscr{L}_2((L^2(\mathcal{O}))^6,$ $(\mathbf{H}_0^1(\mathcal{O}))^6)$. Then under the zero boundary condition, the phase flow of (4.16) preserves the symplectic structure, i.e.,*

$$\varpi_2(t) = \int_{\mathcal{O}} d\mathbf{E}(t) \wedge d\mathbf{H}(t) dx = \int_{\mathcal{O}} d\mathbf{e} \wedge d\mathbf{h} dx = \varpi_2(0), \quad t \in (0, T].$$

Proof Utilizing the formula of change of variables in differential forms yields

$$\varpi_2(t) = \int_{\mathcal{O}} d\mathbf{e} \wedge \left(\frac{\partial \mathbf{E}(t)}{\partial \mathbf{e}}\right)^{\top} \frac{\partial \mathbf{H}(t)}{\partial \mathbf{e}} d\mathbf{e} dx + \int_{\mathcal{O}} d\mathbf{h} \wedge \left(\frac{\partial \mathbf{E}(t)}{\partial \mathbf{h}}\right)^{\top} \frac{\partial \mathbf{H}(t)}{\partial \mathbf{h}} d\mathbf{h} dx$$

$$+ \int_{\mathcal{O}} d\mathbf{e} \wedge \left(\left(\frac{\partial \mathbf{E}(t)}{\partial \mathbf{e}}\right)^{\top} \frac{\partial \mathbf{H}(t)}{\partial \mathbf{h}} - \left(\frac{\partial \mathbf{H}(t)}{\partial \mathbf{e}}\right)^{\top} \frac{\partial \mathbf{E}(t)}{\partial \mathbf{h}}\right) d\mathbf{h} dx,$$

where $t \in [0, T]$. For the sake of simplicity, we set $\mathbf{E_e} = \frac{\partial \mathbf{E}}{\partial \mathbf{e}}, \mathbf{E_h} = \frac{\partial \mathbf{E}}{\partial \mathbf{h}}, \mathbf{H_e} = \frac{\partial \mathbf{H}}{\partial \mathbf{e}}$ and $\mathbf{H_h} = \frac{\partial \mathbf{H}}{\partial \mathbf{h}}$. According to the differentiability with respect to initial value (see [85, Chapter 9]), we derive

$$\begin{cases} d\mathbf{E_e} = \nabla \times \mathbf{H_e} dt, & \mathbf{E_e}(0) = I, \\ d\mathbf{H_e} = -\nabla \times \mathbf{E_e} dt, & \mathbf{H_e}(0) = 0, \\ d\mathbf{E_h} = \nabla \times \mathbf{H_h} dt, & \mathbf{E_h}(0) = 0, \\ d\mathbf{H_h} = -\nabla \times \mathbf{E_h} dt, & \mathbf{H_h}(0) = I. \end{cases}$$

As a consequence, we obtain

$$\frac{d\varpi_2(t)}{dt} = \int_{\mathcal{O}} \left(d\mathbf{e} \wedge \left((\nabla \times \mathbf{H_e})^{\top} \mathbf{H_e} - \mathbf{E_e}^{\top} \nabla \times \mathbf{E_e}\right) d\mathbf{e}\right) dx$$

$$+ \int_{\mathcal{O}} \left(d\mathbf{h} \wedge \left((\nabla \times \mathbf{H_h})^{\top} \mathbf{H_h} - \mathbf{E_h}^{\top} \nabla \times \mathbf{E_h}\right) d\mathbf{h}\right) dx$$

$$+ \int_{\mathcal{O}} \left(d\mathbf{e} \wedge \left((\nabla \times \mathbf{H_e})^{\top} \mathbf{H_h} - \mathbf{E_e}^{\top} \nabla \times \mathbf{E_h}\right) d\mathbf{h}\right) dx$$

$$+ \int_{\mathcal{O}} \left(d\mathbf{e} \wedge \left((\nabla \times \mathbf{E_e})^{\top} \mathbf{E_h} - \mathbf{H_e}^{\top} \nabla \times \mathbf{H_h}\right) d\mathbf{h}\right) dx.$$

Then a direct calculation leads to

$$\frac{d\varpi_2(t)}{dt} = \int_{\mathcal{O}} (d(\nabla \times \mathbf{H}) \wedge d\mathbf{H}) + (d(\nabla \times \mathbf{E}) \wedge d\mathbf{E}) dx$$

$$= \int_{\mathcal{O}} \left(\frac{\partial}{\partial x_1} \left(d\tilde{\mathbf{H}}_2 \wedge d\tilde{\mathbf{H}}_3 \right) + \frac{\partial}{\partial x_2} \left(d\tilde{\mathbf{H}}_3 \wedge d\tilde{\mathbf{H}}_1 \right) + \frac{\partial}{\partial x_3} \left(d\tilde{\mathbf{H}}_1 \wedge d\tilde{\mathbf{H}}_2 \right) \right) dx$$

$$+ \int_{\mathcal{O}} \left(\frac{\partial}{\partial x_1} \left(d\tilde{\mathbf{E}}_2 \wedge d\tilde{\mathbf{E}}_3 \right) + \frac{\partial}{\partial x_2} \left(d\tilde{\mathbf{E}}_3 \wedge d\tilde{\mathbf{E}}_1 \right) + \frac{\partial}{\partial x_3} \left(d\tilde{\mathbf{E}}_1 \wedge d\tilde{\mathbf{E}}_2 \right) \right) dx,$$

where $\mathbf{E} = (\tilde{\mathbf{E}}_1, \tilde{\mathbf{E}}_2, \tilde{\mathbf{E}}_3)^\top$ and $\mathbf{H} = (\tilde{\mathbf{H}}_1, \tilde{\mathbf{H}}_2, \tilde{\mathbf{H}}_3)^\top$. From the zero boundary condition it follows that (4.17) holds. □

Lemma 4.2.6 (See [50]) *Assume that* $\mathcal{O} = [0, 1]^3$, *and* $\mathbf{Q}^{\frac{1}{2}} \in \mathscr{L}_2((L^2(\mathcal{O}))^6, (H_0^1(\mathcal{O}))^6)$. *Then the averaged energy of the exact solution satisfies the trace formula*

$$\mathbb{E}\left[\Phi^{exact}(t)\right] = \mathbb{E}\left[\Phi^{exact}(0)\right] + 6\lambda^2 \mathrm{Tr}(\mathbf{Q})t \quad \forall\, t \in (0, T], \tag{4.17}$$

where $\Phi^{exact}(t) := \int_{\mathcal{O}} \left(\|\mathbf{E}(t)\|^2 + \|\mathbf{H}(t)\|^2 \right) dx$ *denotes the energy of the system. Moreover, the solution to Eq. (4.16) preserves the averaged divergence*

$$\mathbb{E}\left[\mathrm{div}(\mathbf{E}(t))\right] = \mathbb{E}\left[\mathrm{div}(\mathbf{e})\right], \quad \mathbb{E}\left[\mathrm{div}(\mathbf{H}(t))\right] = \mathbb{E}\left[\mathrm{div}(\mathbf{h})\right] \quad \forall\, t \in (0, T].$$

Recently, there have been a number of papers devoted especially to constructing structure-preserving numerical methods for stochastic Maxwell equations (see e.g., [50, 51, 125, 127, 246] and references therein). Below we shall introduce the stochastic exponential integrator and stochastic symplectic Runge–Kutta discretization.

4.2.1 Stochastic Exponential Integrator

The stochastic exponential integrator has been widely utilized to approximate stochastic ordinary differential equations, stochastic heat equations, stochastic wave equations, stochastic Schrödinger equations and so on (see e.g., [11, 14, 15, 36, 65–67, 72, 75, 145, 154, 156, 178, 219, 236, 261] and references therein). This subsection is concerned with the stochastic exponential integrator for the stochastic Maxwell equation.

Approximating the integrals of the exact solution

$$\mathbb{U}(t_{n+1}) = E(h)\mathbb{U}(t_n) + \int_{t_n}^{t_{n+1}} E(t_{n+1} - s)\mathbb{F}(\mathbb{U}(s))ds + \int_{t_n}^{t_{n+1}} E(t_{n+1} - s)\mathbb{G}(\mathbb{U}(s))dW(s)$$

by the left rectangular formula over $[t_n, t_{n+1}]$ for $n \in \{0, 1, \ldots, N-1\}$, one obtains the stochastic exponential integrator

$$\mathbb{U}_{n+1} = E(h)\mathbb{U}_n + E(h)\mathbb{F}(\mathbb{U}_n)h + E(h)\mathbb{G}(\mathbb{U}_n)\Delta_n W, \quad \mathbb{U}_0 = \mathbb{U}(0), \qquad (4.18)$$

where $\Delta_n W = W(t_{n+1}) - W(t_n)$ and $h = \frac{T}{N}$. It can be observed that (4.18) is an explicit and effective numerical method of (4.9).

Stochastic Exponential Integrator for (4.16) Applying (4.18) to (4.16), we obtain

$$\mathbb{U}_{n+1} = \exp(h\tilde{A})\mathbb{U}_n + \exp(h\tilde{A})\mathbb{G}\Delta_n W, \quad n = 0, 1, \ldots, N-1, \qquad (4.19)$$

where $\mathbb{G}$ is the Nemytskij operator of $\mathrm{diag}(\lambda, \lambda, \lambda, \lambda, \lambda, \lambda)$ and $\tilde{A} = \begin{bmatrix} 0 & \nabla\times \\ -\nabla\times & 0 \end{bmatrix}$. The stochastic exponential integrator possesses the same long-time behavior as the exact solution to the stochastic linear Maxwell equation (4.16). Now we focus on the discrete symplecticity of (4.19).

Proposition 4.2.1 (See [68]) *The stochastic exponential integrator* (4.19) *possesses the discrete stochastic symplectic conservation law almost surely, i.e.,*

$$\varpi_2^n = \int_{\mathcal{O}} d\mathbf{E}_n \wedge d\mathbf{H}_n dx = \int_{\mathcal{O}} d\mathbf{e} \wedge d\mathbf{h} dx = \varpi_2^0, \quad a.s.,$$

where $n = 0, 1, \ldots, N$.

Proof Taking the differential of (4.19) implies

$$d\mathbb{U}_{n+1} = d\big(\exp(h\tilde{A})\mathbb{U}_n\big), \quad n = 0, 1, \ldots, N-1.$$

Therefore, showing the symplecticity of the stochastic exponential integrator is equivalent to proving the symplecticity of the phase flow of the linear deterministic Maxwell equation, which is a well-known fact (see e.g., [30, Section 6.7] or [186, Section 4] and references therein). □

Proposition 4.2.2 (See [68]) *The exponential integrator* (4.19) *exactly preserves the following discrete averaged divergence*

$$\mathbb{E}\left[\mathrm{div}(\mathbf{E}_n)\right] = \mathbb{E}\left[\mathrm{div}(\mathbf{E}_{n-1})\right], \quad \mathbb{E}\left[\mathrm{div}(\mathbf{H}_n)\right] = \mathbb{E}\left[\mathrm{div}(\mathbf{H}_{n-1})\right]$$

for all $n \in \{1, \ldots, N\}$.

Proof Fix $n \in \{1, \ldots, N\}$ and let $(\mathrm{div}, \mathrm{div})(\mathbf{E}^\top, \mathbf{H}^\top)^\top := (\mathrm{div}(\mathbf{E}), \mathrm{div}(\mathbf{H}))^\top$. Taking the divergence and expectation, we arrive at

$$\mathbb{E}\left[(\mathrm{div}, \mathrm{div})\mathbb{U}_n\right] = \mathbb{E}\left[(\mathrm{div}, \mathrm{div})(\exp(h\tilde{A})\mathbb{U}_{n-1})\right]. \qquad (4.20)$$

Notice that $\exp(h\tilde{A})\mathbb{U}_{n-1}$ is the solution of the deterministic Maxwell equation

$$\begin{cases} d\mathbf{E}(t) - \nabla \times \mathbf{H}(t)dt = 0, \\ d\mathbf{H}(t) + \nabla \times \mathbf{E}(t)dt = 0, \\ (\mathbf{E}^\top, \mathbf{H}^\top)^\top(0) = \mathbb{U}_{n-1} \end{cases}$$

at time $t = h$. Making use of the property $\mathrm{div}(\nabla \times \cdot) = 0$ and similar arguments as in [50, Theorem 2.2], we derive

$$(\mathrm{div}, \mathrm{div})(\exp(h\tilde{A})\mathbb{U}_{n-1}) = (\mathrm{div}, \mathrm{div})(\mathbb{U}_{n-1}). \qquad (4.21)$$

Finally, combining (4.20) and (4.21) leads to the desired result. □

From the Itô isometry, it follows that the exact trace formula for the energy also holds for the numerical solution given by (4.19), which is illustrated in the following proposition.

Proposition 4.2.3 (See [68]) *The stochastic exponential integrator* (4.19) *satisfies*

$$\mathbb{E}[\Phi(t_n)] = \mathbb{E}[\Phi(0)] + 6\lambda^2 \mathrm{Tr}(\mathbf{Q})t_n,$$

where $\Phi(t_n) := \int_{\mathcal{O}} \left(\|\mathbf{E}_n\|^2 + \|\mathbf{H}_n\|^2 \right) dx$ *with* $t_n = nh$, $n \in \{1, \ldots, N\}$.

Below we shall perform computations for the stochastic exponential integrator (4.19) for the stochastic Maxwell equation with additive noise, which is denoted by SEXP, and compare it with semi-implicit Euler–Maruyama method as follows,

$$\mathbb{U}_{n+1} = \mathbb{U}_n + \tilde{A}\mathbb{U}_{n+1}h + \mathbb{G}\Delta_n W. \qquad \text{(SEM)}$$

Here, we take into consideration the stochastic Maxwell equation (4.16) with TM polarization (see e.g., [167, 191] and references therein) on the domain $[0, 1] \times [0, 1]$. In this setting, the electric and magnetic fields are $\mathbf{E} = (0, 0, \tilde{E}_3)^\top$ and $\mathbf{H} = (\tilde{H}_1, \tilde{H}_2, 0)^\top$, respectively. In order to check the zero divergence property, we let the initial condition be

$$\tilde{E}_3(x_1, x_2, 0) = 0.1 \exp(-50((x_1 - 0.5)^2 + (x_2 - 0.5)^2)),$$

$$\tilde{H}_1(x_1, x_2, 0) = \mathrm{rand}_{x_2}, \quad \tilde{H}_2(x_1, x_2, 0) = \mathrm{rand}_{x_1},$$

where rand_{x_1} and rand_{x_2} are random values in one direction whereas the other direction is kept constant. Moreover, we take $\mu_{j,k} \sim (j + k)^{-2-\epsilon}$ for some $\epsilon > 0$ with $\mu_{j,k}$ being the eigenvalue of $\mathbf{Q}$ for $j, k \in \mathbb{N}_+$. The trace formula for the energy of SEXP, as stated in Proposition 4.2.3, can be observed in Fig. 4.1, which shows a perfect alignment between the SEXP method and the exact solution. This is in

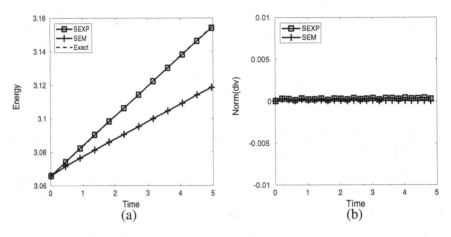

Fig. 4.1 (a) Averaged energy on a longer time, (b) Averaged divergence

contrast with the wrong behavior of the SEM method. Moreover, it can be observed that the averaged divergence of magnetic field is preserved along the numerical solution given by both the SEXP and SEM method. The illustration of the averaged divergence of the SEXP method confirms theoretical findings in Proposition 4.2.2.

Strong Convergence of Stochastic Exponential Integrator In this part, we turn to the convergence analysis in strong convergence sense of the stochastic exponential integrator (4.18). At first, we denote $\mathbb{Z}_{N+1} = \{0, 1, \ldots, N\}$ and introduce the a priori estimate of (4.18).

Theorem 4.2.1 (See [68]) *Under Assumptions 4.2.1-4.2.3, for all $p \geq 1$, there exists a constant $C := C(p, T, \mathbf{Q}, \mathbb{U}(0), \mathbb{F}, \mathbb{G}) > 0$ such that (4.18) satisfies*

$$\sup_{n \in \mathbb{Z}_{N+1}} \mathbb{E}\left[\|\mathbb{U}_n\|_{\mathrm{Dom}(A)}^{2p}\right] \leq C.$$

Proof Fix $n \in \{0, 1, \ldots, N\}$. Then the numerical approximation given by (4.18) can be rewritten as

$$\mathbb{U}_n = E(t_n)\mathbb{U}(0) + h \sum_{j=0}^{n-1} E(t_n - t_j)\mathbb{F}(\mathbb{U}_j) + \sum_{j=0}^{n-1} E(t_n - t_j)\mathbb{G}(\mathbb{U}_j)\Delta_j W,$$

which yields

$$
\mathbb{E}\left[\|\mathbb{U}_n\|_{\mathrm{Dom}(A)}^{2p}\right] \le C\mathbb{E}\left[\|E(t_n)\mathbb{U}(0)\|_{\mathrm{Dom}(A)}^{2p}\right] + C\mathbb{E}\left[\left\|h\sum_{j=0}^{n-1}E(t_n-t_j)\mathbb{F}(\mathbb{U}_j)\right\|_{\mathrm{Dom}(A)}^{2p}\right]
$$

$$
+ C\mathbb{E}\left[\left\|\sum_{j=0}^{n-1}E(t_n-t_j)\mathbb{G}(\mathbb{U}_j)\varDelta_j W\right\|_{\mathrm{Dom}(A)}^{2p}\right]
$$

for $p \ge 1$. For the first term, making use of the definition of the graph norm and (4.11), we arrive at

$$
\|E(t_n)\mathbb{U}(0)\|_{\mathrm{Dom}(A)}^{2p} = (\|E(t_n)\mathbb{U}(0)\|_V + \|E(t_n)A\mathbb{U}(0)\|_V)^{2p} = \|\mathbb{U}(0)\|_{\mathrm{Dom}(A)}^{2p},
$$

which means $\mathbb{E}\left[\|E(t_n)\mathbb{U}(0)\|_{\mathrm{Dom}(A)}^{2p}\right] = \mathbb{E}\left[\|\mathbb{U}(0)\|_{\mathrm{Dom}(A)}^{2p}\right]$. Based on the linear growth property of $\mathbb{F}$ and the Hölder inequality,

$$
\mathbb{E}\left[\left\|h\sum_{j=0}^{n-1}E(t_n-t_j)\mathbb{F}(\mathbb{U}_j)\right\|_{\mathrm{Dom}(A)}^{2p}\right] \le C + Ch\mathbb{E}\left[\sum_{j=0}^{n-1}\|\mathbb{U}_j\|_{\mathrm{Dom}(A)}^{2p}\right].
$$

Combining (4.11) with the Burkholder–Davis–Gundy inequality in [85, Theorem 4.37] leads to

$$
\mathbb{E}\left[\left\|\sum_{j=0}^{n-1}E(t_n-t_j)\mathbb{G}(\mathbb{U}_j)\varDelta_j W\right\|_{\mathrm{Dom}(A)}^{2p}\right]
$$

$$
= \mathbb{E}\left[\left\|\int_0^{t_n}E\left(t_n-\left[\frac{s}{h}\right]h\right)\mathbb{G}(\mathbb{U}_{[\frac{s}{h}]})dW(s)\right\|_{\mathrm{Dom}(A)}^{2p}\right]
$$

$$
\le C + C\mathbb{E}\left[\left(h\sum_{j=0}^{n-1}\|\mathbb{U}_j\|_{\mathrm{Dom}(A)}^2\right)^p\right] \tag{4.22}
$$

with $\left[\frac{s}{h}\right]$ being the integer part of $\frac{s}{h}$. Based on the Jensen inequality, we obtain

$$
\mathbb{E}\left[\left\|\sum_{j=0}^{n-1}E(t_n-t_j)\mathbb{G}(\mathbb{U}_j)\varDelta_j W\right\|_{\mathrm{Dom}(A)}^{2p}\right] \le C + Ch\sum_{j=0}^{n-1}\mathbb{E}\left[\|\mathbb{U}_j\|_{\mathrm{Dom}(A)}^{2p}\right].
$$

As a result, we get

$$\mathbb{E}\left[\|\mathbb{U}_n\|^{2p}_{\text{Dom}(A)}\right] \leq C + Ch\mathbb{E}\left[\sum_{j=0}^{n-1}\|\mathbb{U}_j\|^{2p}_{\text{Dom}(A)}\right].$$

A discrete Grönwall's inequality concludes the proof.   □

Corollary 4.2.1 (See [68]) *Under the same assumptions as in Theorem 4.2.1, for all $p \geq 1$, there exists a constant $C := C(p, T, \mathbf{Q}, \mathbb{U}(0), \mathbb{F}, \mathbb{G}) > 0$ such that*

$$\mathbb{E}\left[\sup_{n\in\mathbb{Z}_{N+1}}\|\mathbb{U}_n\|^{2p}_{\text{Dom}(A)}\right] \leq C. \tag{4.23}$$

Proof For the sake of simplicity, we only estimate the stochastic integral

$$\mathbb{E}\left[\sup_{n\in\mathbb{Z}_{N+1}}\left\|\sum_{j=0}^{n-1}E(t_n - t_j)\mathbb{G}(\mathbb{U}_j)\Delta_j W\right\|^{2p}_{\text{Dom}(A)}\right]$$

$$= \mathbb{E}\left[\sup_{n\in\mathbb{Z}_{N+1}}\left\|\int_0^{t_n}E\left(t_n - \left[\frac{s}{h}\right]h\right)\mathbb{G}(\mathbb{U}_{[\frac{s}{h}]})dW(s)\right\|^{2p}_{\text{Dom}(A)}\right]$$

for (4.23), and the estimations of the other terms can be obtained by means of similar arguments. Based on (4.11), the Burkholder–Davis–Gundy inequality and the Hölder inequality, the right hand side (RHS) of the above formula meets

$$\text{RHS} \leq C\mathbb{E}\left[\left(\int_0^T\left\|\mathbb{G}(\mathbb{U}_{[\frac{s}{h}]})\right\|^2_{\mathscr{L}_2(U_0,\text{Dom}(A))}ds\right)^p\right]$$

$$\leq C + Ch\sum_{j=0}^{N-1}\mathbb{E}\left[\|\mathbb{U}_j\|^{2p}_{\text{Dom}(A)}\right] \leq C,$$

where we have used Theorem 4.2.1.   □

Now we show the strong convergence of the stochastic exponential integrator for the stochastic Maxwell equation (4.9) driven by additive noise.

Theorem 4.2.2 (See [68]) *Suppose that Assumptions 4.2.1–4.2.3 hold, $\mathbb{F} \in C_b^2(V, V)$ and $\mathbb{G}$ is independent of $\mathbb{U}$. Then for all $p \geq 1$, there exists a constant $C := C(p, T, \mathbf{Q}, \mathbb{U}(0), \mathbb{F}, \mathbb{G}) > 0$ such that the exponential integrator (4.18)* *satisfies*

$$\mathbb{E}\left[\sup_{n\in\mathbb{Z}_{N+1}}\|\mathbb{U}(t_n)-\mathbb{U}_n\|_V^{2p}\right]\leq Ch^{2p}.$$

Proof Letting $\epsilon_n = \mathbb{U}(t_n)-\mathbb{U}_n$ for $n = 0, 1, \ldots, N$, we obtain

$$\epsilon_{n+1} = \sum_{j=0}^n \int_{t_j}^{t_{j+1}} \left(E(t_{n+1}-s)\mathbb{F}(\mathbb{U}(s)) - E(t_{n+1}-t_j)\mathbb{F}(\mathbb{U}_j)\right)ds$$

$$+ \sum_{j=0}^n \int_{t_j}^{t_{j+1}} (E(t_{n+1}-s)-E(t_{n+1}-t_j))\mathbb{G}dW(s)$$

$$=: Err_1^n + Err_2^n. \tag{4.24}$$

Rewrite the term Err_1^n as

$$Err_1^n = \sum_{j=0}^n \int_{t_j}^{t_{j+1}} E(t_{n+1}-s)(\mathbb{F}(\mathbb{U}(s))-\mathbb{F}(\mathbb{U}(t_j)))ds$$

$$+ \sum_{j=0}^n \int_{t_j}^{t_{j+1}} \left(E(t_{n+1}-s)-E(t_{n+1}-t_j)\right)\mathbb{F}(\mathbb{U}(t_j))ds$$

$$+ \sum_{j=0}^n \int_{t_j}^{t_{j+1}} E(t_{n+1}-t_j)(\mathbb{F}(\mathbb{U}(t_j))-\mathbb{F}(\mathbb{U}_j))ds =: I_1^n + I_2^n + I_3^n.$$

Applying the mild formulation of the exact solution, we achieve $I_1^n = \mathscr{A}_1^n + \mathscr{A}_2^n$, where

$$\mathscr{A}_1^n = \sum_{j=0}^n \int_{t_j}^{t_{j+1}} E(t_{n+1}-s)\mathbb{F}_u(\mathbb{U}(t_j))(E(s-t_j)-I)\mathbb{U}(t_j)ds$$

$$+ \sum_{j=0}^n \int_{t_j}^{t_{j+1}} E(t_{n+1}-s)\mathbb{F}_u(\mathbb{U}(t_j)) \int_{t_j}^s E(s-r)\mathbb{F}(\mathbb{U}(r))drds$$

$$+ \sum_{j=0}^n \int_{t_j}^{t_{j+1}} E(t_{n+1}-s)\mathbb{F}_u(\mathbb{U}(t_j)) \int_{t_j}^s E(s-r)\mathbb{G}dW(r)ds$$

$$=: II_1^n + II_2^n + II_3^n,$$

$$\mathscr{A}_2^n = \sum_{j=0}^n \int_{t_j}^{t_{j+1}} E(t_{n+1}-s) \int_0^1 \theta\mathbb{F}_{uu}(\mathbb{U}(s)+\theta(\mathbb{U}(s)-\mathbb{U}(t_j)))$$

$$(\mathbb{U}(s)-\mathbb{U}(t_j))(\mathbb{U}(s)-\mathbb{U}(t_j))d\theta ds.$$

The assumption that $\mathbb{F} \in \mathbf{C}_b^2(V, V)$ and the Hölder continuity of the exact solution $\mathbb{U}$ in Lemma 4.2.4 lead to $\mathbb{E}\left[\|\mathscr{A}_2\|_V^{2p}\right] \leq Ch^{2p}$. For II_1^n, it follows from (4.11) and Lemma 4.2.1 that

$$\|\mathrm{II}_1^n\|_V \leq \sum_{j=0}^{n} \int_{t_j}^{t_{j+1}} \|\mathbb{F}_u(\mathbb{U}(t_j))(E(s - t_j) - I)\mathbb{U}(t_j)\|_V \, ds$$

$$\leq Ch\left(\sup_{j \in \mathbb{Z}_{n+1}} \|\mathbb{U}(t_j)\|_{\mathrm{Dom}(A)}^{2p}\right)^{\frac{1}{2p}},$$

which indicates

$$\mathbb{E}\left[\sup_{n \in \mathbb{Z}_N} \|\mathrm{II}_1^n\|_V^{2p}\right] \leq Ch^{2p}\mathbb{E}\left[\sup_{j \in \mathbb{Z}_{N+1}} \|\mathbb{U}(t_j)\|_{\mathrm{Dom}(A)}^{2p}\right] \leq Ch^{2p}.$$

Employing Lemma 4.2.1 and the Hölder inequality, we derive

$$\|\mathrm{II}_2^n\|_V \leq C\sum_{j=0}^{n} \int_{t_j}^{t_{j+1}} \int_{t_j}^{s} (1 + \|\mathbb{U}(r)\|_V)drds \leq Ch + Ch\left(\sup_{0 \leq t \leq T} \|\mathbb{U}(t)\|_V^{2p}\right)^{\frac{1}{2p}},$$

which means

$$\mathbb{E}\left[\sup_{n \in \mathbb{Z}_N} \|\mathrm{II}_2^n\|_V^{2p}\right] \leq Ch^{2p} + Ch^{2p}\mathbb{E}\left[\sup_{0 \leq t \leq T} \|\mathbb{U}(t)\|_V^{2p}\right] \leq Ch^{2p}.$$

Notice that stochastic Fubini's theorem yields

$$\mathrm{II}_3^n = \int_0^{t_{n+1}} \int_r^{([\frac{r}{h}]+1)h} E(t_{n+1} - s)\mathbb{F}_u(\mathbb{U}([\tfrac{s}{h}]h))E(s - r)\mathbb{G}dsdW(r)$$

and the integrand in the above equation is $\mathscr{F}_r$-adaptive. Then making use of (4.11), the Burkholder–Davis–Gundy inequality and the assumption $\mathbb{F} \in \mathbf{C}_b^2(V, V)$ leads to

$$\mathbb{E}\left[\sup_{n \in \mathbb{Z}_N} \|\mathrm{II}_3^n\|_V^{2p}\right]$$

$$\leq C\mathbb{E}\left[\left(\int_0^T \left\|\int_r^{([\frac{r}{h}]+1)h} E(-s)\mathbb{F}_u(\mathbb{U}([\tfrac{s}{h}]h))E(s - r)\mathbb{G}ds\right\|_{\mathscr{L}_2(U_0, V)}^2 dr\right)^p\right]$$

$$\leq C\mathbb{E}\left[\left(\sum_{j=0}^{N-1}\int_{t_j}^{t_{j+1}}\left(\int_r^{t_{j+1}}\left\|E(-s)\mathbb{F}_u(\mathbb{U}(t_j))E(s-r)\mathbb{G}\right\|_{\mathscr{L}_2(U_0,V)}ds\right)^2 dr\right)^p\right]$$

$$\leq C\mathbb{E}\left[\left(\sum_{j=0}^{N-1}\int_{t_j}^{t_{j+1}}\left(\int_r^{t_{j+1}}\left\|\mathbb{G}\mathbb{Q}^{\frac{1}{2}}\right\|_{\mathscr{L}_2(U,V)}ds\right)^2 dr\right)^p\right]\leq Ch^{2p}.$$

As a consequence, $\mathbb{E}\left[\sup_{n\in\mathbb{Z}_N}\|\mathscr{A}_1^n\|_V^{2p}\right]\leq Ch^{2p}$ and $\mathbb{E}\left[\sup_{n\in\mathbb{Z}_N}\|\mathrm{I}_1^n\|_V^{2p}\right]\leq Ch^{2p}$.
Exploiting (4.11), Lemma 4.2.1 and the linear growth property of $\mathbb{F}$, we deduce

$$\|\mathrm{I}_2^n\|_V\leq\sum_{j=0}^n\int_{t_j}^{t_{j+1}}\left\|(E(s-t_j)-I)\mathbb{F}(\mathbb{U}(t_j))\right\|_V ds\leq Ch+Ch^2\sum_{j=0}^n\|\mathbb{U}(t_j)\|_{\mathrm{Dom}(A)},$$

which indicates

$$\mathbb{E}\left[\sup_{n\in\mathbb{Z}_N}\|\mathrm{I}_2^n\|_V^{2p}\right]\leq Ch^{2p}+Ch^{2p}\mathbb{E}\left[\sup_{0\leq t\leq T}\|\mathbb{U}(t)\|_{\mathrm{Dom}(A)}^{2p}\right]\leq Ch^{2p}.$$

Similarly, $\mathbb{E}\left[\sup_{n\in\mathbb{Z}_N}\|\mathrm{I}_3^n\|_V^{2p}\right]\leq Ch\sum_{j=0}^{N-1}\mathbb{E}\left[\sup_{l\in\mathbb{Z}_{j+1}}\|\epsilon_l\|_V^{2p}\right]$. The last term Err_2^n can
be bounded as follows

$$\mathbb{E}\left[\sup_{n\in\mathbb{Z}_N}\|Err_2^n\|_V^{2p}\right]=\mathbb{E}\left[\sup_{n\in\mathbb{Z}_N}\left\|\int_0^{t_{n+1}}\left(E\left(t_{n+1}-\left[\frac{s}{h}\right]h\right)-E(t_{n+1}-s)\right)\mathbb{G}dW(s)\right\|_V^{2p}\right]$$

$$\leq\mathbb{E}\left[\sup_{0\leq t\leq T}\left\|\int_0^t E(t-s)\left(E\left(s-\left[\frac{s}{h}\right]h\right)-I\right)\mathbb{G}dW(s)\right\|_V^{2p}\right].$$

According to (4.11), the Burkholder–Davis–Gundy inequality and Lemma 4.2.1, we
obtain

$$\mathbb{E}\left[\sup_{n\in\mathbb{Z}_N}\|Err_2^n\|_V^{2p}\right]\leq C\mathbb{E}\left[\left(\sum_{j=0}^{N-1}\int_{t_j}^{t_{j+1}}\left\|(E(s-t_j)-I)\mathbb{G}\right\|_{\mathscr{L}_2(U_0,V)}^2 ds\right)^p\right]\leq Ch^{2p},$$

where we have used the linear growth property of $\mathbb{G}$ in $\mathscr{L}_2(U_0,\mathrm{Dom}(A))$. Collecting
all the above estimates gives

$$\mathbb{E}\left[\sup_{n\in\mathbb{Z}_N}\|\epsilon_{n+1}\|_V^{2p}\right]\leq Ch^{2p}+Ch\sum_{j=0}^{N-1}\mathbb{E}\left[\sup_{l\in\mathbb{Z}_{j+1}}\|\epsilon_l\|_V^{2p}\right].$$

The Grönwall inequality completes the proof. □

Now we introduce the strong error estimate of (4.18) applied to stochastic Maxwell equation (4.9) driven by the multiplicative noise. In this case, the assumption on $\mathbb{F}$ in Theorem 4.2.2 could be weakened, since the numerical method without simulating Lévy's area is in general limited to $\frac{1}{2}$ in strong convergence sense. As a result, taking the Taylor expansion up to the first term is enough.

Theorem 4.2.3 (See [68]) *Let Assumptions 4.2.1–4.2.3 hold. The strong error of the exponential integrator (4.18) when applied to the stochastic Maxwell equation (4.9) satisfies*

$$\mathbb{E}\left[\sup_{n \in \mathbb{Z}_{N+1}} \|\mathbb{U}(t_n) - \mathbb{U}_n\|_V^{2p} \right] \le Ch^p$$

for $p \ge 1$, where $C := C(p, T, \mathbf{Q}, \mathbb{U}(0), \mathbb{F}, \mathbb{G}) > 0$.

Proof When the noise is multiplicative, the term Err_2^n in (4.24) reads

$$Err_2^n = \sum_{j=0}^{n} \int_{t_j}^{t_{j+1}} E(t_{n+1} - s)(\mathbb{G}(\mathbb{U}(s)) - \mathbb{G}(\mathbb{U}(t_j)))dW(s)$$

$$+ \sum_{j=0}^{n} \int_{t_j}^{t_{j+1}} \left(E(t_{n+1} - s) - E(t_{n+1} - t_j) \right) \mathbb{G}(\mathbb{U}(t_j))dW(s)$$

$$+ \sum_{j=0}^{n} \int_{t_j}^{t_{j+1}} E(t_{n+1} - t_j)(\mathbb{G}(\mathbb{U}(t_j)) - \mathbb{G}(\mathbb{U}_j))dW(s)$$

$$=: \mathrm{III}_1^n + \mathrm{III}_2^n + \mathrm{III}_3^n.$$

By (4.11) and the Burkholder–Davis–Gundy inequality, one obtains

$$\mathbb{E}\left[\sup_{n \in \mathbb{Z}_N} \|\mathrm{III}_1^n\|_V^{2p} \right] \le C\mathbb{E}\left[\left(\int_0^T \left\| \mathbb{G}(\mathbb{U}(s)) - \mathbb{G}(\mathbb{U}(\left[\frac{s}{h}\right]h)) \right\|_{\mathscr{L}_2(U_0,V)}^2 ds \right)^p \right]$$

$$\le C\mathbb{E}\left[\left(\int_0^T \left\| \mathbb{U}(s) - \mathbb{U}(\left[\frac{s}{h}\right]h) \right\|_V^2 ds \right)^p \right].$$

From the Hölder inequality and the continuity of $\mathbb{U}$ in Lemma 4.2.4 it follows that

$$\mathbb{E}\left[\sup_{n \in \mathbb{Z}_N} \|\mathrm{III}_1^n\|_V^{2p} \right] \le C\mathbb{E}\left[\int_0^T \left\| \mathbb{U}(s) - \mathbb{U}(\left[\frac{s}{h}\right]h) \right\|_V^{2p} ds \right]$$

$$\le C \sum_{j=0}^{N-1} \int_{t_j}^{t_{j+1}} |s - t_j|^p ds \le Ch^p.$$

Analogously, we derive

$$
\mathbb{E}\left[\sup_{n\in\mathbb{Z}_N}\|\text{III}_2^n\|_V^{2p}\right]
$$

$$
\leq \mathbb{E}\left[\sup_{0\leq t\leq T}\left\|\int_0^t\left(E(t-s)-E\left(t-\left[\frac{s}{h}\right]h\right)\right)\mathbb{G}(\mathbb{U}(\left[\frac{s}{h}\right]h))dW(s)\right\|_V^{2p}\right]
$$

$$
\leq C\sum_{j=0}^{N-1}\int_{t_j}^{t_{j+1}}\left|s-\left[\frac{s}{h}\right]h\right|^{2p}\mathbb{E}\left[\left\|\mathbb{G}(\mathbb{U}(\left[\frac{s}{h}\right]h))\right\|_{\mathscr{L}_2(U_0,\text{Dom}(A))}^{2p}\right]ds \leq Ch^{2p}.
$$

For III_3^n, taking advantage of Assumption 4.2.3, we deduce

$$
\mathbb{E}\left[\sup_{n\in\mathbb{Z}_N}\|\text{III}_3^n\|_V^{2p}\right] \leq C\mathbb{E}\left[\left(\int_0^T\left\|\mathbb{G}(\mathbb{U}(\left[\frac{s}{h}\right]h))-\mathbb{G}(\mathbb{U}_{[\frac{s}{h}]})\right\|_{\mathscr{L}_2(U_0,V)}^2 ds\right)^p\right]
$$

$$
\leq Ch\sum_{j=0}^{N-1}\mathbb{E}\left[\sup_{l\in\mathbb{Z}_{j+1}}\|\mathbb{U}(t_l)-\mathbb{U}_l\|_V^{2p}\right].
$$

Altogether, we obtain

$$
\mathbb{E}\left[\sup_{n\in\mathbb{Z}_N}\|Err_2^n\|_V^{2p}\right] \leq Ch^p + Ch\sum_{j=0}^{N-1}\mathbb{E}\left[\sup_{l\in\mathbb{Z}_{j+1}}\|\epsilon_l\|_V^{2p}\right],
$$

where $\epsilon_l = \mathbb{U}(t_l)-\mathbb{U}_l$. Employing (4.11) and Assumption 4.2.2 gives

$$
\|\mathbb{I}_1^n\|_V^{2p} \leq \left(\sum_{j=0}^n\int_{t_j}^{t_{j+1}}\|E(t_{n+1}-s)(\mathbb{F}(\mathbb{U}(s))-\mathbb{F}(\mathbb{U}(t_j)))\|_V ds\right)^{2p}
$$

$$
\leq C\sum_{j=0}^n\int_{t_j}^{t_{j+1}}\|\mathbb{U}(s)-\mathbb{U}(t_j)\|_V^{2p}ds.
$$

Based on Lemma 4.2.4, we deduce

$$
\mathbb{E}\left[\sup_{n\in\mathbb{Z}_N}\|\mathbb{I}_1^n\|_V^{2p}\right] \leq C\sum_{j=0}^n\int_{t_j}^{t_{j+1}}|s-t_j|^p ds \leq Ch^p.
$$

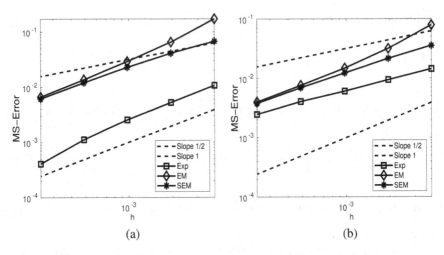

Fig. 4.2 Strong convergence orders of (4.18) for the stochastic Maxwell equation with (a) $\mathbb{F}(\mathbb{U}) = \mathbb{U} + \cos(\mathbb{U})$ and $\mathbb{G}(\mathbb{U}) = \sin(\mathbb{U})$ and (b) $\mathbb{F}(\mathbb{U}) = \mathbb{U}$ and $\mathbb{G}(\mathbb{U}) = (1, 1, 1, 1, 1, 1)^\top$

Putting all estimates together leads to

$$
\mathbb{E}\left[\sup_{n\in\mathbb{Z}_N} \|\epsilon_{n+1}\|_V^{2p}\right] \le Ch^p + Ch \sum_{j=0}^{N-1} \mathbb{E}\left[\sup_{l\in\mathbb{Z}_{j+1}} \|\epsilon_l\|_V^{2p}\right].
$$

An application of the Grönwall inequality completes the proof.   □

Figure 4.2 shows the error $\mathbb{E}\left[\|\mathbb{U}_N - \mathbb{U}(T)\|_V^2\right]$ of stochastic exponential integrator, semi-implicit Euler–Maruyama method and Euler–Maruyama method for stochastic Maxwell equation with TM polarization on $[0, 1] \times [0, 1]$, where $T = 0.5$ and $h = 2^{-i}$, $i \in \{8, \ldots, 12\}$. Expected values are approximated by computing averages over 500 samples. The strong convergence orders of the stochastic exponential integrator stated in Theorems 4.2.2 and 4.2.3 can be observed from these plots.

4.2.2 Stochastic Symplectic Runge–Kutta Discretizations

In previous subsection, it is shown that the stochastic exponential integrator preserves the symplecticity for the linear stochastic Maxwell equation with additive noise. Below we shall introduce a general class of stochastic symplectic Runge–Kutta discretizations for the nonlinear stochastic Maxwell equation

$$
\begin{cases}
d\mathbb{U}(t) = A\mathbb{U}(t)dt + \mathbb{F}(\mathbb{U}(t))dt + \mathbb{G}dW(t), \\
\mathbb{U}(0) = (\mathbf{e}^\top, \mathbf{h}^\top)^\top,
\end{cases}
\tag{4.25}
$$

where $\mathbb{F}\colon U \to U$ is Fréchet differentiable, and $\mathbb{G}$ is independent of $\mathbb{U}$. Moreover, set $\tilde{\mathbb{F}} = \mathbb{JF}$ with $\mathbb{J} = \begin{bmatrix} 0 & I \\ -I & 0 \end{bmatrix}$, and assume that $D\tilde{\mathbb{F}}(\mathbb{U}) \in \mathscr{L}(U, U)$ is a symmetric operator, i.e.,

$$\langle D\tilde{\mathbb{F}}(\mathbb{U})\phi, \psi\rangle_U = \langle \phi, D\tilde{\mathbb{F}}(\mathbb{U})\psi\rangle_U \quad \forall \, \phi, \psi \in U,$$

It can be verified that the phase flow preserves the symplecticity, i.e.,

$$\varpi_2(t) = \int_{\mathscr{O}} d\mathbf{E}(t) \wedge d\mathbf{H}(t) dx = \int_{\mathscr{O}} d\mathbf{e} \wedge d\mathbf{h} dx = \varpi_2(0), \quad a.s.,$$

for $t \in (0, T]$.

Applying the s-stage stochastic Runge–Kutta method with $s \geq 1$, which only relies on the increment of the $\mathbf{Q}$-Wiener process, to (4.25) in the temporal direction, we arrive at

$$U_{ni} = \mathbb{U}_n + \sum_{j=1}^{s} m_{ij} \left(AU_{nj} + \mathbb{F}\left(U_{nj}\right)\right) h + \sum_{j=1}^{s} \tilde{m}_{ij} \mathbb{G}\Delta_n W,$$

$$U_{n+1} = \mathbb{U}_n + \sum_{i=1}^{s} b_i \left(AU_{ni} + \mathbb{F}\left(U_{ni}\right)\right) h + \sum_{i=1}^{s} \tilde{b}_i \mathbb{G}\Delta_n W,$$

(4.26)

where $i = 1, \ldots, s$, and $n = 0, 1, \ldots, N - 1$. In what follows, $M = \left(m_{ij}\right)_{s \times s}$ and $\tilde{M} = \left(\tilde{m}_{ij}\right)_{s \times s}$ are $s \times s$ matrices of real elements while $b = (b_1, \ldots, b_s)^\top$ and $\tilde{b} = \left(\tilde{b}_1, \ldots, \tilde{b}_s\right)^\top$ are real vectors. To present more clearly the stochastic Runge–Kutta discretization (4.26), we give a concrete example as follows.

Example 4.2.3 (Midpoint Method) If we make use of the midpoint method to discretize the stochastic Maxwell equation (4.25), we obtain the recursion

$$U_{n1} = \mathbb{U}_n + \frac{1}{2} (AU_{n1} + \mathbb{F}(U_{n1})) h + \frac{1}{2} \mathbb{G}\Delta_n W,$$

$$\mathbb{U}_{n+1} = \mathbb{U}_n + (AU_{n1} + \mathbb{F}(U_{n1})) h + \mathbb{G}\Delta_n W$$

with $n = 0, 1, \ldots, N - 1$. Clearly, we derive $U_{n1} = (\mathbb{U}_{n+1} + \mathbb{U}_n)/2$ and hence we rewrite the midpoint method compactly as

$$\mathbb{U}_{n+1} = \mathbb{U}_n + \frac{1}{2} A \left(\mathbb{U}_{n+1} + \mathbb{U}_n\right) h + \mathbb{F}\left(\frac{\mathbb{U}_{n+1} + \mathbb{U}_n}{2}\right) h + \mathbb{G}\Delta_n W, \quad n = 0, 1, \ldots, N - 1.$$

By introducing operators $E_h^{\text{Mid}} := \left(I - \frac{h}{2}A\right)^{-1}\left(I + \frac{h}{2}A\right)$ and $T_h^{\text{Mid}} := \left(I - \frac{h}{2}A\right)^{-1}$, we have the equivalent form of the midpoint method as follows

$$U_{n+1} = E_h^{\text{Mid}}U_n + T_h^{\text{Mid}}\mathbb{F}\left(\frac{U_{n+1} + U_n}{2}\right)h + T_h^{\text{Mid}}G\Delta_n W, \quad n = 0, 1, \ldots, N-1.$$

Now we turn to the condition of symplecticity for (4.26).

Theorem 4.2.4 (See [52]) *Assume that the phase flow of the solution associated with (4.31) is twice continuously differentiable with respect to the initial value, and the coefficients m_{ij}, b_i of the stochastic Runge–Kutta discretization (4.26) meet*

$$b_i m_{ij} + b_j m_{ji} - b_i b_j = 0$$

for $i, j = 1, \ldots, s$, with $s \geq 1$. Then under the zero boundary condition, (4.26) is stochastically symplectic with the discrete symplectic conservation law, i.e.,

$$\varpi_2^{n+1} = \int_{\mathcal{O}} d\mathbf{E}_{n+1} \wedge d\mathbf{H}_{n+1}dx = \int_{\mathcal{O}} d\mathbf{E}_n \wedge d\mathbf{H}_n dx = \varpi_2^n, \quad a.s.,$$

where $n = 0, 1, \ldots, N-1$.

Proof Let $\tilde{H}_0(\mathbb{U}) = -\int_0^1 \langle \mathbb{U}, \tilde{\mathbb{F}}(\lambda\mathbb{U})\rangle_V d\lambda + C$, where $C \in \mathbb{R}$. Thus, it can be verified that $\mathbb{F} = \mathbb{J}\frac{\delta\tilde{H}_0}{\delta u}$. Based on these notations, we obtain

$$d\mathbb{U}_{n+1} \wedge \mathbb{J}d\mathbb{U}_{n+1} - d\mathbb{U}_n \wedge \mathbb{J}d\mathbb{U}_n$$

$$= h\sum_{i=1}^s b_i \left(d\mathbb{U}_n \wedge \mathbb{J}AdU_{ni} + AdU_{ni} \wedge \mathbb{J}d\mathbb{U}_n\right)$$

$$+ h\sum_{i=1}^s b_i \left(d\mathbb{U}_n \wedge \mathbb{J}^2\frac{\delta^2\tilde{H}_0}{\delta u^2}dU_{ni} + \mathbb{J}\frac{\delta^2\tilde{H}_0}{\delta u^2}dU_{ni} \wedge d\mathbb{U}_n\right)$$

$$+ h^2\sum_{i,j=1}^s b_i b_j \left(AdU_{ni} \wedge \mathbb{J}AdU_{nj} + \mathbb{J}\frac{\delta^2\tilde{H}_0}{\delta u^2}dU_{ni} \wedge \mathbb{J}^2\frac{\delta^2\tilde{H}_0}{\delta u^2}dU_{nj}\right)$$

$$+ h^2\sum_{i,j=1}^s b_i b_j \left(AdU_{ni} \wedge \mathbb{J}^2\frac{\delta^2\tilde{H}_0}{\delta u^2}dU_{nj} + \mathbb{J}\frac{\delta^2\tilde{H}_0}{\delta u^2}dU_{ni} \wedge \mathbb{J}AdU_{nj}\right).$$

Making use of the relation as follows

$$d\mathbb{U}_n = dU_{ni} - h\sum_{j=1}^s m_{ij}AdU_{nj} - h\sum_{j=1}^s m_{ij}\mathbb{J}\frac{\delta^2\tilde{H}_0}{\delta u^2}dU_{nj}$$

leads to

$$d\mathbb{U}_{n+1} \wedge \mathbb{J}d\mathbb{U}_{n+1} - d\mathbb{U}_n \wedge \mathbb{J}d\mathbb{U}_n$$

$$= h \sum_{i=1}^{s} b_i \left(d\mathbb{U}_{ni} \wedge \mathbb{J}Ad\mathbb{U}_{ni} + Ad\mathbb{U}_{ni} \wedge \mathbb{J}d\mathbb{U}_{ni} \right)$$

$$+ h \sum_{i=1}^{s} b_i \left(d\mathbb{U}_{ni} \wedge \mathbb{J}^2 \frac{\delta^2 \tilde{H}_0}{\delta u^2} d\mathbb{U}_{ni} + \mathbb{J} \frac{\delta^2 \tilde{H}_0}{\delta u^2} d\mathbb{U}_{ni} \wedge \mathbb{J}d\mathbb{U}_{ni} \right)$$

$$+ h^2 \sum_{i,j=1}^{s} \left(b_i b_j - b_i m_{ij} - b_j m_{ji} \right) \left(Ad\mathbb{U}_{ni} \wedge \mathbb{J}Ad\mathbb{U}_{nj} \right)$$

$$+ 2h^2 \sum_{i,j=1}^{s} \left(b_i b_j - b_i m_{ij} - b_j m_{ji} \right) \left(Ad\mathbb{U}_{ni} \wedge \mathbb{J}^2 \frac{\delta^2 \tilde{H}_0}{\delta u^2} d\mathbb{U}_{nj} \right)$$

$$+ h^2 \sum_{i,j=1}^{s} \left(b_i b_j - b_i m_{ij} - b_j m_{ji} \right) \left(\mathbb{J} \frac{\delta^2 \tilde{H}_0}{\delta u^2} d\mathbb{U}_{ni} \wedge \mathbb{J} \frac{\delta^2 \tilde{H}_0}{\delta u^2} d\mathbb{U}_{nj} \right).$$

According to the symmetry of $\frac{\delta^2 \tilde{H}_0}{\delta u^2}$ and the symplectic condition, we derive

$$d\mathbb{U}_{n+1} \wedge \mathbb{J}d\mathbb{U}_{n+1} - d\mathbb{U}_n \wedge \mathbb{J}d\mathbb{U}_n = \tau \sum_{i=1}^{s} b_i \left(d\mathbb{U}_{ni} \wedge \mathbb{J}Ad\mathbb{U}_{ni} + Ad\mathbb{U}_{ni} \wedge \mathbb{J}d\mathbb{U}_{ni} \right).$$

Taking advantage of $\mathbb{U} = \left(\mathbf{E}^{\top}, \mathbf{H}^{\top} \right)^{\top}$ yields

$$d\mathbf{E}_{n+1} \wedge d\mathbf{H}_{n+1} - d\mathbf{E}_n \wedge d\mathbf{H}_n$$

$$= \frac{1}{2} \left(d\mathbb{U}_{n+1} \wedge \mathbb{J}d\mathbb{U}_{n+1} - d\mathbb{U}_n \wedge \mathbb{J}d\mathbb{U}_n \right) = h \sum_{i=1}^{s} b_i \left(d\mathbb{U}_{ni} \wedge \mathbb{J}Ad\mathbb{U}_{ni} \right)$$

$$= -h \sum_{i=1}^{s} b_i \left(\mu^{-1} d\mathbf{E}_{ni} \wedge (\nabla \times d\mathbf{E}_{ni}) + \varepsilon^{-1} d\mathbf{H}_{ni} \wedge (\nabla \times d\mathbf{H}_{ni}) \right).$$

Thereby, under the zero boundary condition, it holds that

$$\int_{\mathscr{O}} d\mathbf{E}_{n+1} \wedge d\mathbf{H}_{n+1} d\mathbf{x} - \int_{\mathscr{O}} d\mathbf{E}_n \wedge d\mathbf{H}_n dx$$

$$= -h \sum_{i=1}^{s} b_i \int_{\mathscr{O}} \left(\mu^{-1} d\mathbf{E}_{ni} \wedge (\nabla \times d\mathbf{E}_{ni}) + \varepsilon^{-1} d\mathbf{H}_{ni} \wedge (\nabla \times d\mathbf{H}_{ni}) \right) dx = 0.$$

Thus, the proof is completed. $\qquad\qquad\qquad\qquad\qquad\qquad\qquad\qquad\qquad\qquad\square$

The following theorem states the error analysis in the mean-square sense of stochastic Runge–Kutta discretization (4.26).

Theorem 4.2.5 (See [52]) *Assume that* $\mathbb{F}\colon \mathrm{Dom}(A^2) \to \mathrm{Dom}(A^2)$ *is a* $\mathbf{C}^2$ *function with bounded derivatives up to order 2 and* $\mathbb{G} \in \mathscr{L}_2(U_0, \mathrm{Dom}(A^3))$. *In addition suppose that the initial data* $\mathbb{U}_0 \in \mathbf{L}^2(\Omega, \mathrm{Dom}(A^2))$ *and* $\sum_{i=1}^{s} b_i = \sum_{i=1}^{s} \tilde{b}_i = 1$. *Then there exists a positive constant* $C := C(T, \mathbb{U}(0), \mathbb{F}, \mathbb{G})$ *such that (4.26) applied to (4.25) satisfies*

$$\sup_{n \in \mathbb{Z}_{n+1}} \mathbb{E}\left[\|\mathbb{U}(t_n) - \mathbb{U}_n\|_V^2\right] \leq Ch^2.$$

From the above theorem, we achieve immediately that the mean-square convergence order of stochastic symplectic Runge–Kutta discretization applied to the stochastic Maxwell equation with additive noises is 1.

4.3 Stochastic Schrödinger Equation

The stochastic Schrödinger equations have been widely employed to model the propagation of nonlinear dispersive waves in random or inhomogenous media in quantum physics (see e.g., [19, 20, 221] and references therein). In this section, we study symplectic Runge–Kutta discretizations in the temporal direction and the probabilistic superiority of the spatial symplectic discretization based on the spectral Galerkin method for the stochastic Schrödinger equation.

4.3.1 Stochastic Symplectic Runge–Kutta Discretizations

Recently, much attention has been paid to the stochastic cubic Schrödinger equation (4.4) from both theoretical and numerical views (see e.g., [47, 74–76, 89] and references therein). For instance, [89] investigates the local and global existence of the solution, and the charge conservation law, i.e.,

$$\|u(t)\|_{\mathbf{L}^2(\mathcal{O},\mathbb{C})}^2 := \int_{\mathcal{O}} |u(t,x)|^2 dx = \|u(0)\|_{\mathbf{L}^2(\mathcal{O},\mathbb{C})}^2 \tag{4.27}$$

for all $t \in (0, T]$. Moreover, the equivalent equation of (4.4) is

$$\begin{cases} du(t) = \left(\mathrm{i}\Delta u(t) + \mathrm{i}\lambda |u(t)|^2 u(t) + \dfrac{1}{2} F_Q u(t)\right)dt + \mathrm{i}u(t)dW(t), \\ u(0) = u_0 \end{cases} \tag{4.28}$$

with $F_Q := \sum_{k=1}^{\infty} (\mathbf{Q}^{\frac{1}{2}} e_k)^2$ and $\mathscr{O} = [0, 1]$. Following the same arguments as in [74], we can obtain the uniform boundedness for moments of high order derivatives of the exact solution as follows.

Theorem 4.3.1 (See [74]) *Let $p = 2$ or $p \geq 4$ and $\mathbf{s} \geq 2$. Assume that*

$$u_0 \in \bigcap_{m=2}^{\mathbf{s}} \mathbf{L}^{3^{\mathbf{s}-m}p}(\Omega, \mathbf{H}^m(\mathscr{O}, \mathbb{C})) \cap \bigcap_{m=0}^{1} \mathbf{L}^{3^{\mathbf{s}-m-1}5p}(\Omega, \mathbf{H}^m(\mathscr{O}, \mathbb{C})) \qquad (4.29)$$

and $\mathbf{Q}^{\frac{1}{2}} \in \mathscr{L}_2(\mathbf{L}^2(\mathscr{O}, \mathbb{C}), \mathbf{H}^{\mathbf{s}}(\mathscr{O}, \mathbb{C}))$. Then there exists a constant $C := C(T, p, u_0, \mathbf{Q}) > 0$ such that the solution of (4.4) or (4.28) satisfies

$$\mathbb{E}\Big[\sup_{t \in [0,T]} \|u(t)\|_{\mathbf{H}^{\mathbf{s}}(\mathscr{O}, \mathbb{C})}^p \Big] \leq C. \qquad (4.30)$$

Remark 4.3.1 One can follow the same approach as in [89] to construct the unique local mild solution with continuous $\mathbf{H}^{\mathbf{s}}(\mathscr{O}, \mathbb{C})$-valued paths by a contraction argument, and then show that it is global by the a priori estimate (4.30) with $\mathbf{s} \geq 2$. According to [176], it can also be proved that this mild solution is also a strong one of (4.28).

Since the stochastic cubic Schrödinger equation (4.4) is an infinite-dimensional stochastic Hamiltonian system, it is significant to design a numerical method preserving the symplecticity. Below we shall study the stochastic symplectic Runge–Kutta discretization for (4.4). The s-stage Runge–Kutta method with $s \in \mathbb{N}_+$ applied to (4.4) in the temporal direction reads

$$p_i = P_n - h \sum_{j=1}^{s} a_{ij}^{(0)} \left(\Delta q_j + \lambda \left(p_j^2 + q_j^2 \right) q_j \right) - \Delta_n W \sum_{j=1}^{s} a_{ij}^{(1)} q_j,$$

$$q_i = Q_n + h \sum_{j=1}^{s} a_{ij}^{(0)} \left(\Delta p_j + \lambda \left(p_j^2 + q_j^2 \right) p_j \right) + \Delta_n W \sum_{j=1}^{s} a_{ij}^{(1)} p_j,$$

$$P_{n+1} = P_n - h \sum_{i=1}^{s} b_i^{(0)} \left(\Delta q_i + \lambda \left(p_i^2 + q_i^2 \right) q_i \right) - \Delta_n W \sum_{i=1}^{s} b_i^{(1)} q_i,$$

$$Q_{n+1} = Q_n + h \sum_{i=1}^{s} b_i^{(0)} \left(\Delta p_i + \lambda \left(p_i^2 + q_i^2 \right) p_i \right) + \Delta_n W \sum_{i=1}^{s} b_i^{(1)} p_i,$$

$$(4.31)$$

where $i = 1, \ldots, s$, $n = 0, 1, \ldots, N - 1$, and $\Delta_n W = W(t_{n+1}) - W(t_n)$ denotes the increment of the $\mathbf{Q}$-Wiener process. It is noticeable that two classes of parameters $\left\{a_{ij}^{(0)}, b_i^{(0)}\right\}_{i,j=1,\dots,s}$ and $\left\{a_{ij}^{(1)}, b_i^{(1)}\right\}_{i,j=1,\dots,s}$ may be different. Denoting $U_i := p_i + iq_i$ and $u_j := P_j + iQ_j$ for $i \in \{1,\dots,s\}$ and $j \in \{0, 1, \dots, N\}$, we obtain an approximation for the original system (4.4) as follows

$$U_i = u_n + ih \sum_{j=1}^{s} a_{ij}^{(0)} \left(\Delta U_j + \lambda |U_j|^2 U_j\right) + i\Delta_n W \sum_{j=1}^{s} a_{ij}^{(1)} U_j, \tag{4.32}$$

$$u_{n+1} = u_n + ih \sum_{i=1}^{s} b_i^{(0)} \left(\Delta U_i + \lambda |U_i|^2 U_i\right) + i\Delta_n W \sum_{i=1}^{s} b_i^{(1)} U_i \tag{4.33}$$

with $n = 0, 1, \dots, N - 1$.

Theorem 4.3.2 (See [47]) *Assume that the phase flow of the solution associated with (4.31) is twice continuously differentiable with respect to the initial value, and the coefficients satisfy*

$$b_i^{(0)} b_j^{(0)} = b_i^{(0)} a_{ij}^{(0)} + b_j^{(0)} a_{ji}^{(0)},$$

$$b_i^{(0)} b_j^{(1)} = b_i^{(0)} a_{ij}^{(1)} + b_j^{(1)} a_{ji}^{(0)},$$

$$b_i^{(1)} b_j^{(1)} = b_i^{(1)} a_{ij}^{(1)} + b_j^{(1)} a_{ji}^{(1)}$$

for all $i, j = 1, \dots, s$. Then the temporal semi-discretization (4.31) preserves both the symplecticity

$$\int_{\mathcal{O}} dP_{n+1} \wedge dQ_{n+1} dx = \int_{\mathcal{O}} dP_n \wedge dQ_n dx \quad a.s.,$$

and the discrete charge conservation law

$$\|u_{n+1}\|_{L^2(\mathcal{O},\mathbb{C})}^2 = \|u_n\|_{L^2(\mathcal{O},\mathbb{C})}^2,$$

where $n = 0, 1, \dots, N - 1$.

Proof The symplecticity of (4.31) can be received by similar arguments as in the proof of Theorem 4.2.4. For the preservation of charge conservation law of (4.31), we first denote $h_i = \Delta U_i + \lambda |U_i|^2 U_i$ and $\tilde{h}_i = U_i$ for $i \in \{1, \dots, s\}$. Multiplying (4.33) with $\bar{u}_{n+1} + \bar{u}_n$ and taking the real part, we derive

$$|u_{n+1}|^2 = |u_n|^2 + \Re\left(ih \sum_{i=1}^{s} b_i^{(0)} h_i (\bar{u}_{n+1} + \bar{u}_n) + i\Delta_n W \sum_{i=1}^{s} b_i^{(1)} \tilde{h}_i (\bar{u}_{n+1} + \bar{u}_n)\right).$$

where $n = 0, 1, \ldots, N - 1$. From (4.32) it follows that

$$u_n = U_i + \mathrm{i}h \sum_{j=1}^{s} a_{ij}^{(0)} h_j + \mathrm{i} \Delta_n W \sum_{j=1}^{s} a_{ij}^{(1)} \tilde{h}_j$$

with $i = 1, \ldots, s$, which leads to

$$\bar{u}_{n+1} + \bar{u}_n = 2\bar{U}_i + \mathrm{i}2h \sum_{j=1}^{s} a_{ij}^{(0)} \bar{h}_j + \mathrm{i}2 \Delta_n W \sum_{j=1}^{s} a_{ij}^{(1)} \bar{\tilde{h}}_j + \mathrm{i}h \sum_{j=1}^{s} b_j^{(0)} \bar{h}_j + \mathrm{i} \Delta_n W \sum_{j=1}^{s} b_j^{(1)} \bar{\tilde{h}}_j.$$

As a consequence,

$$\Re \left(\mathrm{i}h \sum_{i=1}^{s} b_i^{(0)} h_i (\bar{u}_{n+1} + \bar{u}_n) + \mathrm{i} \Delta_n W \sum_{i=1}^{s} b_i^{(1)} \tilde{h}_i (\bar{u}_{n+1} + \bar{u}_n) \right)$$

$$= -\Im \left(2h \sum_{i=1}^{s} b_i^{(0)} \Delta U_i \bar{U}_i \right) - \Im \left(2 \Delta_n W \sum_{i=1}^{s} b_i^{(1)} U_i \bar{U}_i \right)$$

$$- h^2 \sum_{ij=1}^{s} \left(b_i^{(0)} b_j^{(0)} - b_i^{(0)} a_{ij}^{(0)} - b_j^{(0)} a_{ji}^{(0)} \right) \Re \left(h_i \bar{h}_j \right)$$

$$- 2h \Delta_n W \sum_{ij=1}^{s} \left(b_i^{(0)} b_j^{(1)} - b_i^{(0)} a_{ij}^{(1)} - b_j^{(1)} a_{ji}^{(0)} \right) \Re \left(h_i \bar{\tilde{h}}_j \right)$$

$$- (\Delta_n W)^2 \sum_{ij=1}^{s} \left(b_i^{(1)} b_j^{(1)} - b_i^{(1)} a_{ij}^{(1)} - b_j^{(1)} a_{ji}^{(1)} \right) \Re \left(\tilde{h}_i \bar{\tilde{h}}_j \right).$$

Integrating the above equation with respect to x and taking advantage of the symplectic condition yield the preservation of charge conservation law. □

Remark 4.3.2 For the general stochastic Runge–Kutta semi-discretizations in the temporal direction of the stochastic cubic Schrödinger equation (4.4), the mean-square convergence order is still an open problem as far as we have known.

4.3.2 Approximating Large Deviation Rate Functions via Spectral Galerkin Method

In this subsection, we consider the following stochastic linear Schrödinger equation

$$\begin{cases} du(t) = \mathrm{i} \Delta u(t) dt + \mathrm{i} \alpha dW(t), & t \in (0, T], \\ u(0) = u_0 \in \mathbf{H}_0^1([0, \pi], \mathbb{C}), \end{cases} \tag{4.34}$$

where $\alpha > 0$, Δ is the Laplace operator with the Dirichlet boundary condition, and

$$W(t) = \sum_{k \in \mathbb{N}_+} \sqrt{\eta_k} \beta_k(t) e_k$$

is an $\mathbf{L}^2([0, \pi], \mathbb{R})$-valued $\mathbf{Q}$-Wiener process with $e_k(x) = \sqrt{\frac{2}{\pi}} \sin(kx)$. Here, $\mathbf{Q}$ is a nonnegative symmetric operator on $\mathbf{L}^2([0, \pi], \mathbb{R})$ with $\mathbf{Q}e_k = \eta_k e_k$ for some non-increasing sequence $\{\eta_k\}_{k \in \mathbb{N}_+}$, and $\{e_k\}_{k \in \mathbb{N}_+}$ also forms an orthonormal basis of $\mathbf{L}^2([0, \pi], \mathbb{R})$. Noting that $\Delta e_k = -k^2 e_k$, $k \in \mathbb{N}_+$, we have $\Delta \mathbf{Q} = \mathbf{Q}\Delta$ and $\|\Delta^{-1}\|_{\mathscr{L}(\mathbf{L}^2([0,\pi],\mathbb{C}))} = 1$. Assume that $\mathbf{Q}^{\frac{1}{2}} \in \mathscr{L}_2(\mathbf{L}^2([0, \pi], \mathbb{R}), \mathbf{H}^1([0, \pi], \mathbb{R}))$, then (4.34) admits a unique mild solution in $\mathbf{H}_0^1([0, \pi], \mathbb{C})$ (see [11]):

$$u(t) = E(t)u_0 - \alpha \int_0^t \sin((t-s)\Delta) dW(s) + i\alpha \int_0^t \cos((t-s)\Delta) dW(s)$$

$$=: E(t)u_0 - \alpha W_{\sin}(t) + i\alpha W_{\cos}(t),$$

where $t \in (0, T]$, $E(t) = \exp(it\Delta)$ and

$$W_{\sin}(t) \sim \mathscr{N}\left(0, \int_0^t \sin^2((t-s)\Delta)\mathbf{Q} ds\right),$$

$$W_{\cos}(t) \sim \mathscr{N}\left(0, \int_0^t \cos^2((t-s)\Delta)\mathbf{Q} ds\right).$$

Based on Proposition 4.1.1, it is known that

$$\mathbb{COR}\,(W_{\sin}(t), W_{\cos}(t)) = \int_0^t \sin((t-s)\Delta)\cos((t-s)\Delta)\mathbf{Q} ds, \quad t > 0.$$

Denote the real inner product by $\langle f, g \rangle_{\mathbb{R}} = \mathfrak{R}\big(\int_0^\pi f(x)\bar{g}(x)dx\big)$, where $f, g \in \mathbf{L}^2([0, \pi], \mathbb{C})$. For $\lambda = \mathfrak{R}(\lambda) + i\mathfrak{I}(\lambda) \in \mathbf{L}^2([0, \pi], \mathbb{C})$, we obtain

$$\langle u(t), \lambda \rangle_{\mathbb{R}} = \langle E(t)u_0, \lambda \rangle_{\mathbb{R}} - \alpha\langle W_{\sin}(t), \mathfrak{R}(\lambda) \rangle_{\mathbb{R}} + \alpha\langle W_{\cos}(t), \mathfrak{I}(\lambda) \rangle_{\mathbb{R}}. \tag{4.35}$$

Hence, $\mathbb{E}\big[\langle u(t), \lambda \rangle_{\mathbb{R}}\big] = \langle E(t)u_0, \lambda \rangle_{\mathbb{R}}$. The fact that $\Delta\mathbf{Q} = \mathbf{Q}\Delta$ leads to

$$\mathbb{VAR}\,[\langle u(t), \lambda \rangle_{\mathbb{R}}] = \alpha^2 \big\langle \int_0^t \sin^2((t-s)\Delta)\mathbf{Q} ds \mathfrak{R}(\lambda), \mathfrak{R}(\lambda) \big\rangle_{\mathbb{R}}$$

$$+ \alpha^2 \big\langle \int_0^t \cos^2((t-s)\Delta)\mathbf{Q} ds \mathfrak{I}(\lambda), \mathfrak{I}(\lambda) \big\rangle_{\mathbb{R}}$$

$$- 2\alpha^2 \big\langle \int_0^t \sin((t-s)\Delta)\cos((t-s)\Delta)\mathbf{Q} ds \mathfrak{R}(\lambda), \mathfrak{I}(\lambda) \big\rangle_{\mathbb{R}}.$$

According to

$$\int_0^t \sin^2((t-s)\varDelta)ds = \frac{1}{2}\int_0^t (I - \cos(2(t-s)\varDelta))\,ds = \frac{tI}{2} - \frac{\varDelta^{-1}}{4}\sin(2t\varDelta),$$

$$\int_0^t \cos^2((t-s)\varDelta)ds = \frac{1}{2}\int_0^t (I + \cos(2(t-s)\varDelta))\,ds = \frac{tI}{2} + \frac{\varDelta^{-1}}{4}\sin(2t\varDelta),$$

$$\int_0^t \sin(2(t-s)\varDelta)ds = \frac{\varDelta^{-1}}{2}(I - \cos(2t\varDelta)),$$

we achieve

$$\mathrm{VAR}\,[\langle u(t), \lambda \rangle_{\mathbb{R}}]$$

$$= \frac{t\alpha^2}{2}(\langle \mathbf{Q}\Re(\lambda), \Re(\lambda)\rangle_{\mathbb{R}} + \langle \mathbf{Q}\Im(\lambda), \Im(\lambda)\rangle_{\mathbb{R}})$$

$$- \frac{\alpha^2}{4}\Big(\langle \varDelta^{-1}\sin(2t\varDelta)\mathbf{Q}\Re(\lambda), \Re(\lambda)\rangle_{\mathbb{R}} - \langle \varDelta^{-1}\sin(2t\varDelta)\mathbf{Q}\Im(\lambda), \Im(\lambda)\rangle_{\mathbb{R}}\Big)$$

$$- \frac{\alpha^2}{2}\langle \varDelta^{-1}(I - \cos(2t\varDelta))\mathbf{Q}\Re(\lambda), \Im(\lambda)\rangle_{\mathbb{R}}.$$

When $\alpha = 0$, the charge $\|u\|_{\mathbf{L}^2([0,\pi],\mathbb{C})}^2 = \int_0^\pi |u|^2 dx$ of (4.34) is conservative. However, in the stochastic setting, $\|u\|_{\mathbf{L}^2([0,\pi],\mathbb{C})}^2$ grows linearly in the mean sense, i.e.,

$$\mathbb{E}\left[\|u(t)\|_{\mathbf{L}^2([0,\pi],\mathbb{C})}^2\right] = \|u_0\|_{\mathbf{L}^2([0,\pi],\mathbb{C})}^2 + \alpha^2\mathrm{Tr}(\mathbf{Q})t, \quad t \in [0, T].$$

By the Markov inequality, one has that for each $R > 0$ and a sufficiently large T,

$$\mathbb{P}\left(\|u(T)\|_{\mathbf{L}^2([0,\pi],\mathbb{C})}^2 \geq T^2 R^2\right) \leq \frac{\mathbb{E}[\|u(T)\|_{\mathbf{L}^2([0,\pi],\mathbb{C})}^2]}{T^2 R^2} \leq \frac{C}{T} \tag{4.36}$$

for some constant $C := C(R, \alpha)$. Let $B_T := \frac{u(T)}{T}$. A direct calculation leads to

$$\lim_{T \to \infty} \mathbb{P}\left(\|B(T)\|_{\mathbf{L}^2([0,\pi],\mathbb{C})} \geq R\right) = 0. \tag{4.37}$$

To characterize the speed of convergence or give an exponential tail estimate, we turn to the LDP of $\{B_T\}_{T>0}$ on $\mathbf{L}^2([0, \pi], \mathbb{C})$ by using Theorem A.2.2 and the concept of the reproducing kernel Hilbert space.

Denote by μ a centered Gaussian measure on a separable Banach space E. For any $\varphi \in E^*$, it can be identified with an element of the Hilbert space $\mathbf{L}^2(\mu) :=$

$\mathbf{L}^2(\mathrm{E}, \mathscr{B}(\mathrm{E}), \mu; \mathbb{R})$. Let $\overline{\mathrm{E}^*} := \overline{\mathrm{E}^*}^{\mathbf{L}^2(\mu)}$ be the closure of E^* in $\mathbf{L}^2(\mu)$. Define a mapping $J : \overline{\mathrm{E}^*} \to \mathrm{E}$ by setting

$$J(\varphi) = \int_{\mathrm{E}} x\varphi(x)\mu(dx) \quad \forall \varphi \in \overline{\mathrm{E}^*}.$$

Then the image $\mathscr{H}_\mu = J(\overline{\mathrm{E}^*})$ of J is the *reproducing kernel Hilbert space* of μ with the scalar product

$$\langle J(\varphi), J(\psi) \rangle_{\mathscr{H}_\mu} = \int_{\mathrm{E}} \varphi(x)\psi(x)\mu(dx).$$

Further, if $\mu = \mathscr{N}(0, \widetilde{\mathbf{Q}})$ is a Gaussian measure on some Hilbert space $\mathbb{H}$, where $\widetilde{\mathbf{Q}}$ is a nonnegative symmetric operator with finite trace, then the reproducing kernel Hilbert space $\mathscr{H}_\mu$ of μ is $\mathscr{H}_\mu = \widetilde{\mathbf{Q}}^{\frac{1}{2}}(\mathbb{H})$ with the norm $\|x\|_{\mathscr{H}_\mu} = \|\widetilde{\mathbf{Q}}^{-\frac{1}{2}}x\|_{\mathbb{H}}$. We refer to [85, Section 2.2.2] for more details of the reproducing kernel Hilbert space.

Theorem 4.3.3 (See [56]) $\{B_T\}_{T>0}$ *satisfies an LDP on* $\mathbf{L}^2([0, \pi], \mathbb{C})$ *with the good rate function*

$$\mathbf{I}(x) = \begin{cases} \frac{1}{\alpha^2}\left\|\mathbf{Q}^{-\frac{1}{2}}x\right\|^2_{\mathbf{L}^2([0,\pi],\mathbb{C})}, & \text{if } x \in \mathbf{Q}^{\frac{1}{2}}(\mathbf{L}^2([0, \pi], \mathbb{C})); \\ +\infty, & \text{otherwise}, \end{cases}$$

where $\mathbf{Q}^{-\frac{1}{2}}$ *is the pseudo inverse of* $\mathbf{Q}^{\frac{1}{2}}$.

Proof The proof is divided into three steps.

Step 1: The Logarithmic Moment Generating Function of $\{B_T\}_{T>0}$ For any $\lambda' : \mathbf{L}^2([0, \pi], \mathbb{C}) \to \mathbb{R}$, by the Riesz representation theorem, there is a unique $\lambda \in \mathbf{L}^2([0, \pi], \mathbb{C})$ such that $\lambda'(x) = \langle x, \lambda \rangle_{\mathbb{R}}$, $x \in \mathbf{L}^2([0, \pi], \mathbb{C})$. Based on the fact that $\langle u(t), \lambda \rangle_{\mathbb{R}}$ is Gaussian, we have

$$\Lambda(\lambda') = \lim_{T \to \infty} \frac{1}{T} \log(\mathbb{E}[\exp(T \langle B_T, \lambda \rangle_{\mathbb{R}})])$$

$$= \lim_{T \to \infty} \frac{1}{T}\left(\mathbb{E}[\langle u(T), \lambda \rangle_{\mathbb{R}}] + \frac{1}{2}\mathbb{VAR}[\langle u(T), \lambda \rangle_{\mathbb{R}}] \right)$$

$$= \frac{\alpha^2}{4}\left\|\mathbf{Q}^{\frac{1}{2}}\lambda\right\|^2_{\mathbf{L}^2([0,\pi],\mathbb{C})}, \tag{4.38}$$

where we have used $\|\sin(t\Delta)\|_{\mathscr{L}(\mathbf{L}^2([0,\pi],\mathbb{C}))} \leq 1$, $\|\cos(t\Delta)\|_{\mathscr{L}(\mathbf{L}^2([0,\pi],\mathbb{C}))} \leq 1$ and $\|\Delta^{-1}\|_{\mathscr{L}(\mathbf{L}^2([0,\pi],\mathbb{C}))} = 1$. Thus, $\Lambda(\lambda')$ is finite valued and Fréchet differentiable, and its Fréchet derivative is $D\Lambda(\lambda')(\cdot) = \frac{\alpha^2}{2}\langle \mathbf{Q}\lambda, \cdot \rangle_{\mathbb{R}}$.

Step 2: Exponential Tightness of $\{B_T\}_{T>0}$ To obtain the exponential tightness of $\{B_T\}_{T>0}$ (see Definition A.2.4), it suffices to show that there exists a family of pre-compact sets $\{K_L\}_{L>0}$ such that

$$\lim_{L\to\infty} \limsup_{T\to\infty} \frac{1}{T} \log(\mathbb{P}\left(B_T \in K_L^c\right)) = -\infty. \tag{4.39}$$

Denoting $K_L := \{f \in \mathbf{H}^1([0,\pi],\mathbb{C}) \mid \|f\|_{\mathbf{H}^1([0,\pi],\mathbb{C})} \le L\}$, which is a pre-compact set of $\mathbf{L}^2([0,\pi],\mathbb{C})$, we arrive at

$$\mathbb{P}\left(B_T \in K_L^c\right) = \mathbb{P}\left(\|u(T)\|_{\mathbf{H}^1([0,\pi],\mathbb{C})} > LT\right)$$

$$\le \mathbb{P}\left(\|E(T)u_0\|_{\mathbf{H}^1([0,\pi],\mathbb{C})} > \frac{TL}{3}\right) + \mathbb{P}\left(\alpha\|W_{\sin}(T)\|_{\mathbf{H}^1([0,\pi],\mathbb{R})} > \frac{TL}{3}\right)$$

$$+ \mathbb{P}\left(\alpha\|W_{\cos}(T)\|_{\mathbf{H}^1([0,\pi],\mathbb{R})} > \frac{TL}{3}\right) =: \mathrm{I} + \mathrm{II} + \mathrm{III}. \tag{4.40}$$

Since I is 0 for sufficiently large T, we only need to estimate II and III in (4.40).

Denoting $m_k = \frac{1}{\sqrt{1+k^2}}e_k$ for $\mathbb{N}_+$, one obtains that $\{m_k\}_{k\in\mathbb{N}}$ is an orthonormal basis of $\mathbf{H}^1([0,\pi],\mathbb{R})$. Since $\mathbf{Q}^{\frac{1}{2}} \in \mathscr{L}_2(\mathbf{L}^2([0,\pi],\mathbb{R}), \mathbf{H}^1([0,\pi],\mathbb{R}))$, we have

$$\sum_{k\in\mathbb{N}_+} \eta_k(1+k^2) = \|\mathbf{Q}^{\frac{1}{2}}\|^2_{\mathscr{L}_2(\mathbf{L}^2([0,\pi],\mathbb{R}),\mathbf{H}^1([0,\pi],\mathbb{R}))} < \infty, \tag{4.41}$$

which implies that $\mathbb{Q} \in \mathscr{L}(\mathbf{H}^1([0,\pi],\mathbb{R}))$ defined by $\mathbb{Q}m_k = \eta_k(1+k^2)m_k$ for $k \in \mathbb{N}_+$ is a finite trace operator. Moreover, $\mathbb{Q}$ is a nonnegative symmetric operator on $\mathbf{H}^1([0,\pi],\mathbb{R})$. Notice that for any $t \ge 0$,

$$W(t) = \sum_{k\in\mathbb{N}_+} \sqrt{\eta_k}e_k\beta_k(t) = \sum_{k\in\mathbb{N}_+} \sqrt{\eta_k(1+k^2)}m_k\beta_k(t) = \sum_{k\in\mathbb{N}_+} \mathbb{Q}^{\frac{1}{2}}m_k\beta_k(t).$$

Therefore, the $\mathbf{L}^2([0,\pi],\mathbb{R})$-valued $\mathbf{Q}$-Wiener process $W(\cdot)$ can be considered as an $\mathbf{H}^1([0,\pi],\mathbb{R})$-valued $\mathbb{Q}$-Wiener process. Then it follows Proposition 4.1.1 that

$$W_{\sin}(T) \sim \mathscr{N}\left(0, \int_0^T \sin^2((t-s)\varDelta)\mathbb{Q}ds\right) = \mathscr{N}\left(0, \left(\frac{TI}{2} - \frac{\varDelta^{-1}\sin(2T\varDelta)}{4}\right)\mathbb{Q}\right)$$

on $\mathbf{H}^1([0,\pi],\mathbb{R})$, which implies

$$\frac{W_{\sin}(T)}{\sqrt{T}} \sim \mathscr{N}\left(0, \left(\frac{I}{2} - \frac{\varDelta^{-1}\sin(2T\varDelta)}{4T}\right)\mathbb{Q}\right). \tag{4.42}$$

By the Markov inequality, for each $\varepsilon > 0$,

$$\mathbb{P}\big(\alpha\|W_{\sin}(T)\|_{\mathbf{H}^1([0,\pi],\mathbb{R})} > \frac{TL}{3}\big) \le \exp\Big(-\frac{\varepsilon TL^2}{9\alpha^2}\Big)\mathbb{E}\Big[\exp\Big(\varepsilon \Big\|\frac{W_{\sin}(T)}{\sqrt{T}}\Big\|^2_{\mathbf{H}^1([0,\pi],\mathbb{R})}\Big)\Big].$$

Notice that the kth eigenvalue of $\big(\frac{I}{2} - \frac{\Delta^{-1}\sin(2T\Delta)}{4T}\big)\mathbb{Q}$ with $k \in \mathbb{N}_+$ satisfies

$$\nu_k\Big(\Big(\frac{I}{2} - \frac{\Delta^{-1}\sin(2T\Delta)}{4T}\Big)\mathbb{Q}\Big) = \Big(\frac{1}{2} - \frac{\sin(2Tk^2)}{4Tk^2}\Big)\eta_k(1+k^2) < C(\mathbb{Q}),$$

where $C(\mathbb{Q})$ is a positive constant. It follows from Proposition 4.1.2 that for each $\varepsilon \in \big(0, \frac{1}{2C(\mathbb{Q})}\big)$,

$$\mathbb{E}\Big[\exp\Big(\varepsilon\Big\|\frac{W_{\sin}(T)}{\sqrt{T}}\Big\|^2_{\mathbf{H}^1([0,\pi],\mathbb{R})}\Big)\Big] < (\det(I - 2\varepsilon\mathbb{Q}))^{-\frac{1}{2}} = C(\varepsilon, \mathbb{Q}),$$

which yields

$$\limsup_{T\to\infty}\frac{1}{T}\log\Big(\mathbb{P}\Big(\alpha\|W_{\sin}(T)\|_{\mathbf{H}^1([0,\pi],\mathbb{R})} > \frac{TL}{3}\Big)\Big) \le -\frac{\varepsilon L^2}{9\alpha^2}.$$

Analogously, we arrive that for each $\varepsilon \in \big(0, \frac{1}{2C(\mathbb{Q})}\big)$,

$$\limsup_{T\to\infty}\frac{1}{T}\log\Big(\mathbb{P}\Big(\alpha\|W_{\cos}(T)\|_{\mathbf{H}^1([0,\pi],\mathbb{R})} > \frac{TL}{3}\Big)\Big) \le -\frac{\varepsilon L^2}{9\alpha^2}.$$

Making use of Proposition A.2.1, we obtain

$$\limsup_{T\to\infty}\frac{1}{T}\log(\mathbb{P}\big(B_T \in K_L^c\big)) \le -\frac{\varepsilon L^2}{9\alpha^2}.$$

As a consequence, we have (4.39) which means the exponential tightness of $\{B_T\}_{T>0}$. According to Theorem A.2.2, $\{B_T\}_{T>0}$ satisfies an LDP on $\mathbf{L}^2([0,\pi],\mathbb{C})$ with the good rate function Λ^*.

Step 3: The Explicit Expression of Λ^* As we known, $\mu := \mathcal{N}(0, \mathbf{Q})$ is a centered Gaussian measure on $\mathbf{L}^2(\mu)$. Define $\mathbf{J}\colon \overline{(\mathbf{L}^2([0,\pi],\mathbb{C}))^*}^{\mathbf{L}^2(\mu)} \to \mathbf{L}^2([0,\pi],\mathbb{C})$ by $\mathbf{J}(\tilde{h}) = \int_{\mathbf{L}^2([0,\pi],\mathbb{C})} z\tilde{h}(z)\mu(dz)$. Then the associated reproducing kernel Hilbert space is $\mathscr{H}_\mu = \mathbf{J}\Big(\overline{(\mathbf{L}^2([0,\pi],\mathbb{C}))^*}^{\mathbf{L}^2(\mu)}\Big) = \mathbf{Q}^{\frac{1}{2}}(\mathbf{L}^2([0,\pi],\mathbb{C}))$, and we obtain

$$\int_{\mathbf{L}^2([0,\pi],\mathbb{C})} \langle\lambda, x\rangle^2_\mathbb{R}\mu(dx) = \big\|\mathbf{Q}^{\frac{1}{2}}\lambda\big\|^2_{\mathbf{L}^2([0,\pi],\mathbb{C})}.$$

Thus, $\Lambda(\lambda') = \frac{\alpha^2}{4}\left\|\mathbf{Q}^{\frac{1}{2}}\lambda\right\|^2_{L^2([0,\pi],\mathbb{C})} = \frac{\alpha^2}{4}\left\|\lambda'\right\|^2_{L^2(\mu)}$. For a given $x \in \mathbf{L}^2([0,\pi],\mathbb{C})$, if

$$\Lambda^*(x) = \sup_{\lambda' \in (\mathbf{L}^2([0,\pi],\mathbb{C}))^*} \{\lambda'(x) - \Lambda(\lambda')\} < +\infty,$$

then there exists a positive constant $C(x) < +\infty$ such that $\lambda'(x) \leq \frac{\alpha^2}{4}\left\|\lambda'\right\|^2_{L^2(\mu)} + C(x)$. Define the linear functional x^{**} on

$$\left((\mathbf{L}^2([0,\pi],\mathbb{C}))^*, \|\cdot\|_{L^2(\mu)}\right) \subseteq \overline{(\mathbf{L}^2([0,\pi],\mathbb{C}))^*}^{L^2(\mu)}$$

by setting $x^{**}(\lambda') = \lambda'(x)$ for every $\lambda' \in (\mathbf{L}^2([0,\pi],\mathbb{C}))^*$. Then we obtain

$$\sup_{\lambda' \in (\mathbf{L}^2([0,\pi],\mathbb{C}))^*, \|\lambda'\|_{L^2(\mu)} \leq 1} x^{**}(\lambda') \leq \frac{\alpha^2}{4} + C(x).$$

It implies that x^{**} is a bounded linear functional on $\left((\mathbf{L}^2([0,\pi],\mathbb{C}))^*, \|\cdot\|_{L^2(\mu)}\right)$. From the Hahn–Banach theorem and the fact that $\left((\mathbf{L}^2([0,\pi],\mathbb{C}))^*, \|\cdot\|_{L^2(\mu)}\right)$ is dense in $\overline{(\mathbf{L}^2([0,\pi],\mathbb{C}))^*}^{L^2(\mu)}$ it follows that x^{**} can be uniquely extended to $\overline{(\mathbf{L}^2([0,\pi],\mathbb{C}))^*}^{L^2(\mu)}$. For each $\lambda' \in \overline{(\mathbf{L}^2([0,\pi],\mathbb{C}))^*}^{L^2(\mu)}$, take $\lambda'_n \in (\mathbf{L}^2([0,\pi],\mathbb{C}))^*$ such that $\lambda'_n \to \lambda'$ in the $\|\cdot\|_{L^2(\mu)}$-norm and then the extended functional is $x^{**}(\lambda') = \lim_{n\to\infty} x^{**}(\lambda'_n)$. Denote the extended functional still by x^{**}. As a result, for every $x \in \mathbf{L}^2([0,\pi],\mathbb{C})$ satisfying $\Lambda^*(x) < +\infty$, we arrive at a bounded linear functional on $\overline{(\mathbf{L}^2([0,\pi],\mathbb{C}))^*}^{L^2(\mu)}$ such that $x^{**}(\lambda') = \lambda'(x)$ for each $\lambda' \in (\mathbf{L}^2([0,\pi],\mathbb{C}))^*$. By the Riesz representation theorem, there is some $\tilde{h} \in \overline{(\mathbf{L}^2([0,\pi],\mathbb{C}))^*}^{L^2(\mu)}$ such that $x^{**}(\lambda') = \langle\lambda', \tilde{h}\rangle_{L^2(\mu)}$ for every $\lambda' \in \overline{(\mathbf{L}^2([0,\pi],\mathbb{C}))^*}^{L^2(\mu)}$. Therefore, $\lambda'(x) = \langle\lambda', \tilde{h}\rangle_{L^2(\mu)}$ for each $\lambda' \in (\mathbf{L}^2([0,\pi],\mathbb{C}))^*$. Further, we obtain

$$\lambda'(x) = \int_{L^2([0,\pi],\mathbb{C})} \tilde{h}(z)\lambda'(z)\mu(dz) = \lambda'\left(\int_{L^2([0,\pi],\mathbb{C})} z\tilde{h}(z)\mu(dz)\right) = \lambda'(\mathbf{J}(\tilde{h}))$$

for $\lambda' \in (\mathbf{L}^2([0,\pi],\mathbb{C}))^*$. Since λ' is arbitrary, then $x = \mathbf{J}(\tilde{h})$. As a consequence, $\Lambda^*(x) < +\infty$ means that $x \in \mathscr{H}_\mu = \mathbf{J}\left(\overline{(\mathbf{L}^2([0,\pi],\mathbb{C}))^*}^{L^2(\mu)}\right) = Im(\mathbf{Q}^{\frac{1}{2}})$, where $Im(\mathbf{Q}^{\frac{1}{2}})$ is the image of $\mathbf{Q}^{\frac{1}{2}}$.

On the other hand, if $x = \mathbf{J}(\tilde{h}) \in \mathbf{L}^2([0,\pi],\mathbb{C})$ for some $\tilde{h} \in \overline{(\mathbf{L}^2([0,\pi],\mathbb{C}))^*}^{\mathbf{L}^2(\mu)}$, then

$$\Lambda^*(x) = \Lambda^*(\mathbf{J}(\tilde{h})) = \sup_{\lambda' \in (\mathbf{L}^2([0,\pi],\mathbb{C}))^*} \left\{ \langle \lambda', \tilde{h} \rangle_{\mathbf{L}^2(\mu)} - \frac{\alpha^2}{4} \left\| \lambda' \right\|^2_{\mathbf{L}^2(\mu)} \right\}.$$

Noticing that $\langle \lambda', \tilde{h} \rangle_{\mathbf{L}^2(\mu)} - \frac{\alpha^2}{4} \left\| \lambda' \right\|^2_{\mathbf{L}^2(\mu)}$ is continuous with respect to λ' on $\|\cdot\|_{\mathbf{L}^2(\mu)}$, and that the space $\left((\mathbf{L}^2([0,\pi],\mathbb{C}))^*, \|\cdot\|_{\mathbf{L}^2(\mu)} \right)$ is dense in $\overline{(\mathbf{L}^2([0,\pi],\mathbb{C}))^*}^{\mathbf{L}^2(\mu)}$, we have

$$\Lambda^*(x) \leq \sup_{g \in \overline{(\mathbf{L}^2([0,\pi],\mathbb{C}))^*}^{\mathbf{L}^2(\mu)}} \left\{ \frac{1}{2} \left(\frac{\alpha^2}{2} \|g\|^2_{\mathbf{L}^2(\mu)} + \frac{2}{\alpha^2} \left\| \tilde{h} \right\|^2_{\mathbf{L}^2(\mu)} \right) - \frac{\alpha^2}{4} \|g\|^2_{\mathbf{L}^2(\mu)} \right\}$$

$$= \frac{1}{\alpha^2} \left\| \tilde{h} \right\|^2_{\mathbf{L}^2(\mu)}.$$

Taking $g = \frac{2}{\alpha^2}\tilde{h}$ leads to $\Lambda^*(x) \geq \frac{1}{\alpha^2} \left\| \tilde{h} \right\|^2_{\mathbf{L}^2(\mu)}$. Hence, we obtain

$$\Lambda^*(x) = \frac{1}{\alpha^2} \left\| \tilde{h} \right\|^2_{\mathbf{L}^2(\mu)} = \frac{1}{\alpha^2} \|x\|^2_{\mathscr{H}_\mu} = \frac{1}{\alpha^2} \left\| \mathbf{Q}^{-\frac{1}{2}} x \right\|^2_{\mathbf{L}^2([0,\pi],\mathbb{C})}, \tag{4.43}$$

which completes this proof. □

Remark 4.3.3 (See [56]) For a sufficiently large $L > 0$, one can always find R and T such that $T^2 R^2 \leq L$. Then

$$\mathbb{P}\left(\|u(T)\|^2_{\mathbf{L}^2([0,\pi],\mathbb{C})} \geq L \right) \leq \exp\left(-T \left(\inf_{y \geq R} \tilde{\mathbf{I}}(y) - \varepsilon \right) \right),$$

where $\tilde{\mathbf{I}}(y) = \frac{1}{\alpha^2} \inf_{z \in \mathbf{L}^2([0,\pi],\mathbb{C}), \|\mathbf{Q}^{\frac{1}{2}}z\|_{\mathbf{L}^2([0,\pi],\mathbb{C})}=y} \|z\|^2_{\mathbf{L}^2([0,\pi],\mathbb{C})}$. This implies that the probability of the tail event of the mass of (4.34) is exponentially small for a sufficiently large time.

LDP of $\{B_T^M\}_{T>0}$ for Spectral Galerkin Method For $M \in \mathbb{N}_+$, we define the finite-dimensional subspace $(U_M, \langle \cdot, \cdot \rangle_{\mathbb{C}})$ of $(\mathbf{L}^2([0,\pi],\mathbb{C}), \langle \cdot, \cdot \rangle_{\mathbb{C}})$ with $U_M := \mathrm{span}\{e_1, \ldots, e_M\}$ and $\langle f, g \rangle_{\mathbb{C}} = \int_0^\pi f(x)\bar{g}(x)dx$ for $f, g \in \mathbf{L}^2([0,\pi],\mathbb{C})$, and the projection operator $P_M : \mathbf{L}^2([0,\pi],\mathbb{C}) \to U_M$ by

$$P_M x = \sum_{k=1}^M \langle x, e_k \rangle_{\mathbb{C}} e_k \quad \forall x \in \mathbf{L}^2([0,\pi],\mathbb{C}).$$

In fact, P_M is also a projection operator from $(\mathbf{L}^2([0, \pi], \mathbb{R}), \langle \cdot, \cdot \rangle_{\mathbb{R}})$ onto $(H_M, \langle \cdot, \cdot \rangle_{\mathbb{R}})$ such that $P_M x = \sum_{k=1}^{M} \langle x, e_k \rangle_{\mathbb{R}} e_k$ for $x \in \mathbf{L}^2([0, \pi], \mathbb{R})$, where $H_M := \text{span}\{e_1, \ldots, e_M\}$. Denoting $\Delta_M = \Delta P_M$ and using the above notations, we get the spectral Galerkin method as follows

$$\begin{cases} du^M(t) = i\Delta_M u^M(t)dt + i\alpha P_M dW(t), & t > 0, \\ u^M(0) = P_M u_0 \in U_M, \end{cases} \tag{4.44}$$

which admits a unique mild solution on U_M given by

$$u^M(t) = E_M(t)u^M(0) + i\alpha \int_0^t E_M(t - s) P_M dW(s) \tag{4.45}$$

with $E_M(t) = \exp(it\Delta_M)$. Moreover, (4.44) is associated with a finite-dimensional stochastic Hamiltonian system. Now we define $B_T^M = \frac{u^M(T)}{T}$ which can be viewed as a discrete approximation for B_T. Following the ideas of deriving the LDP of $\{B_T\}_{T>0}$, in this part, we present the LDP of $\{B_T^M\}_{T>0}$.

Theorem 4.3.4 (See [56]) *For each fixed $M \in \mathbb{N}_+$, $\{B_T^M\}_{T>0}$ satisfies an LDP on $\mathbf{L}^2([0, \pi], \mathbb{C})$ with the good rate function*

$$\mathbf{I}^M(x) = \begin{cases} \frac{1}{\alpha^2} \left\| \mathbf{Q}_M^{-\frac{1}{2}} x \right\|_{\mathbf{L}^2([0,\pi],\mathbb{C})}^2, & \text{if } x \in \mathbf{Q}_M^{\frac{1}{2}}(\mathbf{L}^2([0, \pi], \mathbb{C})); \\ +\infty, & \text{otherwise,} \end{cases} \tag{4.46}$$

where $\mathbf{Q}_M := \mathbf{Q}P_M$ and $\mathbf{Q}_M^{-\frac{1}{2}}$ is the pseudo inverse of $\mathbf{Q}_M^{\frac{1}{2}}$ on U_M, i.e., $\mathbf{Q}_M^{-\frac{1}{2}} x = \arg\min_{z} \{\|z\|_{\mathbf{L}^2([0,\pi],\mathbb{C})} : z \in U_M, \mathbf{Q}_M^{\frac{1}{2}} z = x\}$ for every $x \in U_M$.

Proof According to $E_M(t) = \cos(t\Delta_M) + i\sin(t\Delta_M)$, we have

$u^M(T)$

$$= E_M(T)u^M(0) - \alpha \int_0^T \sin((T - s)\Delta_M) P_M dW(s) + i\alpha \int_0^T \cos((T - s)\Delta_M) P_M dW(s)$$

$$=: E_M(T)u^M(0) - \alpha W_{\sin}^M(T) + i\alpha W_{\cos}^M(T). \tag{4.47}$$

By means of Proposition 4.1.1,

$$\mathbb{VAR}\left[W_{\sin}^M(T)\right] = \frac{T\mathbf{Q}_M}{2} - \frac{\mathbf{Q}_M \Delta_M^{-1}}{4} \sin(2T\Delta_M),$$

where $W^M_{\sin}(T)$ is a Gaussian random variable taking values on $(H_M, \langle \cdot, \cdot \rangle_{\mathbb{R}})$ for each fixed $T > 0$. Similarly, we have $W^M_{\cos}(T) \sim \mathcal{N}(0, \frac{TQ_M}{2} + \frac{Q_M \Delta_M^{-1}}{4} \sin(2T\Delta_M))$ on H_M, and the correlation operator

$$\text{COR}\left(W^M_{\sin}(T), W^M_{\cos}(T) \right) = \frac{Q_M \Delta_M^{-1}}{4} (I - \cos(2T\Delta_M)).$$

For each $\lambda = \Re(\lambda) + i\Im(\lambda) \in U_M$ with $\Re(\lambda), \Im(\lambda) \in H_M$, by (4.47), we get

$$\langle u^M(T), \lambda \rangle_{\mathbb{R}} = \langle E_M(T)u^M(0), \lambda \rangle_{\mathbb{R}} - \alpha \langle W^M_{\sin}(T), \Re(\lambda) \rangle_{\mathbb{R}} + \alpha \langle W^M_{\cos}(T), \Im(\lambda) \rangle_{\mathbb{R}},$$

which yields

$$\left| \mathbb{E}[\langle u^M(T), \lambda \rangle_{\mathbb{R}}] \right| = \left| \langle E_M(T)u^M(0), \lambda \rangle_{\mathbb{R}} \right| \le K(\lambda). \tag{4.48}$$

Putting all the estimates together, we obtain

$$\begin{aligned}
\text{VAR}&[\langle u^M(T), \lambda \rangle_{\mathbb{R}}] \\
&= \frac{\alpha^2 T}{2} \langle Q_M \Re(\lambda), \Re(\lambda) \rangle_{\mathbb{R}} + \frac{\alpha^2 T}{2} \langle Q_M \Im(\lambda), \Im(\lambda) \rangle_{\mathbb{R}} \\
&\quad - \frac{\alpha^2}{4} \langle Q_M \Delta_M^{-1} \sin(2T\Delta_M) \Re(\lambda), \Re(\lambda) \rangle_{\mathbb{R}} + \frac{\alpha^2}{4} \langle Q_M \Delta_M^{-1} \sin(2T\Delta_M) \Im(\lambda), \Im(\lambda) \rangle_{\mathbb{R}} \\
&\quad - \frac{\alpha^2}{2} \langle Q_M \Delta_M^{-1} (I - \cos(2T\Delta_M)) \Re(\lambda), \Im(\lambda) \rangle_{\mathbb{R}} \\
&=: \frac{\alpha^2 T}{2} \langle Q_M \Re(\lambda), \Re(\lambda) \rangle_{\mathbb{R}} + \frac{\alpha^2 T}{2} \langle Q_M \Im(\lambda), \Im(\lambda) \rangle_{\mathbb{R}} + R(T)
\end{aligned} \tag{4.49}$$

with $|R(T)| \le K(M, \lambda)$. Based on (4.48) and (4.49), we derive

$$\Lambda^M(\lambda) = \frac{1}{2} \left(\frac{\alpha^2}{2} \langle Q_M \Re(\lambda), \Re(\lambda) \rangle_{\mathbb{R}} + \frac{\alpha^2}{2} \langle Q_M \Im(\lambda), \Im(\lambda) \rangle_{\mathbb{R}} \right) = \frac{\alpha^2}{4} \left\| Q_M^{\frac{1}{2}} \lambda \right\|^2_{\mathbf{L}^2([0,\pi],\mathbb{C})}$$

for every $\lambda \in U_M$. Analogous to the proof of (4.43), we obtain

$$(\Lambda^M)^*(x) = \begin{cases} \frac{1}{\alpha^2} \left\| Q_M^{-\frac{1}{2}} x \right\|^2_{\mathbf{L}^2([0,\pi],\mathbb{C})}, & \text{if } x \in Q_M^{\frac{1}{2}}(U_M); \\ +\infty, & \text{otherwise.} \end{cases}$$

Next, we present that $\left\{ B_T^M \right\}_{T>0}$ is exponentially tight. Define $K_L := \Big\{ f \in U_M \big| \ \| f \|_{\mathbf{L}^2([0,\pi],\mathbb{C})} \leq L \Big\}$. According to (4.47), we derive

$$
\mathbb{P}\left(B_T^M \in K_L^c \right) = \mathbb{P}\left(\| u^M(T) \|_{\mathbf{L}^2([0,\pi],\mathbb{C})} > LT \right)
$$

$$
\leq \mathbb{P}\left(\| E_M(T) u_0^M \|_{\mathbf{L}^2([0,\pi],\mathbb{C})} > \frac{TL}{3} \right) + \mathbb{P}\left(\alpha \| W_{\sin}^M(T) \|_{\mathbf{L}^2([0,\pi],\mathbb{R})} > \frac{TL}{3} \right)
$$

$$
+ \mathbb{P}\left(\alpha \| W_{\cos}^M(T) \|_{\mathbf{L}^2([0,\pi],\mathbb{R})} > \frac{TL}{3} \right). \tag{4.50}
$$

It can be derived that

$$
\frac{W_{\sin}^M(T)}{\sqrt{T}} \sim \mathcal{N}\left(0, \left(\frac{I}{2} - \frac{\Delta_M^{-1} \sin(2T\Delta_M)}{4T} \right) \mathbf{Q}_M \right). \tag{4.51}
$$

Hence, we obtain that the kth eigenvalue of $\mathrm{VAR}\left[\frac{W_{\sin}^M(T)}{\sqrt{T}} \right]$ satisfies

$$
\nu_k \left(\mathrm{VAR}\left[\frac{W_{\sin}^M(T)}{\sqrt{T}} \right] \right) = \left(\frac{1}{2} - \frac{\sin(2Tk^2)}{4Tk^2} \right) \eta_k \leq \eta_k
$$

for $k = 1, \ldots, M$. From Proposition 4.1.2, it follows that

$$
\mathbb{E}\left(\exp\left(\varepsilon \left\| \frac{W_{\sin}^M(T)}{\sqrt{T}} \right\|_{\mathbf{L}^2([0,\pi],\mathbb{R})}^2 \right) \right) < (\det(I - 2\varepsilon \mathbf{Q}_M))^{-\frac{1}{2}} = C(\varepsilon, \mathbf{Q}_M)
$$

for every $\varepsilon \in \left(0, \frac{1}{2\eta_1} \right)$. The above formula yields

$$
\mathbb{P}\left(\alpha \| W_{\sin}^M(T) \|_{\mathbf{L}^2([0,\pi],\mathbb{R})} > \frac{TL}{3} \right) \leq \exp\left(-\frac{\varepsilon TL^2}{9\alpha^2} \right) C(\varepsilon, \mathbf{Q}_M).
$$

Similar arguments lead to

$$
\mathbb{P}\left(\alpha \| W_{\cos}^M(T) \|_{\mathbf{L}^2([0,\pi],\mathbb{R})} > \frac{TL}{3} \right) \leq \exp\left(-\frac{\varepsilon TL^2}{9\alpha^2} \right) C(\varepsilon, \mathbf{Q}_M).
$$

Due to Proposition A.2.1, we arrive at

$$
\limsup_{T \to \infty} \frac{1}{T} \log\left(\mathbb{P}\left(B_T^M \in K_L^c \right) \right) \leq -\frac{\varepsilon L^2}{9\alpha^2}, \qquad 0 < \varepsilon < \frac{1}{2\eta_1},
$$

where we have used the fact that $\mathbb{P}\left(\|E_M(T)u_0^M\|_{\mathbf{L}^2([0,\pi],\mathbb{C})} > \frac{TL}{3}\right) = 0$ for sufficiently large T. Then, we arrive at

$$\lim_{L\to\infty} \limsup_{T\to\infty} \frac{1}{T} \log\left(\mathbb{P}\left(B_T^M \in K_L^c\right)\right) = -\infty,$$

which indicates the exponential tightness of $\left\{B_T^M\right\}_{T>0}$.

Notice that $\Lambda^M(\cdot)$ is Fréchet differentiable and $D\Lambda^M(\lambda)(\cdot) = \frac{\alpha^2}{2}\langle\mathbf{Q}_M\lambda, \cdot\rangle$ for any $\lambda \in U_M$. Then it follows from Theorem A.2.2 that $\left\{B_T^M\right\}_{T>0}$ satisfies an LDP on U_M with the good rate function

$$\widetilde{\mathbf{I}}^M(x) = (\Lambda^M)^*(x) = \begin{cases} \frac{1}{\alpha^2}\left\|\mathbf{Q}_M^{-\frac{1}{2}}x\right\|_{\mathbf{L}^2([0,\pi],\mathbb{C})}^2, & \text{if } x \in \mathbf{Q}_M^{\frac{1}{2}}(U_M); \\ +\infty, & \text{if } x \in U_M \setminus \mathbf{Q}_M^{\frac{1}{2}}(U_M). \end{cases}$$

Clearly, U_M is a closed subspace of $\mathbf{L}^2([0,\pi],\mathbb{C})$, and for each $T > 0$, $\mathbb{P}(B_T^M \in U_M) = 1$. Thus, using Lemma A.2.3 and the fact $\mathbf{Q}_M^{\frac{1}{2}}(U_M) = \mathbf{Q}_M^{\frac{1}{2}}(\mathbf{L}^2([0,\pi],\mathbb{C}))$, we conclude the result. $\qquad\square$

Weakly Asymptotical Preservation for LDP of $\{B_T\}_{T>0}$ We have obtained the LDP for $\{B_T^M\}_{T>0}$ of the spectral Galerkin method $\{u^M(T)\}_{T>0}$ with $M \in \mathbb{N}_+$. It is natural to ask whether $\mathbf{I}^M$ converges to $\mathbf{I}$ as M tends to infinity. In Sect. 2.4, the definition of *asymptotical preservation for the LDP* of the original system, i.e., the rate functions associated with numerical methods converging to that of the original system is given in the pointwise sense. In this case, since $\mathbf{Q}_M^{\frac{1}{2}}(\mathbf{L}^2([0,\pi],\mathbb{C})) \subsetneqq \mathbf{Q}^{\frac{1}{2}}(\mathbf{L}^2([0,\pi],\mathbb{C}))$ in general, it can not be assured that $\mathbf{I}^M$ converges to $\mathbf{I}$ pointwise. However, the sequence $\left\{\mathbf{Q}_M^{\frac{1}{2}}(\mathbf{L}^2([0,\pi],\mathbb{C}))\right\}_{M\in\mathbb{N}_+}$ of sets converges to $\mathbf{Q}^{\frac{1}{2}}(\mathbf{L}^2([0,\pi],\mathbb{C}))$ by making use of $\lim_{M\to\infty}\mathbf{Q}_M^{\frac{1}{2}}x = \mathbf{Q}^{\frac{1}{2}}x$ for each $x \in \mathbf{L}^2([0,\pi],\mathbb{C})$. It is possible that $\mathbf{I}^M$ is a good approximation of $\mathbf{I}$ when M is large enough. To this end, we give the following definition.

Definition 4.3.1 (See [56]) For a spatial semi-discretization $\{u^M\}_{M\in\mathbb{N}_+}$ of (4.34), denote $B_T^M = \frac{u^M(T)}{T}$. Assume that $\{B_T^M\}_{T>0}$ satisfies an LDP on $\mathbf{L}^2([0,\pi],\mathbb{C})$ with the rate function $\mathbf{I}^M$ for all sufficiently large M. Then $\{u^M\}_{M\in\mathbb{N}_+}$ is said to weakly asymptotically preserve the LDP of $\{B_T\}_{T>0}$ if for each $x \in \mathbf{Q}^{\frac{1}{2}}(\mathbf{L}^2([0,\pi],\mathbb{C}))$ and $\varepsilon > 0$, there exist $x_0 \in \mathbf{L}^2([0,\pi],\mathbb{C})$ and $M \in \mathbb{N}_+$ such that

$$\|x - x_0\|_{\mathbf{L}^2([0,\pi],\mathbb{C})} < \varepsilon, \qquad \left|\mathbf{I}(x) - \mathbf{I}^M(x_0)\right| < \varepsilon, \tag{4.52}$$

where $\mathbf{I}$ is the rate function of $\{B_T\}_{T>0}$.

Theorem 4.3.5 (See [56]) *Let assumptions of Theorem 4.3.3 hold. For the spectral Galerkin method* (4.44), $\{u^M\}_{M\in\mathbb{N}_+}$ *weakly asymptotically preserves the LDP of* $\{B_T\}_{T>0}$, *i.e.,* (4.52) *holds.*

Proof This problem is discussed in the following two cases.

   Case 1: There are finitely many non-zero entries in $\{\eta_k\}_{k\in\mathbb{N}_+}$, *i.e., for some* $l \in \mathbb{N}_+$, $\eta_l > \eta_{l+1} = \eta_{l+2} = \cdots = 0$.

   For this case, $\mathbf{Q}$ degenerates to a finite-rank operator. If $M \geq l$, then $\mathbf{Q}_M = \mathbf{Q}$. Hence, $\mathbf{I}^M(x) = \mathbf{I}(x)$ for every $x \in \mathbf{L}^2([0, \pi], \mathbb{C})$, which means (4.52). Thus, $\{u^M\}_{M\in\mathbb{N}_+}$ exactly preserves the LDP of $\{B_T\}_{T>0}$ (see [54, Definition 4.1]).

   Case 2: There are finitely many 0 in $\{\eta_k\}_{k\in\mathbb{N}_+}$.

   Notice that for each finite $M \in \mathbb{N}_+$, $\eta_1 \geq \cdots \geq \eta_M > 0$. We denote $y = \mathbf{Q}^{-\frac{1}{2}}x$ with $x \in \mathbf{L}^2([0, \pi], \mathbb{C})$ and define $x_M = \mathbf{Q}_M^{\frac{1}{2}}y$. Further, we have

$$\mathbf{Q}_M^{-\frac{1}{2}}x_M = arg\min_z \left\{ \|z\|_{\mathbf{L}^2([0,\pi],\mathbb{C})} \,\big|\, z \in U_M, \ \mathbf{Q}^{\frac{1}{2}}z = \mathbf{Q}^{\frac{1}{2}}P_My \right\}$$

$$= arg\min_z \left\{ \|z\|_{\mathbf{L}^2([0,\pi],\mathbb{C})} \,\big|\, z \in U_M, \ \sqrt{\eta_k}\langle z, e_k\rangle_\mathbb{C} = \sqrt{\eta_k}\langle y, e_k\rangle_\mathbb{C}, \ k = 1, \dots, M \right\}$$

$$= P_My.$$

The above formula leads to

$$\lim_{M\to\infty} \left| \mathbf{I}^M(x_M) - \mathbf{I}(x) \right| = \frac{1}{\alpha^2} \lim_{M\to\infty} \left| \|P_My\|^2_{\mathbf{L}^2([0,\pi],\mathbb{C})} - \|y\|^2_{\mathbf{L}^2([0,\pi],\mathbb{C})} \right| = 0. \tag{4.53}$$

In addition, it holds that

$$\lim_{M\to\infty} x_M = \lim_{M\to\infty} \mathbf{Q}_M^{\frac{1}{2}}y = \lim_{M\to\infty} P_M\mathbf{Q}^{\frac{1}{2}}y = \mathbf{Q}^{\frac{1}{2}}y = x. \tag{4.54}$$

Therefore, it follows from (4.53) and (4.54) that for each $x \in \mathbf{Q}^{\frac{1}{2}}(\mathbf{L}^2([0, \pi], \mathbb{C}))$ and $\varepsilon > 0$, there exist sufficiently large M and $x_0 = \mathbf{Q}_M^{\frac{1}{2}}(\mathbf{Q}^{-\frac{1}{2}}x)$ such that (4.52) holds.

   Combining *Case 1* and *Case 2*, we complete the proof.   □

Remark 4.3.4 (See [56]) As seen in the proof of Theorem 4.3.5, for every $x \in \mathbf{Q}^{\frac{1}{2}}(\mathbf{L}^2([0, \pi], \mathbb{C}))$ and sufficiently large M, $\mathbf{I}^M(\mathbf{Q}_M^{\frac{1}{2}}\mathbf{Q}^{-\frac{1}{2}}x)$ is a good approximation of $\mathbf{I}(x)$.

4.4 Stochastic Wave Equation with Cubic Nonlinearity

In recent years, the stochastic wave equation has been widely exploited to characterize the sound propagation in the sea, the dynamics of the primary current density vector field within the grey matter of the human brain, heat conduction around a ring, the dilatation of shock waves throughout the sun, the vibration of a string under the action of stochastic forces, etc., (see e.g., [87, 98, 107, 211, 251] and references therein). The stochastic wave equation with Lipschitz and regular coefficients has been investigated theoretically and numerically (see e.g., [14, 86, 144, 159, 217, 220] and references therein). However, in many practical applications, the drift coefficient of the stochastic wave equation fails to satisfy the global Lipschitz condition, which is always considered in the vast majority of research articles concerning the numerical approximation of stochastic partial differential equation. In this section, we turn to the stochastic wave equation (4.3) and its exponentially integrable energy-preserving numerical discretizations.

By denoting $X = (u, v)^\top$, the stochastic wave equation (4.3) can be rewritten in the following abstract form

$$\begin{cases} dX(t) = AX(t)dt + \mathbb{F}(X(t))dt + \mathbb{G}dW(t), & t \in (0, T], \\ X(0) = X_0, \end{cases} \tag{4.55}$$

where

$$X_0 = \begin{bmatrix} u_0 \\ v_0 \end{bmatrix}, \quad A = \begin{bmatrix} 0 & I \\ \Delta & 0 \end{bmatrix}, \quad \mathbb{F}(X(t)) = \begin{bmatrix} 0 \\ -f(u(t)) \end{bmatrix}, \quad \mathbb{G} = \begin{bmatrix} 0 \\ I \end{bmatrix}.$$

Here and below we denote by I the identity operator. The eigenvalues of the operator $-\Delta$ are $0 < \lambda_1 \le \lambda_2 \le \cdots$ and the corresponding eigenfunctions $\{e_i\}_{i=1}^\infty$ satisfying $-\Delta e_i = \lambda_i e_i$ form an orthonormal basis of $\mathbf{L}^2(\mathcal{O})$. Define the interpolation space $\dot{\mathbb{H}}^r := \mathrm{Dom}((-\Delta)^{\frac{r}{2}})$ for $r \in \mathbb{R}$ equipped with the inner product $\langle x, y \rangle_{\dot{\mathbb{H}}^r} := \langle (-\Delta)^{\frac{r}{2}} x, (-\Delta)^{\frac{r}{2}} y \rangle_{\mathbf{L}^2(\mathcal{O})} = \sum_{i=1}^\infty \lambda_i^r \langle x, e_i \rangle_{\mathbf{L}^2(\mathcal{O})} \langle y, e_i \rangle_{\mathbf{L}^2(\mathcal{O})}$ and the corresponding norm $\|x\|_{\dot{\mathbb{H}}^r} := \sqrt{\langle x, x \rangle_{\dot{\mathbb{H}}^r}}$. Furthermore, we introduce the product space

$$\mathbb{H}^r := \dot{\mathbb{H}}^r \times \dot{\mathbb{H}}^{r-1}, \quad r \in \mathbb{R}$$

endowed with the inner product $\langle X_1, X_2 \rangle_{\mathbb{H}^r} := \langle x_1, x_2 \rangle_{\dot{\mathbb{H}}^r} + \langle y_1, y_2 \rangle_{\dot{\mathbb{H}}^{r-1}}$ for any $X_1 = (x_1, y_1)^\top \in \mathbb{H}^r$ and $X_2 = (x_2, y_2)^\top \in \mathbb{H}^r$, and the corresponding norm

$$\|X\|_{\mathbb{H}^r} := \sqrt{\langle X, X \rangle_{\mathbb{H}^r}} = \sqrt{\|u\|_{\dot{\mathbb{H}}^r}^2 + \|v\|_{\dot{\mathbb{H}}^{r-1}}^2}, \quad X \in \mathbb{H}^r.$$

Moreover, we define the domain of operator A by

$$\text{Dom}(A) = \left\{ X \in \mathbb{H} \mid AX = \begin{bmatrix} v \\ \Delta u \end{bmatrix} \in \mathbb{H} = \mathbf{L}^2(\mathcal{O}) \times \dot{\mathbb{H}}^{-1} \right\} = \dot{\mathbb{H}}^1 \times \mathbf{L}^2(\mathcal{O}).$$

Then the operator A generates a unitary group $E(t)$, $t \in \mathbb{R}$ given by

$$E(t) = \exp(tA) = \begin{bmatrix} \cos(t(-\Delta)^{\frac{1}{2}}) & (-\Delta)^{-\frac{1}{2}}\sin(t(-\Delta)^{\frac{1}{2}}) \\ -(-\Delta)^{\frac{1}{2}}\sin(t(-\Delta)^{\frac{1}{2}}) & \cos(t(-\Delta)^{\frac{1}{2}}) \end{bmatrix},$$

where $\cos(t(-\Delta)^{\frac{1}{2}})$ and $\sin(t(-\Delta)^{\frac{1}{2}})$ are cosine and sine operators, respectively.

Lemma 4.4.1 (See [14]) *For all $r \in [0, 1]$, there exists a positive constant C depending on r such that for all $t \geq s \geq 0$,*

$$\|(\sin(t(-\Delta)^{\frac{1}{2}}) - \sin(s(-\Delta)^{\frac{1}{2}}))(-\Delta)^{-\frac{r}{2}}\|_{\mathscr{L}(\mathbf{L}^2(\mathcal{O}))} \leq C(t-s)^r,$$

$$\|(\cos(t(-\Delta)^{\frac{1}{2}}) - \cos(s(-\Delta)^{\frac{1}{2}}))(-\Delta)^{-\frac{r}{2}}\|_{\mathscr{L}(\mathbf{L}^2(\mathcal{O}))} \leq C(t-s)^r,$$

$$\|(E(t) - E(s))X\|_{\mathbf{L}^2(\mathcal{O})} \leq C(t-s)^r\|X\|_{\mathbb{H}^r}.$$

Lemma 4.4.2 (See [14]) *For any $t \in \mathbb{R}$, $\cos(t(-\Delta)^{\frac{1}{2}})$ and $\sin(t(-\Delta)^{\frac{1}{2}})$ satisfy a trigonometric identity in the sense that*

$$\|\sin(t(-\Delta)^{\frac{1}{2}})x\|^2_{\mathbf{L}^2(\mathcal{O})} + \|\cos(t(-\Delta)^{\frac{1}{2}})x\|^2_{\mathbf{L}^2(\mathcal{O})} = \|x\|^2_{\mathbf{L}^2(\mathcal{O})} \quad \forall x \in \mathbf{L}^2(\mathcal{O}).$$

According to the above trigonometric identity, we have $\|E(t)\|_{\mathscr{L}(\mathbb{H})} \leq 1$, $t \in \mathbb{R}$. Define the potential functional F by $\frac{\delta F(u)}{\delta u} = f(u)$, where $\frac{\delta \cdot}{\delta u}$ is the Fréchet derivative (see [119]). Then F can be chosen such that

$$a_1\|u\|^4_{\mathbf{L}^4(\mathcal{O})} - b_1 \leq F(u) \leq a_2\|u\|^4_{\mathbf{L}^4(\mathcal{O})} + b_2 \tag{4.56}$$

for some positive constants a_1, a_2, b_1, b_2. Similar to [61], applying the Itô formula to the Lyapunov energy functional

$$V_1(u, v) = \frac{1}{2}\|u\|^2_{\dot{\mathbb{H}}^1} + \frac{1}{2}\|v\|^2_{\mathbf{L}^2(\mathcal{O})} + F(u) + C_1, \quad C_1 \geq b_1, \tag{4.57}$$

we have the energy evolution law of (4.3) as shown in the following lemma.

Lemma 4.4.3 (See [61, 81]) *Assume that $X_0 \in \mathbb{H}^1$ and $\mathbf{Q}^{\frac{1}{2}} \in \mathscr{L}_2(\mathbf{L}^2(\mathcal{O}))$. Then the stochastic wave equation admits the energy evolution law*

$$\mathbb{E}[V_1(u(t), v(t))] = V_1(u_0, v_0) + \frac{1}{2}\text{Tr}(\mathbf{Q})t, \quad t \in [0, T].$$

4.4.1   Spectral Galerkin Method and Exponential Integrability

In this subsection, we study the spectral Galerkin method for the stochastic wave equation (4.3). For fixed $M \in \mathbb{N}_+$, we define a finite-dimensional subspace U_M of $\mathbf{L}_2(\mathscr{O})$ spanned by $\{e_1, \ldots, e_M\}$, and the projection operator $P_M : \dot{\mathbb{H}}^r \to U_M$ by

$$P_M \zeta = \sum_{i=1}^{M} \langle \zeta, e_i \rangle_{\mathbf{L}^2(\mathscr{O})} e_i \quad \forall \, \zeta \in \dot{\mathbb{H}}^r, \, r \geq -1. \tag{4.58}$$

The definition of P_M immediately implies

$$\|P_M\|_{\mathscr{L}(\mathbf{L}_2(\mathscr{O}))} \leq 1, \quad \Delta P_M \zeta = P_M \Delta \zeta \quad \forall \, \zeta \in \dot{\mathbb{H}}^r, \, r \geq 2.$$

Now we define $\Delta_M : U_M \to U_M$ by

$$\Delta_M \zeta = -\sum_{i=1}^{M} \lambda_i \langle \zeta, e_i \rangle_{\mathbf{L}^2(\mathscr{O})} e_i, \quad \forall \, \zeta \in U_M. \tag{4.59}$$

By denoting $X^M = (u^M, v^M)^\top$, the spectral Galerkin method for (4.55) yields

$$\begin{cases} dX^M(t) = A_M X^M(t)dt + \mathbb{F}_M(X^M(t))dt + \mathbb{G}_M dW(t), \quad t \in (0, T], \\ X^M(0) = X_0^M, \end{cases}$$

$$\tag{4.60}$$

where

$$X_0^M = \begin{bmatrix} u_0^M \\ v_0^M \end{bmatrix}, \quad A_M = \begin{bmatrix} 0 & I \\ \Delta_M & 0 \end{bmatrix}, \quad \mathbb{F}_M(X^M) = \begin{bmatrix} 0 \\ -P_M \left(f(u^M) \right) \end{bmatrix}, \quad \mathbb{G}_M = \begin{bmatrix} 0 \\ P_M \end{bmatrix}.$$

Similarly, the discrete operator A_M generates a unitary group $E_M(t)$, $t \in \mathbb{R}$, given by

$$E_M(t) = \exp(t A_M) = \begin{bmatrix} C_M(t) & (-\Delta_M)^{-\frac{1}{2}} S_M(t) \\ -(-\Delta_M)^{\frac{1}{2}} S_M(t) & C_M(t) \end{bmatrix},$$

where $C_M(t) = \cos(t(-\Delta_M)^{\frac{1}{2}})$ and $S_M(t) = \sin(t(-\Delta_M)^{\frac{1}{2}})$ are discrete cosine and sine operators defined on U_M, respectively. It can be verified straightforwardly that

$$C_M(t)P_M\zeta = \cos(t(-\Delta)^{\frac{1}{2}})P_M\zeta = P_M\cos(t(-\Delta)^{\frac{1}{2}})\zeta,$$

$$S_M(t)P_M\zeta = \sin(t(-\Delta)^{\frac{1}{2}})P_M\zeta = P_M\sin(t(-\Delta)^{\frac{1}{2}})\zeta$$

for any $\zeta \in \dot{\mathbb{H}}^r$, $r \geq -1$.

Since the drift coefficient in (4.60) is locally Lipschitz continuous and A_M is a bounded operator in $U_M \times U_M$, the local existence of the unique mild solution is obtained by making use of the Banach fixed point theorem or the Picard iterations under the $\mathbf{C}([0, T], \mathbf{L}^p(\Omega, \mathbb{H}^1))$-norm for $p \geq 2$. To extend the local solution to a global solution, analogous to the proof of [61, Theorem 4.2], we present the a priori estimate in $\mathbf{L}^p(\Omega, \mathbf{C}([0, T], \mathbb{H}^1))$, by applying the Itô formula to $\left(V_1(u^M, v^M)\right)^p$, $p \in [2, \infty)$ and utilizing the unitary property of $E_M(t)$. Finally, we get the following properties of X^M.

Lemma 4.4.4 (See [81]) *Assume that $X_0 \in \mathbb{H}^1$, $T > 0$ and $\mathbf{Q}^{\frac{1}{2}} \in \mathscr{L}_2(\mathbf{L}^2(\mathcal{O}))$. Then the spectral Galerkin method (4.60) has a unique mild solution given by*

$$X^M(t) = E_M(t)X_0^M + \int_0^t E_M(t-s)\mathbb{F}_M(X^M(s))ds + \int_0^t E_M(t-s)\mathbb{G}_M dW(s)$$

$$(4.61)$$

for $t \in [0, T]$. Moreover, for $p \geq 2$, there exists a positive constant $C := C(X_0, T, \mathbf{Q}, p)$ such that

$$\|X^M\|_{\mathbf{L}^p(\Omega, \mathbf{C}([0,T], \mathbb{H}^1))} \leq C.$$

$$(4.62)$$

Proposition 4.4.1 (See [81]) *Assume that $X_0 \in \mathbb{H}^1$ and $\mathbf{Q}^{\frac{1}{2}} \in \mathscr{L}_2(\mathbf{L}^2(\mathcal{O}))$. The mild solution X^M satisfies the energy evolution law*

$$\mathbb{E}[V_1(u^M(t), v^M(t))] = V_1(u_0^M, v_0^M) + \frac{1}{2}\mathrm{Tr}\left((P_M\mathbf{Q}^{\frac{1}{2}})(P_M\mathbf{Q}^{\frac{1}{2}})^*\right)t \quad \forall t \geq 0.$$

Now we show the exponential integrability property of X^M. It is remarkable that [69, Section 5.4] shows the exponential integrability of the spectral Galerkin method of the 2-dimensional stochastic wave equation driven by the multiplicative noise on a non-empty compact domain.

Lemma 4.4.5 (See [81]) *Assume that $X_0 \in \mathbb{H}^1$ and $\mathbf{Q}^{\frac{1}{2}} \in \mathscr{L}_2(\mathbf{L}^2(\mathcal{O}))$. Then there exist a constant $\alpha \geq \mathrm{Tr}(\mathbf{Q})$ and a positive constant $C := C(X_0, T, \mathbf{Q}, \alpha)$ such that*

$$\sup_{s\in[0,T]} \mathbb{E}\left[\exp\left(\frac{V_1\left(u^M(s), v^M(s)\right)}{\exp(\alpha s)}\right)\right] \leq C.$$

$$(4.63)$$

Proof Denote $\mathbb{G}_M \circ \mathbf{Q}^{\frac{1}{2}} := (0, P_M \mathbf{Q}^{\frac{1}{2}})^\top$ and

$$G_{A_M + \mathbb{F}_M, \mathbb{G}_M}(V_1) := \langle DV_1, A_M + \mathbb{F}_M \rangle_{\mathbf{L}^2(\mathscr{O})} + \frac{1}{2}\mathrm{Tr}\left((D^2 V_1)\mathbb{G}_M \circ \mathbf{Q}^{\frac{1}{2}}\left(\mathbb{G}_M \circ \mathbf{Q}^{\frac{1}{2}}\right)^*\right).$$

A direct calculation similar to (5.43) in [69, Section 5.4] leads to

$$\begin{aligned}
&\left(G_{A_M + \mathbb{F}_M, \mathbb{G}_M}(V_1)\right)\left(u^M, v^M\right) \\
&= \langle f(u^M) - \Delta u^M, v^M \rangle_{\mathbf{L}^2(\mathscr{O})} + \langle v^M, P_M(\Delta u^M - f(u^M)) \rangle_{\mathbf{L}^2(\mathscr{O})} \\
&\quad + \frac{1}{2}\mathrm{Tr}(P_M \mathbf{Q}^{\frac{1}{2}}(P_M \mathbf{Q}^{\frac{1}{2}})^*) \\
&= \frac{1}{2}\mathrm{Tr}(P_M \mathbf{Q}^{\frac{1}{2}}(P_M \mathbf{Q}^{\frac{1}{2}})^*).
\end{aligned}$$

Then we get that for $\alpha > 0$,

$$\begin{aligned}
&\left(G_{A_M + \mathbb{F}_M, \mathbb{G}_M}(V_1)\right)\left(u^M, v^M\right) + \frac{1}{2\exp(\alpha t)}\|(\mathbb{G}_M \circ \mathbf{Q}^{\frac{1}{2}})^*(DV_1)(u^M, v^M)\|^2_{\mathbf{L}^2(\mathscr{O})} \\
&\leq \frac{1}{2}\mathrm{Tr}(\mathbf{Q}) + \frac{1}{2\exp(\alpha t)}\sum_{i=1}^{\infty}\langle v^M, \mathbf{Q}^{\frac{1}{2}}e_i \rangle^2_{\mathbf{L}^2(\mathscr{O})} \leq \frac{1}{2}\mathrm{Tr}(\mathbf{Q}) \\
&\quad + \frac{1}{\exp(\alpha t)}V_1(u^M, v^M)\mathrm{Tr}(\mathbf{Q}).
\end{aligned}$$

Let $\bar{U} = -\frac{1}{2}\mathrm{Tr}(\mathbf{Q}), \alpha \geq \mathrm{Tr}(\mathbf{Q})$. Taking advantage of the exponential integrability lemma in the appendix, we obtain

$$\mathbb{E}\left[\exp\left(\frac{V_1(u^M(t), v^M(t))}{\exp(\alpha t)} + \int_0^t \frac{\bar{U}(s)}{\exp(\alpha s)}ds\right)\right] \leq \exp(V_1(u_0^M, v_0^M)),$$

which implies (4.63).   □

Corollary 4.4.1 (See [81]) *Assume that $X_0 \in \mathbb{H}^1$ and $\mathbf{Q}^{\frac{1}{2}} \in \mathscr{L}_2(\mathbf{L}^2(\mathscr{O}))$. Let $d = 1, 2$. For any positive constant c, it holds that*

$$\mathbb{E}\left[\exp\left(\int_0^T c\|u^M(s)\|^2_{\mathbf{L}^6(\mathscr{O})}ds\right)\right] < \infty.$$

Proof By making use of the Jensen inequality, the Gagliardo–Nirenberg inequality $\|u\|_{\mathbf{L}^6(\mathscr{O})} \le C\|\nabla u\|^a_{\mathbf{L}^2(\mathscr{O})}\|u\|^{1-a}_{\mathbf{L}^2(\mathscr{O})}$ with $a = \frac{d}{3}$, and the Young inequality, we obtain

$$\mathbb{E}\left[\exp\left(\int_0^T c\|u^M(s)\|^2_{\mathbf{L}^6(\mathscr{O})}ds\right)\right] \le \sup_{t\in[0,T]} \mathbb{E}\left[\exp(cT\|u^M(t)\|^2_{\mathbf{L}^6(\mathscr{O})})\right]$$

$$\le \sup_{t\in[0,T]} \mathbb{E}\left[\exp\left(\frac{\|\nabla u^M(t)\|^2_{\mathbf{L}^2(\mathscr{O})}}{2\exp(\alpha t)}\right)\exp\left(\exp(\frac{a}{1-a}\alpha T)\|u^M(t)\|^2_{\mathbf{L}^2(\mathscr{O})}(cCT)^{\frac{1}{1-a}}2^{\frac{a}{1-a}}\right)\right].$$

Then the Hölder inequality and the Young inequality yield that for sufficiently small $\epsilon > 0$,

$$\mathbb{E}\left[\exp\left(\int_0^T c\|u^M(s)\|^2_{\mathbf{L}^6(\mathscr{O})}ds\right)\right]$$

$$\le C(\epsilon, d) \sup_{t\in[0,T]} \mathbb{E}\left[\exp\left(\frac{\|\nabla u^M(t)\|^2_{\mathbf{L}^2(\mathscr{O})}}{2\exp(\alpha t)}\right)\exp(\epsilon\|u^M(t)\|^4_{\mathbf{L}^4(\mathscr{O})})\right]$$

$$\le C(\epsilon, d) \sup_{t\in[0,T]} \mathbb{E}\left[\exp\left(\frac{V_1(u^M(t), v^M(t))}{\exp(\alpha t)}\right)\right].$$

Applying Lemma 4.4.5, we complete the proof. ☐

Remark 4.4.1 (See [81]) When $d = 1$, using the Gagliardo–Nirenberg inequality $\|u\|_{\mathbf{L}^\infty(\mathscr{O})} \le \|\nabla u\|^{\frac{1}{2}}_{\mathbf{L}^2(\mathscr{O})}\|u\|^{\frac{1}{2}}_{\mathbf{L}^2(\mathscr{O})}$, we achieve $\mathbb{E}\left[\exp\left(\int_0^T \|u^M(s)\|^2_{\mathbf{L}^\infty(\mathscr{O})}ds\right)\right] < \infty$.

For the applications of the exponential integrability property, we refer to [32, 36, 71, 72, 77, 137] and references therein. Now we show the higher regularity of the solution of (4.60) in the following proposition.

Proposition 4.4.2 (See [81]) *Let* $p \ge 1$, $d = 1$, $\beta \ge 1$, $X_0 \in \mathbb{H}^\beta$, $\|(-\Delta)^{\frac{\beta-1}{2}}\mathbf{Q}^{\frac{1}{2}}\|_{\mathscr{L}_2(\mathbf{L}^2(\mathscr{O}))} < \infty$ *and* $T > 0$. *Then the mild solution* X^M *of (4.60) satisfies*

$$\|X^M\|_{\mathbf{L}^p(\Omega, \mathbf{C}([0,T], \mathbb{H}^\beta))} \le C(X_0, T, p, \mathbf{Q}).$$

Proof For the stochastic convolution, by the Burkholder–Davis–Gundy inequality, we derive

$$\mathbb{E}\left[\sup_{t\in[0,T]}\left\|\int_0^t E_M(t-s)\mathbb{G}_M dW(s)\right\|^p_{\mathbb{H}^\beta}\right] \le C\left(\int_0^T \|(-\Delta)^{\frac{\beta-1}{2}}\mathbf{Q}^{\frac{1}{2}}\|^2_{\mathscr{L}_2(\mathbf{L}^2(\mathscr{O}))}ds\right)^{\frac{p}{2}} \le C.$$

Now it suffices to estimate $\| \int_0^t E_M(t-s)\mathbb{F}_M(X^M(s))ds \|_{\mathbf{L}^p(\Omega,\mathbf{C}([0,T],\mathbb{H}^\beta))}$. Noticing that

$$E_M(t-s)\mathbb{F}_M(X^M(s)) = \begin{bmatrix} -(-\Delta)^{-\frac{1}{2}}S_M(t-s)f(u^M(s)) \\ -C_M(t-s)f(u^M(s)) \end{bmatrix},$$

thus we only need to estimate $\mathbb{E}\left[\sup_{t\in[0,T]} (\int_0^t \|(-\Delta)^{\frac{\beta-1}{2}} f(u^M(s))\|_{\mathbf{L}^2(\mathscr{O})}ds)^p \right]$. The Sobolev embedding $\dot{\mathbb{H}}^1 \hookrightarrow \mathbf{L}^\infty(\mathscr{O})$ implies

$$\int_0^t \left\| (-\Delta)^{\frac{\beta-1}{2}} f(u^M(s)) \right\|_{\mathbf{L}^2(\mathscr{O})} ds \le C \int_0^t (1 + \|u^M(s)\|_{\dot{\mathbb{H}}^1}^2)\|u^M(s)\|_{\dot{\mathbb{H}}^{\beta-1}}ds.$$

According to the Hölder inequality and the Young inequality, we obtain

$$\mathbb{E}\left[\sup_{t\in[0,T]} \left(\int_0^t \|(-\Delta)^{\frac{\beta-1}{2}} f(u^M(s))\|_{\mathbf{L}^2(\mathscr{O})}ds \right)^p \right]$$

$$\le C\mathbb{E}\left[\int_0^T (1 + \|u^M(s)\|_{\dot{\mathbb{H}}^1}^2)^p \|u^M(s)\|_{\dot{\mathbb{H}}^{\beta-1}}^p ds \right]$$

$$\le C \int_0^T \mathbb{E}[1 + \|u^M(s)\|_{\dot{\mathbb{H}}^1}^{4p}]ds + C \int_0^T \mathbb{E}[\|u^M(s)\|_{\dot{\mathbb{H}}^{\beta-1}}^{2p}]ds,$$

which together with Lemma 4.4.4 indicates the desired result for the case that $\beta \in [1,2)$. For $\beta \in [n, n+1)$, $n \in \mathbb{N}_+$, we complete the proof through induction arguments. □

The following proposition is about the regularity estimate of X^M in the case of $d = 2$. In this case, one needs to use different skills to deal with the cubic nonlinearity since the Sobolev embedding $\dot{\mathbb{H}}^1 \hookrightarrow \mathbf{L}^\infty(\mathscr{O})$ fails.

Proposition 4.4.3 (See [81]) *Let* $d = 2$, $T > 0$, $X_0 \in \mathbf{H}^2$ *and* $\|(-\Delta)^{\frac{1}{2}}\mathbf{Q}^{\frac{1}{2}}\|_{\mathscr{L}_2(\mathbf{L}^2(\mathscr{O}))} < \infty$. *Then for any* $p \ge 2$, *there exists a positive constant* $C := C(X_0, \mathbf{Q}, T, p)$ *such that*

$$\|X^M\|_{\mathbf{L}^p(\Omega,\mathbf{C}([0,T],\mathbb{H}^2))} \le C. \tag{4.64}$$

Proof We only present the proof for $p = 2$, since the proof for general $p > 2$ is similar. Analogous to the proof of Proposition 4.4.2, it suffices to get a uniform bound of X^M under the $\mathbf{C}([0,T], \mathbf{L}^2(\Omega, \mathbb{H}^2))$-norm. We introduce another Lyapunov functional V_2 defined by

$$V_2(u^M, v^M) = \frac{1}{2}\left\| \Delta u^M \right\|_{\mathbf{L}^2(\mathscr{O})}^2 + \frac{1}{2}\left\| \nabla v^M \right\|_{\mathbf{L}^2(\mathscr{O})}^2 + \frac{1}{2}\langle (-\Delta)u^M, f(u^M)\rangle_{\mathbf{L}^2(\mathscr{O})}.$$

By applying the Itô formula to V_2 and the commutativity between Δ and P_M, we get

$$
\begin{aligned}
dV_2 &= \langle \Delta u^M, \Delta v^M \rangle_{\mathbf{L}^2(\mathcal{O})} dt + \langle \nabla v^M, \nabla(\Delta_M u^M) \rangle_{\mathbf{L}^2(\mathcal{O})} dt \\
&\quad + \frac{1}{2} \mathrm{Tr}\left((\nabla P_M \mathbf{Q}^{\frac{1}{2}})(\nabla P_M \mathbf{Q}^{\frac{1}{2}})^* \right) dt \\
&\quad + \langle \nabla v^M, -\nabla\left(P_M f(u^M) \right) dt + \nabla P_M dW(t) \rangle_{\mathbf{L}^2(\mathcal{O})} \\
&\quad + \frac{1}{2} \langle (-\Delta) v^M, f(u^M) \rangle_{\mathbf{L}^2(\mathcal{O})} dt + \frac{1}{2} \langle (-\Delta) u^M, f'(u^M) v^M \rangle_{\mathbf{L}^2(\mathcal{O})} dt \\
&= I_1 dt + \langle \nabla v^M, \nabla P_M dW(t) \rangle_{\mathbf{L}^2(\mathcal{O})} + \frac{1}{2} \mathrm{Tr}\left((\nabla P_M \mathbf{Q}^{\frac{1}{2}})(\nabla P_M \mathbf{Q}^{\frac{1}{2}})^* \right) dt,
\end{aligned}
$$

where

$$
I_1 = \frac{1}{2} \langle \nabla u^M, f''(u^M) \nabla u^M v^M \rangle_{\mathbf{L}^2(\mathcal{O})}.
$$

By the Hölder inequality and the Gagliardo–Nirenberg inequality $\|\nabla u\|_{\mathbf{L}^4(\mathcal{O})} \le C \|\Delta u\|_{\mathbf{L}^2(\mathcal{O})}^{\frac{1}{2}} \|\nabla u\|_{\mathbf{L}^2(\mathcal{O})}^{\frac{1}{2}}$, we have

$$
\begin{aligned}
I_1 &\le C \|\nabla u^M\|_{\mathbf{L}^4(\mathcal{O})}^2 (1 + \|u^M\|_{\mathbf{L}^\infty(\mathcal{O})}) \|v^M\|_{\mathbf{L}^2(\mathcal{O})} \\
&\le C \|\Delta u^M\|_{\mathbf{L}^2(\mathcal{O})} \|\nabla u^M\|_{\mathbf{L}^2(\mathcal{O})} (1 + \|u^M\|_{\mathbf{L}^\infty(\mathcal{O})}) \|v^M\|_{\mathbf{L}^2(\mathcal{O})}.
\end{aligned}
$$

Applying the Gagliardo–Nirenberg inequality $\|u\|_{\mathbf{L}^\infty(\mathcal{O})} \le C \|\Delta u\|_{\mathbf{L}^2(\mathcal{O})}^{\frac{1}{4}} \|u\|_{\mathbf{L}^6(\mathcal{O})}^{\frac{3}{4}}$ and the Young inequality, we derive

$$
\begin{aligned}
I_1 &\le C \|\Delta u^M\|_{\mathbf{L}^2(\mathcal{O})} \|\nabla u^M\|_{\mathbf{L}^2(\mathcal{O})} \left(1 + \|\Delta u^M\|_{\mathbf{L}^2(\mathcal{O})}^{\frac{1}{4}} \|u^M\|_{\mathbf{L}^6(\mathcal{O})}^{\frac{3}{4}} \right) \|v^M\|_{\mathbf{L}^2(\mathcal{O})} \\
&\le C \left(\|\nabla u^M\|_{\mathbf{L}^2(\mathcal{O})}^2 \|v^M\|_{\mathbf{L}^2(\mathcal{O})}^2 + \|\nabla u^M\|_{\mathbf{L}^2(\mathcal{O})}^{\frac{8}{3}} \|v^M\|_{\mathbf{L}^2(\mathcal{O})}^{\frac{8}{3}} \|u^M\|_{\mathbf{L}^6(\mathcal{O})}^2 + \|\Delta u^M\|_{\mathbf{L}^2(\mathcal{O})}^2 \right).
\end{aligned}
$$

Making use of the Cauchy–Schwarz inequality and the Young inequality and the fact that $\dot{\mathbb{H}}^1 \hookrightarrow \mathbf{L}^6(\mathcal{O})$, we deduce

$$
\begin{aligned}
\left| \langle (-\Delta) u^M, (u^M)^3 \rangle_{\mathbf{L}^2(\mathcal{O})} \right| &\le \|(-\Delta) u^M\|_{\mathbf{L}^2(\mathcal{O})} \|(u^M)^3\|_{\mathbf{L}^2(\mathcal{O})} \\
&\le \frac{1}{2} \|\Delta u^M\|_{\mathbf{L}^2(\mathcal{O})}^2 + \frac{1}{2} \|u^M\|_{\mathbf{L}^6(\mathcal{O})}^6 \le \frac{1}{2} \|\Delta u^M\|_{\mathbf{L}^2(\mathcal{O})}^2 + \frac{\tilde{C}}{2} \|u^M\|_{\dot{\mathbb{H}}^1}^6,
\end{aligned}
$$

where $\tilde{C}$ is a constant dependent on the Sobolev embedding coefficient in $\dot{\mathbb{H}}^1 \hookrightarrow \mathbf{L}^6(\mathcal{O})$. The above inequality leads to $V_2(u^M, v^M) \ge \frac{1}{4} \|\Delta u^M\|_{\mathbf{L}^2(\mathcal{O})}^2 - \frac{\tilde{C}}{4} \|u^M\|_{\dot{\mathbb{H}}^1}^6$, which yields

$$dV_2 \leq C\left(V_2 + \|u^M\|_{\dot{\mathbb{H}}^1}^6\right) dt + C(1 + \|u^M\|_{\dot{\mathbb{H}}^1}^8 + \|v^M\|_{\mathbf{L}^2(\mathcal{O})}^8) dt$$

$$+ \langle \nabla v^M, \nabla P_M dW \rangle_{\mathbf{L}^2(\mathcal{O})} + \frac{1}{2}\text{Tr}\left((\nabla P_M \mathbf{Q}^{\frac{1}{2}})(\nabla P_M \mathbf{Q}^{\frac{1}{2}})^*\right) dt.$$

Taking expectation and applying the Grönwall inequality, we have

$$\mathbb{E}\left[V_2(u^M(t), v^M(t))\right] \leq C\exp(Ct)\left(\|X_0\|_{\mathbb{H}^2}^2 + \frac{1}{2}\text{Tr}\left((\nabla P_M \mathbf{Q}^{\frac{1}{2}})(\nabla P_M \mathbf{Q}^{\frac{1}{2}})^*\right)t\right.$$

$$\left. + \int_0^t \mathbb{E}[1 + \|v^M(s)\|_{\mathbf{L}^2(\mathcal{O})}^8 + \|u^M(s)\|_{\dot{\mathbb{H}}^1}^8] ds\right),$$

where $t \in [0, T]$, which combined with Lemma 4.4.4 shows the desired result. □

Next we derive the Hölder continuity in time for the numerical solution u^M and X^M under the $\mathbf{L}^p(\Omega, \dot{\mathbb{H}})$- and $\mathbf{L}^p(\Omega, \mathbb{H})$-norm, respectively.

Lemma 4.4.6 (See [81]) *Assume that conditions in Lemma 4.4.4 hold. Then there exists a positive constant $C := C(X_0, p, T, \mathbf{Q})$ such that for each $0 \leq s \leq t \leq T$,*

$$\|u^M(t) - u^M(s)\|_{\mathbf{L}^p(\Omega, \dot{\mathbb{H}})} \leq C|t - s|,$$

$$\|X^M(t) - X^M(s)\|_{\mathbf{L}^p(\Omega, \mathbb{H})} \leq C|t - s|^{\frac{1}{2}}.$$

Proof From (4.60), we have

$$u^M(t) - u^M(s) = (C_M(t) - C_M(s))P_M(u_0) + (-\Delta_M)^{-\frac{1}{2}}(S_M(t) - S_M(s))P_M(v_0)$$

$$+ \int_0^s (-\Delta_M)^{-\frac{1}{2}}(S_M(t - r) - S_M(s - r))P_M(f(u^M(r)))dr$$

$$+ \int_s^t (-\Delta_M)^{-\frac{1}{2}}S_M(t - r)P_M(f(u^M(r)))dr$$

$$+ \int_0^s (-\Delta_M)^{-\frac{1}{2}}(S_M(t - r) - S_M(s - r))P_M dW(r)$$

$$+ \int_s^t (-\Delta_M)^{-\frac{1}{2}}S_M(t - r)P_M dW(r).$$

Therefore, using the properties of $C_M(t)$ and $S_M(t)$ and the Burkholder–Davis–Gundy inequality,

$$\|u^M(t) - u^M(s)\|_{\mathbf{L}^p(\Omega, \dot{\mathbb{H}})}$$

$$\leq C|t - s| \left(\|u_0\|_{\mathbf{L}^p(\Omega, \dot{\mathbb{H}}^1)} + \|v_0\|_{\mathbf{L}^p(\Omega, \dot{\mathbb{H}})} \right)$$

$$+ C \int_0^s (t - s)\|f(u^M(s))\|_{\mathbf{L}^p(\Omega, \dot{\mathbb{H}})} ds + C \int_s^t \|f(u^M(s))\|_{\mathbf{L}^p(\Omega, \dot{\mathbb{H}}^{-1})} ds$$

$$+ \left(\int_0^s \|(-\Delta_M)^{-\frac{1}{2}}(S_M(t - r) - S_M(s - r))P_M\|^2_{\mathscr{L}_2(\dot{\mathbb{H}})} dr \right)^{\frac{1}{2}}$$

$$+ \left(\int_s^t \|(-\Delta_M)^{-\frac{1}{2}} S_M(t - r)P_M\|^2_{\mathscr{L}_2(\dot{\mathbb{H}})} dr \right)^{\frac{1}{2}}$$

$$\leq C|t - s| \left(1 + \|u_0\|_{\mathbf{L}^p(\Omega, \dot{\mathbb{H}}^1)} + \|v_0\|_{\mathbf{L}^p(\Omega, \dot{\mathbb{H}})} + \sup_{0 \leq t \leq T} \|u^M(t)\|^3_{\mathbf{L}^{3p}(\Omega, \dot{\mathbb{H}}^1)} \right)$$

$$\leq C|t - s|,$$

which is the claim for u^M. For X^M, the proof is similar. $\qquad\square$

Below, we prove that the discrete solution given by (4.61) converges to the solution of (4.55) in strong convergence sense based on Lemma 4.4.5.

Proposition 4.4.4 (See [81]) *Assume that $d = 1$, $\beta \geq 1$ or that $d = 2$, $\beta = 2$ and in addition suppose that $X_0 \in \mathbb{H}^\beta$, $\left\| (-\Delta)^{\frac{\beta-1}{2}} Q^{\frac{1}{2}} \right\|_{\mathscr{L}_2(\mathbf{L}^2(\mathscr{O}))} < \infty$. Then for any $p \geq 2$, (4.60) satisfies*

$$\|X^M - X\|_{\mathbf{L}^p(\Omega, \mathbf{C}([0,T], \mathbb{H}))} = O\left(\lambda_M^{-\frac{\beta}{2}} \right).$$

Proof For the sake of simplicity, we consider the strong convergence of u^M whose proof is similar to X^M as example.

Step 1: Strong Convergence and Limit of u^M We claim that $\{u^M\}_{M \in \mathbb{N}_+}$ is a Cauchy sequence in $\mathbf{L}^p(\Omega, \mathbf{C}([0, T], \mathbf{L}^2(\mathscr{O})))$. Notice that

$$u^M(t) - u^{M'}(t) = \left(u^M(t) - P_M u^{M'}(t) \right) + \left((P_M - I)u^{M'}(t) \right),$$

where $M, M' \in \mathbb{N}_+$. Without loss of generality, we assume that $M' > M$. Based on the expression of both u^M and $u^{M'}$, we have

$$\|(P_M - I)u^{M'}(t)\|^2_{\mathbf{L}^2(\mathscr{O})} = \sum_{i=M+1}^{\infty} \lambda_i^{-\beta} \langle u^{M'}(t), \lambda_i^{\frac{\beta}{2}} e_i \rangle^2_{\mathbf{L}^2(\mathscr{O})} \leq \lambda_M^{-\beta} \|u^{M'}(t)\|^2_{\dot{\mathbb{H}}^\beta}$$

with $\beta \geq 1$. For $u^M(t) - P_M u^{M'}(t)$, we have

$$u^M(t) - P_M u^{M'}(t) = \int_0^t (-\Delta)^{-\frac{1}{2}} \sin((t-s)(-\Delta)^{\frac{1}{2}}) P_M \left(f(u^{M'}(s)) - f(u^M(s)) \right) ds.$$

Using the Sobolev embedding $\mathbf{L}^{\frac{6}{5}}(\mathscr{O}) \hookrightarrow \dot{\mathbb{H}}^{-1}$ and the Hölder inequality, we obtain

$$\|u^M(t) - P_M u^{M'}(t)\|_{\mathbf{L}^2(\mathscr{O})}$$

$$\leq \int_0^t \left\| (-\Delta)^{-\frac{1}{2}} \sin((t-s)(-\Delta)^{\frac{1}{2}}) P_M \left(f(u^{M'}(s)) - f(u^M(s)) \right) \right\|_{\mathbf{L}^2(\mathscr{O})} ds$$

$$\leq C \int_0^t (1 + \|u^M(s)\|_{\mathbf{L}^6(\mathscr{O})}^2 + \|u^{M'}(s)\|_{\mathbf{L}^6(\mathscr{O})}^2)$$

$$\times \left(\|u^M(s) - P_M u^{M'}(s)\|_{\mathbf{L}^2(\mathscr{O})} + \|(P_M - I)u^{M'}(s)\|_{\mathbf{L}^2(\mathscr{O})} \right) ds,$$

which implies

$$\|u^M(t) - P_M u^{M'}(t)\|_{\mathbf{L}^2(\mathscr{O})}$$

$$\leq C\lambda_M^{-\frac{\beta}{2}} \exp\left(\int_0^T \left(\|u^M(s)\|_{\mathbf{L}^6(\mathscr{O})}^2 + \|u^{M'}(S)\|_{\mathbf{L}^6(\mathscr{O})}^2 \right) ds \right)$$

$$\times \int_0^t (1 + \|u^M(s)\|_{\mathbf{L}^6(\mathscr{O})}^2 + \|u^{M'}(s)\|_{\mathbf{L}^6(\mathscr{O})}^2) \|u^{M'}(s)\|_{\dot{\mathbb{H}}^\beta} ds$$

due to the Grönwall inequality. Taking the pth moment, and then applying the Hölder and the Young inequalities and Corollary 4.4.1, we arrive at

$$\|u^M - u^{M'}\|_{\mathbf{L}^p(\Omega, \mathbf{C}([0,T], \mathbf{L}^2(\mathscr{O})))}$$

$$\leq C\lambda_M^{-\frac{\beta}{2}} \left\| \exp\left(\int_0^T \left(\|u^M(s)\|_{\mathbf{L}^6(\mathscr{O})}^2 + \|u^{M'}(s)\|_{\mathbf{L}^6(\mathscr{O})}^2 \right) ds \right) \right\|_{\mathbf{L}^{2p}(\Omega, \mathbb{R})}$$

$$\times \left\| \int_0^T (1 + \|u^M(s)\|_{\mathbf{L}^6(\mathscr{O})}^2 + \|u^{M'}(s)\|_{\mathbf{L}^6(\mathscr{O})}^2) \|u^{M'}(s)\|_{\dot{\mathbb{H}}^\beta} ds \right\|_{\mathbf{L}^{2p}(\Omega, \mathbb{R})}$$

$$+ \lambda_M^{-\frac{\beta}{2}} \|u^{M'}(s)\|_{\mathbf{L}^p(\Omega, \mathbf{C}([0,T], \dot{\mathbb{H}}^\beta))},$$

which means

$$\|u^M - u^{M'}\|_{\mathbf{L}^p(\Omega, \mathbf{C}([0,T], \mathbf{L}^2(\mathscr{O})))} \leq C\lambda_M^{-\frac{\beta}{2}}.$$

Similarly, $\{v^{M'}\}_{M' \in \mathbb{N}_+}$ is a Cauchy sequence in $\mathbf{L}^p(\Omega, \mathbf{C}([0,T], \dot{\mathbb{H}}^{-1}))$ which implies that $\{X^{M'}\}_{M' \in \mathbb{N}_+}$ is a Cauchy sequence in $\mathbf{L}^p(\Omega, \mathbf{C}([0,T], \mathbb{H}))$. Denote by $X = (u, v)^\top \in \mathbb{H}$ the limit of $\{X^{M'}\}_{M' \in \mathbb{N}_+}$. Due to Proposition 4.4.2 and Fatou's lemma, we have $\mathbb{E}[\|X\|^p_{\mathbf{C}([0,T],\mathbb{H}^1)}] \le C(X_0, \mathbf{Q}, T, p)$. It follows from the Gagliardo–Nirenberg inequality and the boundedness of X and X^M in $\mathbf{L}^p(\Omega, \mathbf{C}([0, T], \mathbb{H}^1))$ that u^M converges to u in $\mathbf{L}^p(\Omega, \mathbf{C}([0, T], \mathbf{L}^6(\mathcal{O})))$. By means of the Jensen inequality and Fatou's lemma, we deduce

$$\mathbb{E}\left[\exp\left(\int_0^T \|u(s)\|^2_{\mathbf{L}^6(\mathcal{O})} ds\right)\right] \le \frac{1}{T} \int_0^T \mathbb{E}\left[\exp\left(cT\|u(s)\|^2_{\mathbf{L}^6(\mathcal{O})}\right)\right] ds$$

$$\le \lim_{M \to \infty} \frac{1}{T} \int_0^T \mathbb{E}\left[\exp\left(cT\|u^M(s)\|^2_{\mathbf{L}^6(\mathcal{O})}\right)\right] ds.$$

Then the similar procedure as in the proof of Corollary 4.4.1 yields that for any $c > 0$,

$$\mathbb{E}\left(\exp\left(\int_0^T c\|u(s)\|^2_{\mathbf{L}^6(\mathcal{O})} ds\right)\right) < \infty.$$

Step 2: Existence and Uniqueness of the Mild Solution In order to present that the strong limit X is the mild solution of (4.3), it suffices to prove

$$X(t) = E(t)X_0 + \int_0^t E(t - s)\mathbb{F}(X(s)) ds + \int_0^t E(t - s)\mathbb{G} dW(s) \qquad (4.65)$$

for any $t \in [0, T]$. We take the convergence of $\{u^M\}_{M \in \mathbb{N}_+}$ as an example, i.e., to show that

$$u(t) = \cos(t(-\Delta)^{\frac{1}{2}})u_0 + (-\Delta)^{-\frac{1}{2}}\sin(t(-\Delta)^{\frac{1}{2}})v_0$$

$$- \int_0^t (-\Delta)^{-\frac{1}{2}}\sin((t - s)(-\Delta)^{\frac{1}{2}})(f(u(s))ds - dW(s)).$$

To this end, we present that the mild form of the exact solution u^M is convergent to that of u. The assumption on X leads to

$$\|\cos(t(-\Delta)^{\frac{1}{2}})(I - P_M)u_0\|_{\mathbf{L}^2(\mathcal{O})} + \|(-\Delta)^{-\frac{1}{2}}\sin(t(-\Delta)^{\frac{1}{2}})(I - P_M)v_0\|_{\mathbf{L}^2(\mathcal{O})}$$

$$\le C\lambda_M^{-\frac{\beta}{2}}(\|u_0\|_{\dot{\mathbb{H}}^\beta} + \|v_0\|_{\dot{\mathbb{H}}^{\beta-1}}).$$

By the Sobolev embedding $\mathbf{L}^{\frac{6}{5}}(\mathcal{O}) \hookrightarrow \dot{\mathbb{H}}^{-1}$, we obtain

$$\left\| \int_0^t (-\Delta)^{-\frac{1}{2}} \sin((t-s)(-\Delta)^{\frac{1}{2}}) P_M \left(f(u(s)) - f(u^M(s)) \right) ds \right\|_{L^p(\Omega, \mathbf{C}([0,T], \mathbf{L}^2(\mathcal{O})))}$$

$$\leq C \int_0^T \|(1 + \|u(s)\|^2_{\mathbf{L}^6(\mathcal{O})} + \|u^M(s)\|^2_{\mathbf{L}^6(\mathcal{O})})\|_{\mathbf{L}^{2p}(\Omega, \mathbb{R})} \|u(s) - u^M(s)\|_{\mathbf{L}^{2p}(\Omega, \mathbf{L}^2(\mathcal{O}))} ds$$

$$\leq C \lambda_M^{-\frac{\beta}{2}}.$$

For the stochastic term, by the Burkholder–Davis–Gundy inequality, we have that for $p \geq 2$,

$$\left\| \int_0^t (-\Delta)^{-\frac{1}{2}} \sin((t-s)(-\Delta)^{\frac{1}{2}})(I - P_M) dW(s) \right\|_{L^p(\Omega, \mathbf{C}([0,T], \mathbf{L}^2(\mathcal{O})))}$$

$$\leq C \left\| \int_0^t (-\Delta)^{-\frac{1}{2}} \cos(s(-\Delta)^{\frac{1}{2}})(I - P_M) dW(s) \right\|_{L^p(\Omega, \mathbf{C}([0,T], \mathbf{L}^2(\mathcal{O})))}$$

$$+ C \left\| \int_0^t (-\Delta)^{-\frac{1}{2}} \sin(s(-\Delta)^{\frac{1}{2}})(I - P_M) dW(s) \right\|_{L^p(\Omega, \mathbf{C}([0,T], \mathbf{L}^2(\mathcal{O})))} \leq C \lambda_M^{-\frac{\beta}{2}}.$$

Combining the above estimates, we complete the proof.   □

From the proof of Proposition 4.4.4, we have the exponential integrability of the exact solution.

Theorem 4.4.1 (See [81]) *Let $d = 1, 2$, $X_0 \in \mathbb{H}^1$ and $\mathbf{Q}^{\frac{1}{2}} \in \mathscr{L}_2(\mathbf{L}^2(\mathcal{O}))$. For any $c \in \mathbb{R}$,*

$$\mathbb{E}\left(\exp\left(\int_0^T c\|u(s)\|^2_{\mathbf{L}^6(\mathcal{O})} ds \right) \right) < \infty.$$

4.4.2 Exponentially Integrable Full Discretization

Below we introduce an energy-preserving exponentially integrable numerical method for the stochastic wave equation (4.3) by applying the splitting AVF method to (4.60), and finally obtain a strong convergence theorem for the full discrete numerical method.

Let $N \in \mathbb{N}_+$ and $T = Nh$. Denote $\mathbb{Z}_{N+1} = \{0, 1, \ldots, N\}$. For any $T > 0$, we partition the time domain $[0, T]$ uniformly with nodes $t_n = nh$, $n = 0, 1, \ldots, N$ for simplicity. We first decompose (4.3) into a deterministic system on $[t_n, t_{n+1}]$,

$$du^{M,D}(t) = v^{M,D}(t)dt, \quad dv^{M,D}(t) = \Delta_M u^{M,D}(t)dt - P_M(f(u^{M,D}(t)))dt, \tag{4.66}$$

and a stochastic system on $[t_n, t_{n+1}]$,

$$du^{M,S}(t) = 0, \quad dv^{M,S}(t) = P_M dW(t), \tag{4.67}$$

$$u^{M,S}(t_n) = u^{M,D}(t_{n+1}), \quad v^{M,S}(t_n) = v^{M,D}(t_{n+1}).$$

Then on each subinterval $[t_n, t_{n+1}]$, $u^{M,S}(t)$ starting from $u^{M,S}(t_n) = u^{M,D}(t_{n+1})$ and $v^{M,S}(t)$ starting from $v^{M,S}(t_n) = v^{M,D}(t_{n+1})$ can be viewed as approximations of u^M with $u^M(t_n) = u^{M,D}(t_n)$ and v^M with $v^M(t_n) = v^{M,D}(t_n)$ in (4.3), respectively. By further using the explicit solution of (4.67) and the AVF method to discretize (4.66), we obtain the splitting AVF method

$$u_{n+1}^M = u_n^M + h\bar{v}_{n+\frac{1}{2}}^M,$$

$$\bar{v}_{n+1}^M = v_n^M + h\Delta_M u_{n+\frac{1}{2}}^M - hP_M\left(\int_0^1 f(u_n^M + \theta(u_{n+1}^M - u_n^M))d\theta\right), \tag{4.68}$$

$$v_{n+1}^M = \bar{v}_{n+1}^M + P_M(\Delta_n W),$$

where $u_0^M = P_M u_0$, $v_0^M = P_M v_0$, $\bar{v}_{n+\frac{1}{2}}^M = \frac{1}{2}(\bar{v}_{n+1}^M + v_n^M)$, $u_{n+\frac{1}{2}}^M = \frac{1}{2}(u_{n+1}^M + u_n^M)$ and the increment

$$\Delta_n W := W(t_{n+1}) - W(t_n) = \sum_{k=1}^{\infty}(\beta_k(t_{n+1}) - \beta_k(t_n))\mathbf{Q}^{\frac{1}{2}}e_k.$$

Denote

$$\mathbb{A}(t) := \begin{bmatrix} I & \frac{t}{2}I \\ \Delta_M\frac{t}{2} & I \end{bmatrix}, \quad \mathbb{B}(t) := \begin{bmatrix} I & -\frac{t}{2}I \\ -\Delta_M\frac{t}{2} & I \end{bmatrix}$$

and $\mathbb{M}(t) = I - \Delta_M\frac{t^2}{4}$. Then we have

$$\mathbb{B}^{-1}(t)\mathbb{A}(t) = \begin{bmatrix} \mathbb{M}^{-1}(t) & 0 \\ 0 & \mathbb{M}^{-1}(t) \end{bmatrix}\mathbb{A}^2(t) = \begin{bmatrix} 2\mathbb{M}^{-1}(t) - I & \mathbb{M}^{-1}(t)t \\ \mathbb{M}^{-1}(t)\Delta_M t & 2\mathbb{M}^{-1}(t) - I \end{bmatrix}.$$

This formula yields that (4.68) can be rewritten as

$$\begin{bmatrix} u_{n+1}^M \\ v_{n+1}^M \end{bmatrix} = \mathbb{B}^{-1}(h)\mathbb{A}(h)\begin{bmatrix} u_n^M \\ v_n^M \end{bmatrix} \tag{4.69}$$

$$+ \mathbb{B}^{-1}(h)\begin{bmatrix} 0 \\ hP_M\left(\int_0^1 f(u_n^M + \theta(u_{n+1}^M - u_n^M))d\theta\right) \end{bmatrix} + \begin{bmatrix} 0 \\ P_M(\Delta_n W) \end{bmatrix}.$$

For convenience, we assume that there exists a small enough $h_0 > 0$ depending on $X_0, \mathbf{Q}, T$ such that the numerical solution of (4.68) exists and is unique. Throughout this subsection, we always require that the time-step size $h \leq h_0$. In order to study the strong convergence of the numerical method, we first present some estimates of the matrix $\mathbb{B}^{-1}(\cdot)\mathbb{A}(\cdot)$.

Lemma 4.4.7 (See [81]) *For any* $r \geq 0$, $t \geq 0$ *and* $w \in \mathbb{H}^r$, *one has*

$$\|\mathbb{B}^{-1}(t)\mathbb{A}(t)w\|_{\mathbb{H}^r} = \|w\|_{\mathbb{H}^r}.$$

Following [37, Theorem 3], we give the following lemma which will be applied to the error estimate for (4.68).

Lemma 4.4.8 (See [81]) *For any* $r \geq 0$ *and* $h \geq 0$, *there exists a positive constant* $C := C(r)$ *such that*

$$\|(E_M(h) - \mathbb{B}^{-1}(h)\mathbb{A}(h))w\|_{\mathbb{H}^r} \leq Ch^2\|w\|_{\mathbb{H}^{r+2}},$$
$$\|(E_M(h) - \mathbb{B}^{-1}(h))w\|_{\mathbb{H}^r} \leq Ch\|w\|_{\mathbb{H}^{r+1}} \tag{4.70}$$

for any $w \in \mathbb{H}^{r+2}$.

Proposition 4.4.5 (See [81]) *Assume that* $T > 0$, $p \geq 1$, $X_0 \in \mathbb{H}^1$ *and* $\mathbf{Q}^{\frac{1}{2}} \in \mathscr{L}_2(\mathbf{L}^2(\mathcal{O}))$. *Then the solution of (4.68) satisfies*

$$\sup_{n \in \mathbb{Z}_{N+1}} \mathbb{E}[V_1^p(u_n^M, v_n^M)] \leq C, \tag{4.71}$$

where $C = C(X_0, \mathbf{Q}, T, p) > 0$ *and* $V_1(u_n^M, v_n^M) := \frac{1}{2}\|u_n^M\|_{\mathbb{H}^1}^2 + \frac{1}{2}\|v_n^M\|_{\mathbf{L}^2(\mathcal{O})}^2 + F(u_n^M) + C_1$ *with* $C_1 > b_1$.

Proof Fix $t \in T_n := [t_n, t_{n+1}]$ with $n \in \mathbb{Z}_N$. Since on the interval T_n, $u^{M,S}(t) = u_{n+1}^M$, $v^{M,S}(t_n) = \bar{v}_{n+1}^M$, we have

$$V_1(u_{n+1}^M, v_{n+1}^M) = V_1(u_{n+1}^M, \bar{v}_{n+1}^M) + \int_{t_n}^{t_{n+1}} \langle v^{M,S}(s), P_M dW(s) \rangle_{\mathbf{L}^2(\mathcal{O})}$$

$$+ \int_{t_n}^{t_{n+1}} \frac{1}{2} \mathrm{Tr}\left((P_M \mathbf{Q}^{\frac{1}{2}})(P_M \mathbf{Q}^{\frac{1}{2}})^*\right) ds.$$

Applying the Itô formula to $V_1^p(u^{M,S}(t), v^{M,S}(t))$ for $p \geq 2$, we obtain

$$V_1^p(u^{M,S}(t), v^{M,S}(t)) - V_1^p(u^{M,S}(t_n), v^{M,S}(t_n))$$

$$= \frac{p}{2} \int_{t_n}^{t} V_1^{p-1}(u^{M,S}(s), v^{M,S}(s)) \mathrm{Tr}\left((P_M \mathbf{Q}^{\frac{1}{2}})(P_M \mathbf{Q}^{\frac{1}{2}})^*\right) ds$$

$$+ p \int_{t_n}^{t} V_1^{p-1}(u^{M,S}(s), v^{M,S}(s)) \langle v^{M,S}(s), P_M dW(s) \rangle_{\mathbf{L}^2(\mathscr{O})}$$

$$+ \frac{p(p-1)}{2} \sum_{i=1}^{M} \int_{t_n}^{t} V_1^{p-2}(u^{M,S}(s), v^{M,S}(s)) \langle v^{M,S}(s), \mathbf{Q}^{\frac{1}{2}} e_i \rangle^2 ds.$$

Taking expectation, using the martingality of the stochastic integral, the Hölder inequality and the Young inequality, we arrive at

$$\mathbb{E}[V_1^p(u^{M,S}(t), v^{M,S}(t))] \leq \mathbb{E}[V_1^p(u_{n+1}^M, \bar{v}_{n+1}^M)]$$

$$+ C \int_{t_n}^{t} (1 + \mathbb{E}[V_1^p(u^{M,S}(s), v^{M,S}(s))]) ds,$$

which, together with the Grönwall inequality and the property that

$$V_1(u_{n+1}^M, \bar{v}_{n+1}^M) = V_1(u_n^M, v_n^M), \tag{4.72}$$

leads to

$$\mathbb{E}[V_1^p(u_{n+1}^M, v_{n+1}^M)] \leq \exp(Ch) \left(\mathbb{E}[V_1^p(u_n^M, v_n^M)] + Ch \right).$$

Since $Nh = T$, iteration arguments yield

$$\sup_{n \in \mathbb{Z}_N} \mathbb{E}[V_1^p(u_{n+1}^M, v_{n+1}^M)] \leq \exp(CT) \mathbb{E}[V_1^p(u_0^M, v_0^M)] + \exp(CT)CT,$$

which indicates the estimate (4.71). □

From the proof of Proposition 4.4.5, we obtain the following theorem which presents that the proposed method preserves the evolution law of the energy V_1 in (4.57).

Theorem 4.4.2 (See [81]) *Assume that $T > 0$, $p \geq 1$, $X_0 \in \mathbb{H}^1$ and $\mathbf{Q}^{\frac{1}{2}} \in \mathscr{L}_2(\mathbf{L}^2(\mathscr{O}))$. Then the solution of (4.68) satisfies*

$$\mathbb{E}[V_1(u_n^M, v_n^M)] = V_1(u_0^M, v_0^M) + \frac{1}{2} \mathrm{Tr} \left((P_M \mathbf{Q}^{\frac{1}{2}})(P_M \mathbf{Q}^{\frac{1}{2}})^* \right) t_n,$$

where $V_1(u_n^M, v_n^M) := \frac{1}{2} \|u_n^M\|_{\mathbb{H}^1}^2 + \frac{1}{2} \|v_n^M\|_{\mathbf{L}^2(\mathscr{O})}^2 + F(u_n^M) + C_1$ with $C_1 > b_1$.

Besides the energy-preserving property, the proposed numerical method also inherits the exponential integrability property of the original system.

Proposition 4.4.6 (See [81]) *Let $d = 1, 2$, $X_0 \in \mathbb{H}^1$, $T > 0$ and $\|Q^{\frac{1}{2}}\|_{\mathscr{L}_2(L^2(\mathcal{O}))} < \infty$. Then the solution of (4.68) satisfies*

$$\mathbb{E}\left[\exp\left(ch\sum_{i=0}^{n}\|u_i^M\|_{\mathbf{L}^6(\mathcal{O})}^2\right)\right] \leq C \tag{4.73}$$

for every $c > 0$, where $C := C(X_0, Q, T, c) > 0$, $N \in \mathbb{N}_+$, $n \in \mathbb{Z}_{N+1}$ and $Nh = T$.

Proof Notice that $V_1(u_{n+1}^M, v_{n+1}^M) = V_1(u^{M,S}(t_{n+1}), v^{M,S}(t_{n+1}))$, where $v^{M,S}(t_{n+1})$ is the solution of (4.67) defined on $[t_n, t_{n+1}]$ with $v^{M,S}(t_n) = \bar{v}_{n+1}^M$ and $u^{M,S}(t_n) = u_{n+1}^M$. Denote $\widetilde{\mathbb{F}}_M := (0, 0)^\top$, $\widetilde{\mathbb{G}}_M := (0, P_M)^\top$ and $\widetilde{\mathbb{G}}_M \circ Q^{\frac{1}{2}} := (0, P_M Q^{\frac{1}{2}})^\top$. Then we have for $\alpha > 0$,

$$G_{\widetilde{\mathbb{F}}_M, \widetilde{\mathbb{G}}_M}(V_1)(u^{M,S}, v^{M,S}) + \frac{1}{2\exp(\alpha t)}\|(\widetilde{\mathbb{G}}_N \circ Q^{\frac{1}{2}})^*(DV_1)(u^{M,S}, v^{M,S})\|^2$$

$$\leq \frac{1}{2}\mathrm{Tr}(Q) + \frac{1}{\exp(\alpha t)}V_1(u^{M,S}, v^{M,S})\mathrm{Tr}(Q).$$

Let $\bar{U} = -\frac{1}{2}\mathrm{Tr}(Q)$, $\alpha \geq \mathrm{Tr}(Q)$. Applying the Itô formula and taking conditional expectation, we obtain

$$\mathbb{E}\left[\exp\left(\frac{V_1(u^{M,S}(t), v^{M,S}(t))}{\exp(\alpha t)} + \int_{t_n}^t \frac{\bar{U}(s)}{\exp(\alpha s)}ds\right)\right]$$

$$= \mathbb{E}\left[\mathbb{E}\left[\exp\left(\frac{V_1(u^{M,S}(t), v^{M,S}(t))}{\exp(\alpha t)} + \int_{t_n}^t \frac{\bar{U}(s)}{\exp(\alpha s)}ds\right)\Big|\mathscr{F}_{t_n}\right]\right]$$

$$= \mathbb{E}\left[\exp\left(\frac{V_1(u_{n+1}^M, \bar{v}_{n+1}^M)}{\exp(\alpha t_n)}\right)\right] = \mathbb{E}\left[\exp\left(\frac{V_1(u_n^M, v_n^M)}{\exp(\alpha t_n)}\right)\right],$$

where we have used the facts that on $[t_n, t_{n+1}]$, $v^{M,S}(t_n) = \bar{v}_{n+1}^M$ and $u^{M,S}(t_n) = u_{n+1}^M$ and that the energy preservation of the AVF method, $V_1(u_{n+1}^M, \bar{v}_{n+1}^M) = V_1(u_n^M, v_n^M)$.

Repeating the above arguments on every subinterval $[t_l, t_{l+1}]$, $l \leq n - 1$, we get

$$\mathbb{E}\left[\exp\left(\frac{V_1(u_{n+1}^M, v_{n+1}^M)}{\exp(\alpha t_{n+1})}\right)\right] \leq \exp\left(V_1(u_0^M, v_0^M)\right)\exp\left(\int_0^{t_{n+1}} \frac{-\bar{U}(s)}{\exp(\alpha s)}ds\right). \tag{4.74}$$

Now, we are in a position to show (4.73). By using the Jensen inequality, the Gagliardo–Nirenberg inequality $\|u\|_{\mathbf{L}^6(\mathcal{O})} \leq C\|\nabla u\|_{\mathbf{L}^2(\mathcal{O})}^a\|u\|_{\mathbf{L}^2(\mathcal{O})}^{1-a}$ with $a = \frac{d}{3}$, and the Young inequality, we have

$$\mathbb{E}\left[\exp\left(ch\sum_{i=0}^{n}\|u_i^M\|_{\mathbf{L}^6(\mathscr{O})}^2\right)\right] \leq \sup_{i\in\mathbb{Z}_{N+1}}\mathbb{E}\left[\exp(cT\|u_i^M\|_{\mathbf{L}^6(\mathscr{O})}^2)\right]$$

$$\leq \sup_{i\in\mathbb{Z}_{N+1}}\mathbb{E}\left[\exp\left(\frac{\|\nabla u_i^M\|_{\mathbf{L}^2(\mathscr{O})}^2}{2\exp(\alpha t_i)}\right)\exp\left(\exp(\frac{a}{1-a}\alpha T)\|u_i^M\|_{\mathbf{L}^2(\mathscr{O})}^2(cCT)^{\frac{1}{1-a}}2^{\frac{a}{1-a}}\right)\right].$$

Then the Hölder and the Young inequalities imply that for small enough $\epsilon > 0$,

$$\mathbb{E}\left[\exp\left(ch\sum_{i=0}^{n}\|u_i^M\|_{\mathbf{L}^6(\mathscr{O})}^2\right)\right]$$

$$\leq C(\epsilon,d)\sup_{i\in\mathbb{Z}_{N+1}}\mathbb{E}\left[\exp\left(\frac{\|\nabla u_i^M\|_{\mathbf{L}^2(\mathscr{O})}^2}{2\exp(\alpha t_i)}\right)\exp(\epsilon\|u_i^M\|_{\mathbf{L}^4(\mathscr{O})}^4)\right]$$

$$\leq C(\epsilon,d)\sup_{i\in\mathbb{Z}_{N+1}}\mathbb{E}\left[\exp\left(\frac{V_1(u_i^M,v_i^M)}{\exp(\alpha t_i)}\right)\right].$$

By applying (4.74), we complete the proof.   □

Based on the exponential integrability property of $\{u_i^M\}_{1\leq i\leq N}$ and Lemma 4.4.6, we obtain the strong convergence rate in the temporal direction.

Remark 4.4.2 (See [81]) Let $d = 2$. Assume that $X_0 \in \mathbb{H}^2$, $T > 0$ and $\|(-\Delta)^{\frac{1}{2}}\mathbf{Q}^{\frac{1}{2}}\|_{\mathscr{L}_2(\mathbf{L}^2(\mathscr{O}))} < \infty$. By introducing the Lyapunov functional

$$V_2(u_n^M,v_n^M) = \frac{1}{2}\left\|\Delta u_n^M\right\|_{\mathbf{L}^2(\mathscr{O})}^2 + \frac{1}{2}\left\|\nabla v_n^M\right\|_{\mathbf{L}^2(\mathscr{O})}^2 + \frac{1}{2}\langle(-\Delta)u_n^M, f(u_n^M)\rangle_{\mathbf{L}^2(\mathscr{O})},$$

using similar arguments as in the proof of [76, Lemma 3.3] yield that for any $p \geq 1$, there is a constant $C = C(X_0, \mathbf{Q}, T, p) > 0$ such that $\mathbb{E}\left[\sup_{n\in\mathbb{Z}_{N+1}}\|u_n^M\|_{\mathbb{H}^2}^p\right] \leq C$.

Proposition 4.4.7 (See [81]) *Let $d = 1$, $\beta \geq 1$ (or $d = 2$, $\beta = 2$), $\gamma = \min(\beta,2)$ and $T > 0$. Assume that $X_0 \in \mathbb{H}^\beta$, $\|(-\Delta)^{\frac{\beta-1}{2}}\mathbf{Q}^{\frac{1}{2}}\|_{\mathscr{L}_2(\mathbf{L}^2(\mathscr{O}))} < \infty$. Then there exists $h_0 > 0$ such that for $h \leq h_0$ and $p \geq 1$,*

$$\sup_{n\in\mathbb{Z}_{N+1}}\mathbb{E}\left[\|X^M(t_n) - X_n^M\|_{\mathbb{H}}^{2p}\right] \leq Ch^{\gamma p}, \tag{4.75}$$

where $C := C(p, X_0, \mathbf{Q}, T) > 0$.

The convergence result in Proposition 4.4.7, together with Proposition 4.4.4, means the following strong convergence theorem.

Theorem 4.4.3 (See [81]) *Let* $d = 1$, $\beta \geq 1$ *(or* $d = 2$, $\beta = 2$*),* $\gamma = \min(\beta, 2)$ *and* $T > 0$. *Assume that* $X_0 \in \mathbb{H}^\beta$, $\|(-\Delta)^{\frac{\beta-1}{2}} Q^{\frac{1}{2}}\|_{\mathscr{L}_2(\mathbf{L}^2(\mathcal{O}))} < \infty$. *Then there exists* $h_0 > 0$ *such that for* $h \leq h_0$ *and* $p \geq 1$,

$$\sup_{n \in \mathbb{Z}_{N+1}} \mathbb{E}\left[(\|u(t_n) - u_n^M\|_{\mathbf{L}^2(\mathcal{O})}^2 + \|v(t_n) - v_n^M\|_{\mathbb{H}^{-1}}^2)^p \right] \leq C\left(h^{\gamma p} + \lambda_M^{-\beta p} \right),$$

where $C := C(X_0, Q, T, p) > 0$, $M \in \mathbb{N}_+$, $N \in \mathbb{N}_+$, $Nh = T$.

4.5   End Notes

This chapter is devoted to structure-preserving methods for stochastic Maxwell equations, stochastic Schrödinger equations and stochastic wave equations whose phase flows preserve the infinite-dimensional symplectic geometric structure. Especially, a kind of symplectic discretizitions is introduced to approximate the large deviation rate function of an observable for stochastic linear Schrödinger equations. This may indicate the superiority of symplectic discretizations for infinite-dimensional stochastic Hamiltonian systems.

The infinite-dimensional stochastic Hamiltonian system, as a stochastic Hamiltonian partial differential equation, possesses the multi-symplectic conservation law (see e.g., [50, 75, 146] and references therein). Here, the multi-symplecticity is referred to the local conservation of differential 2-forms in both space and time (see Appendix A.4). When taking damping effect into consideration, [134, 239] study the conformal multi-symplecticity, namely the exponential attenuation of differential 2-forms along the phase flow associated with the original system. To inherit these properties, (conformal) multi-symplectic methods have been considered in [50, 75, 129, 134, 239] and references therein. Beyond that it is also of vital significance to design numerical methods preserving the energy evolution law for infinite-dimensional stochastic Hamiltonian systems (see e.g., [50, 66, 67, 75, 81, 146] and references therein). In general, stochastic structure-preserving methods in temporal direction for nonlinear stochastic Hamiltonian systems are implicit, although several efforts have been made to design explicit stochastic structure-preserving methods for some specific equations. What is more, it is far from understand whether stochastic structure-preserving methods could asymptotically preserve the LDP of observables associated with general stochastic Hamiltonian partial differential equations.

The error estimates of stochastic structure-preserving methods in this chapter are studied in the strong convergence sense by using the semigroup framework. Indeed, it is also important to investigate the weak convergence order, which is expected to be higher than the strong one. There have been fruitful results concerning strong and weak error estimates of numerical methods for stochastic partial differential equations with globally Lipschitz continuous coefficients (see e.g., [14, 26, 42, 43, 51] and references therein). When non-globally Lipschitz continuous but monotone coefficients are involved, strong and weak convergence rates of numerical approximations are derived in [25, 36, 72, 79, 177, 262] based on the variational framework and the factorization method in [85]. Unfortunately, the monotonicity assumption may be too restrictive in some physic models, such as the stochastic cubic Schrödinger equation, stochastic wave equation, stochastic Navier–Stokes equation [102, 103], stochastic Burgers equation [27, 28], Cahn–Hilliard–Cook equation [44, 84], etc. At this time, one may need to develop novel numerical techniques and exploit certain exponential integrability to analyze the strong and weak error (see e.g., [74, 76, 78, 95, 139] and references therein). Up to now, there have been still some problems on the convergence analysis of structure-preserving methods for the infinite-dimensional stochastic Hamiltonian system.

- Does the temporal symplectic Runge–Kutta discretization possess the algebraic strong convergence order for stochastic cubic Schrödinger equation and stochastic Klein–Gordon equation?
- How can we analyze the weak convergence order of the temporal Runge–Kutta discretization for infinite-dimensional stochastic Hamiltonian systems under the non-globally monotone condition?

As shown in [113], stochastic numerical methods can be utilized to compute quantities expressed in terms of the probability distribution of stochastic processes. A nature problem is whether the numerical solution could preserve probability properties of the exact solution of stochastic partial differential equations. More and more attentions have been paid on the numerical aspect of ergodicity, density function, central limit theorem and intermittency, which can be used to characterize relevant probabilistic information of stochastic partial differential equations (see e.g., [57–59, 73] and references therein). However, the numerical probability for infinite-dimensional stochastic Hamiltonian systems is still in its infancy and needs novel mathematical tools (see [71, 129]).

Appendix A

A.1 Stochastic Stratonovich Integral

Let $W(\cdot)$ be a 1-dimensional standard Wiener process. From the view of numerical approximation for the stochastic integral $\int_a^b W(t)dW(t)$ with $0 \le a \le b$, we need to replace $W(t)$ by a sequence $\{\Phi_n(t)\}_{n \ge 0}$, which are reasonably smooth so that we can use the ordinary calculus. For the sake of simplicity, we fix an approximating sequence $\{\Phi_n(t)\}_{n \ge 0}$ satisfying the following conditions (see [164]):

(1) for each n and almost all ω, $\Phi_n(\cdot, \omega)$ is a continuous function of bounded variation on $[a, b]$;
(2) for each $t \in [a, b]$, $\Phi_n(t) \to B(t)$ almost surely as $n \to \infty$;
(3) for almost all ω, the sequence $\{\Phi_n(\cdot, \omega)\}$ is uniformly bounded, i.e.,

$$\sup_{n \ge 1} \sup_{a \le t \le b} |\Phi_n(t, \omega)| < \infty.$$

Then we have

$$\int_a^b \Phi_n(t)d\Phi_n(t) = \frac{1}{2}\left(\Phi_n(b)^2 - \Phi_n(a)^2\right) \to \frac{1}{2}\left(W(b)^2 - W(a)^2\right).$$

It is known that the stochastic Itô integral

$$\int_a^b W(t)dW(t) = \frac{1}{2}\left(W(b)^2 - W(a)^2 - (b-a)\right).$$

Hence we see that the Riemann–Stieltjes integral $\int_a^b \Phi_n(t)d\Phi_n(t)$ does not converge to the Itô integral $\int_a^b W(t)dW(t)$.

© The Author(s), under exclusive license to Springer Nature Singapore Pte Ltd. 2022
J. Hong, L. Sun, *Symplectic Integration of Stochastic Hamiltonian Systems*,
Lecture Notes in Mathematics 2314, https://doi.org/10.1007/978-981-19-7670-4

Now we introduce the stochastic Stratonovich integral. The corresponding stochastic Stratonovich integral $\int_0^T W(t) \circ dW(t)$ which is limit of $\int_0^T \Phi_n(t) d\Phi_n(t)$. is defined as

$$\int_0^T W(t) \circ dW(t) := \lim_{h \to 0} \sum_{n=0}^{N-1} \frac{W(t_n) + W(t_{n+1})}{2} (W(t_{n+1}) - W(t_n)) = \frac{W^2(T)}{2}.$$

It also can be verified that

$$\int_0^T W(t) \circ dW(t) = \lim_{h \to 0} \sum_{n=0}^{N-1} W\left(\frac{t_n + t_{n+1}}{2}\right) (W(t_{n+1}) - W(t_n)).$$

Therefore for this case the stochastic Stratonovich integral corresponds to a Riemann sum approximation where we evaluate the integrand at the midpoint of each subinterval. We generalize this example and so introduce

Definition A.1.1 Denote an n-dimensional standard Wiener process by $\mathbf{W}(\cdot)$. Let $\mathbf{X}(\cdot)$ be a stochastic process with values in $\mathbb{R}^n$, and $\mathbf{B} : \mathbb{R}^n \times [0, T] \to \mathbb{R}^{n \times n}$ be a $\mathbf{C}^1$ function such that

$$\mathbb{E}\left(\int_0^T |\mathbf{B}(\mathbf{X}, t)|^2 dt\right) < \infty.$$

We define

$$\int_0^T \mathbf{B}(\mathbf{X}, t) \circ d\mathbf{W}(t) := \lim_{h \to 0} \sum_{n=0}^{N-1} \mathbf{B}\left(\frac{\mathbf{X}(t_{n+1}) + \mathbf{X}(t_n)}{2}, t_n\right) (\mathbf{W}(t_{n+1}) - \mathbf{W}(t_n))$$

provided this limit exists in $\mathbf{L}^2(\Omega, \mathbb{R}^n)$ with $h \to 0$.

Suppose that the process $\mathbf{X}(\cdot)$ solves the Stratonovich integral equation

$$\mathbf{X}(t) = \mathbf{X}(0) + \int_0^t \mathbf{b}(\mathbf{X}(s), s) ds + \int_0^t \mathbf{B}(\mathbf{X}(s), s) \circ d\mathbf{W}(s), \quad t \in [0, T]$$

for $\mathbf{b} \colon \mathbb{R}^n \times [0, T] \to \mathbb{R}^n$ and $\mathbf{B} \colon \mathbb{R}^n \times [0, T] \to \mathbb{M}^{n \times m}$. We then write

$$d\mathbf{X}(t) = \mathbf{b}(\mathbf{X}(t), t) dt + \mathbf{B}(\mathbf{X}(t), t) \circ d\mathbf{W}(t)$$

the second term on the right being the Stratonovich stochastic differential.

Theorem A.1.1 (Stratonovich Chain Rule) *Assume that*

$$d\mathbf{X}(t) = \mathbf{b}(\mathbf{X}(t), t) dt + \mathbf{B}(\mathbf{X}(t), t) \circ d\mathbf{W}(t),$$

and $u : \mathbb{R}^n \times [0, T] \to \mathbb{R}$ *is smooth. Then*

$$du(\mathbf{X}, t) = \frac{\partial u(\mathbf{X}, t)}{\partial t} dt + \sum_{i=1}^{n} \frac{\partial u(\mathbf{X}, t)}{\partial x_i} \circ d\mathbf{X}_i.$$

A.2 Large Deviation Principle

The large deviation principle (LDP) is devoted to the exponential decay of probabilities of rare events, which can be considered as an extension or refinement of the law of large numbers and the central limit theorem. For instance, assume that $X_1, X_2, \ldots$ are i.i.d. $\mathbb{R}$-valued random variables with $X_1 \sim \mathcal{N}(0, 1)$ and let $S_n = \sum_{k=1}^{n} X_k$. Then the weak laws of large number (LLN) states that for any $x > 0$,

$$\lim_{n \to \infty} \mathbb{P}\left(\left|\frac{1}{n} S_n\right| > x\right) = 0.$$

Due to the fact that $S_n \sim \mathcal{N}(0, n)$, we obtain that for any $x > 0$,

$$\mathbb{P}\left(\left|\frac{1}{n} S_n - 0\right| > x\right) = 2\mathbb{P}\left(\frac{1}{n} S_n > x\right) = 2\mathbb{P}\left(\frac{1}{\sqrt{n}} S_n > \sqrt{n}x\right),$$

and

$$\frac{2}{\sqrt{2\pi}} \frac{\sqrt{n}x}{1 + nx^2} \exp\left(-\frac{nx^2}{2}\right) \le 2\mathbb{P}\left(\frac{1}{\sqrt{n}} S_n > \sqrt{n}x\right) \le \frac{2}{\sqrt{2\pi}} \frac{1}{\sqrt{n}x} \exp\left(-\frac{nx^2}{2}\right),$$

where we have used the statement $\frac{1}{\sqrt{2\pi}} \frac{1}{x+\frac{1}{x}} \exp(-\frac{x^2}{2}) \le \mathbb{P}(Y > x) \le \frac{1}{\sqrt{2\pi}} \frac{1}{x} \exp(-\frac{x^2}{2})$ with $Y \sim \mathcal{N}(0, 1)$. A straight computation yields that for any $x > 0$,

$$\lim_{n \to \infty} \frac{1}{n} \log\left(\mathbb{P}\left(\left|\frac{1}{n} S_n\right| > x\right)\right) = -\frac{x^2}{2},$$

and further the LDP indicates that the probability $\mathbb{P}\left(\left|\frac{1}{n} S_n\right| > x\right)$ decays exponentially and the rate is $\frac{x^2}{2}$. Now we introduce some preliminaries upon the theory of large deviations, which can be found in [91, 150]. Let $\mathscr{X}$ be a locally convex Hausdorff topological vector space and $\mathscr{X}^*$ be its dual space.

Definition A.2.1 A function $\mathbf{I}: \mathscr{X} \to [0, \infty]$ is called a rate function, if it is lower semi-continuous, i.e., for each $a \in [0, \infty)$, the level set $\mathbf{I}^{-1}((0, a])$ is a closed subset of $\mathscr{X}$. If all level sets $\mathbf{I}^{-1}((0, a])$, $a \in [0, \infty)$, are compact, then I is called a good rate function.

Definition A.2.2 Let $\mathbf{I}$ be a rate function and $\{\mu_\epsilon\}_{\epsilon>0}$ be a family of probability measures on $\mathscr{X}$. We say that $\{\mu_\epsilon\}_{\epsilon>0}$ satisfies an LDP on $\mathscr{X}$ with the rate function $\mathbf{I}$ if

(LDP1) $\qquad \liminf_{\epsilon \to 0} \epsilon \log(\mu_\epsilon(U)) \geq - \inf_{y \in U} \mathbf{I}(y), \quad$ for every open set $U \subset \mathscr{X}$,

(LDP2) $\qquad \limsup_{\epsilon \to 0} \epsilon \log(\mu_\epsilon(G)) \leq - \inf_{y \in G} \mathbf{I}(Y), \quad$ for every closed set $G \subset \mathscr{X}$.

Analogously, a family of random variables $\{Z_\epsilon\}_{\epsilon>0}$ valued on $\mathscr{X}$ satisfies an LDP with the rate function $\mathbf{I}$ if its distribution law $(\mathbb{P} \circ Z_\epsilon^{-1})_{\epsilon>0}$ satisfies the lower bound LDP (LDP1) and the upper bound LDP (LDP2) in Definition A.2.2 for the rate function $\mathbf{I}$ (see e.g., [46, 91] and references therein).

Proposition A.2.1 (See [91]) *Let N be a fixed integer. Then, for every $a_\epsilon^i \geq 0$,*

$$\limsup_{\epsilon \to 0} \epsilon \log \left(\sum_{i=1}^{N} a_\epsilon^i \right) = \max_{i=1,\ldots,N} \limsup_{\epsilon \to 0} \epsilon \log a_\epsilon^i.$$

Proposition A.2.1 is an important tool in deriving (LDP1) and (LDP2). A useful property for (good) rate function is as follows.

Lemma A.2.1 *If $\mathbf{I}$ is a rate function (resp. good rate function) and $K \subset \mathscr{X}$ is a nonempty compact (resp. closed) set, then $\mathbf{I}$ assumes its infimum on K.*

The Gärtner–Ellis theorem plays an important role in dealing with the LDPs for a family of random variables. When utilizing this theorem, one needs to examine whether the logarithmic moment generating function is essentially smooth.

Definition A.2.3 A convex function $\Lambda: \mathbb{R}^d \to (-\infty, \infty]$ is essentially smooth if

(1) $\mathscr{D}_\Lambda^\circ$ is non-empty, where $\mathscr{D}_\Lambda^\circ$ is the interior of $\mathscr{D}_\Lambda := \{x \in \mathbb{R}^d : \Lambda(x) < \infty\}$;
(2) $\Lambda(\cdot)$ is differentiable throughout $\mathscr{D}_\Lambda^\circ$;
(3) $\Lambda(\cdot)$ is steep, namely,

$$\lim_{n \to \infty} |\nabla \Lambda(\lambda_n)| = \infty,$$

whenever $\{\lambda_n\}_{n \in \mathbb{N}_+}$ is a sequence in $\mathscr{D}_\Lambda^\circ$ converging to a boundary point of $\mathscr{D}_\Lambda^\circ$.

Theorem A.2.1 (Gärtner–Ellis Theorem in $\mathbb{R}^d$) *Let $\{X_n\}_{n\in\mathbb{N}_+}$ be a sequence of random vectors taking values in $\mathbb{R}^d$. Assume that for each $\lambda \in \mathbb{R}^d$, the logarithmic moment generating function, defined as the limit*

$$\Lambda(\lambda) := \lim_{n\to\infty} \frac{1}{n} \log\left(\mathbb{E}[\exp(n\langle\lambda, X_n\rangle)]\right)$$

exists as an extended real number. Further, assume that the origin belongs to $\mathscr{D}_\Lambda^\circ$. If Λ is an essentially smooth and lower semi-continuous function, then the large deviation principle holds for $\{X_n\}_{n\in\mathbb{N}}$ with the good rate function $\Lambda^*(\cdot)$. Here

$$\Lambda^*(x) = \sup_{\lambda\in\mathbb{R}^d} \{\langle\lambda, x\rangle - \Lambda(\lambda)\}$$

for $x \in \mathbb{R}^d$, is the Fenchel–Legendre transform of $\Lambda(\cdot)$.

It is known that the key point of the Gärtner–Ellis theorem is to study the logarithmic moment generating function. Moreover, we would like to mention that the Gärtner–Ellis theorem is also valid in the case of continuous parameter family $\{X_\epsilon\}_{\epsilon>0}$ (see the remarks of [91, Theorem 2.3.6]). When $\mathscr{X}$ is infinite-dimensional, besides the logarithmic moment generating function, we also need to investigate the exponential tightness of $\{\mu_\epsilon\}_{\epsilon>0}$ to derive the LDP of $\{\mu_\epsilon\}_{\epsilon>0}$.

Definition A.2.4 (See [91]) A family of probability measures $\{\mu_\epsilon\}$ on $\mathscr{X}$ is exponentially tight if for every $\alpha < \infty$, there exists a compact set $K_\alpha \subset \mathscr{X}$ such that

$$\limsup_{\epsilon\to 0} \epsilon \log \mu_\epsilon(K_\alpha^c) < -\alpha. \tag{A.1}$$

Remark A.2.1 If the state space $\mathscr{X}$ is finite dimensional, the existence of logarithmic moment generating function implies the exponential tightness.

Theorem A.2.2 *Let $\{\mu_\epsilon\}_{\epsilon>0}$ be an exponentially tight family of Borel probability measures on $\mathscr{X}$. Suppose the logarithmic moment generating function $\Lambda(\cdot) := \lim_{\epsilon\to 0} \epsilon\Lambda_{\mu_\epsilon}(\cdot/\epsilon)$ is finite valued and Gateaux differentiable, where $\Lambda_{\mu_\epsilon}(\lambda) := \log\int_{\mathscr{X}} \exp(\lambda(x))\mu_\epsilon(dx)$, $\lambda \in \mathscr{X}^*$. Then $\{\mu_\epsilon\}_{\epsilon>0}$ satisfies the LDP in $\mathscr{X}$ with the convex, good rate function*

$$\Lambda^*(x) := \sup_{\lambda\in\mathscr{X}^*} \{\lambda(x) - \Lambda(\lambda)\}.$$

The following two lemmas are useful to derive new LDPs based on a given LDP. The first lemma is also called the contraction principle, which produces a new LDP on another space based on the known LDP via a continuous mapping.

Lemma A.2.2 (See [91]) *Let $\mathscr{Y}$ be another Hausdorff topological space, $f\colon \mathscr{X} \to \mathscr{Y}$ be a continuous function, and $\mathbf{I}\colon \mathscr{X} \to [0, \infty]$ be a good rate function.*

1. *For each $y \in \mathscr{Y}$, define*

$$\tilde{\mathbf{I}}(y) := \inf\{\mathbf{I}(x) \mid x \in \mathscr{X},\, y = f(x)\}.$$

Then $\tilde{\mathbf{I}}(y)$ is a good rate function on $\mathscr{Y}$, where as usual the infimum over the empty set is taken as ∞.

2. *If $\mathbf{I}$ controls the LDP associated with a family of probability measures $\{\mu_\epsilon\}_{\epsilon>0}$ on $\mathscr{X}$, then $\tilde{\mathbf{I}}$ controls the LDP associated with the family of probability measures $\{\mu_\epsilon \circ f^{-1}\}_{\epsilon>0}$ on $\mathscr{Y}$.*

The second one gives the relationship between the LDP of $\{\mu_\epsilon\}_{\epsilon>0}$ on $\mathscr{X}$ and that on the subspaces of $\mathscr{X}$.

Lemma A.2.3 (See [91]) *Let E be a measurable subset of $\mathscr{X}$ such that $\mu_\epsilon(E) = 1$ for all $\epsilon > 0$. Suppose that E is equipped with the topology induced by $\mathscr{X}$. If E is a closed subset of $\mathscr{X}$ and $\{\mu_\epsilon\}_{\epsilon>0}$ satisfies the LDP on E with the rate function $\mathbf{I}$, then $\{\mu_\epsilon\}_{\epsilon>0}$ satisfies the LDP on $\mathscr{X}$ with the rate function $\tilde{\mathbf{I}}$ such that $\tilde{\mathbf{I}}(y) = \mathbf{I}$ on E and $\tilde{\mathbf{I}}(y) = \infty$ on E^c.*

A.3 Exponential Integrability

Now we introduce the exponential integrability for the stochastic differential equation in the following lemma.

Lemma A.3.1 (See [74]) *Let H be a Hilbert space and X be an H-valued adapted stochastic process with continuous sample paths satisfying $\int_0^T \|\mu(X_t)\| + \|\sigma(X_t)\|^2 dt < \infty$ a.s., and for all $t \in [0, T]$, $X_t = X_0 + \int_0^t \mu(X_r)dr + \int_0^t \sigma(X_r)dW_r$, a.s. Assume that there exist two functionals $\overline{V}$ and $V \in \mathbf{C}^2(H, \mathbb{R})$ and an $\mathscr{F}_0$-measurable random variable α such that for almost every $t \in [0, T]$,*

$$DV(X_t)\mu(X_t) + \frac{\mathrm{Tr}\big[D^2 V(X_t)\sigma(X_t)\sigma^*(X_t)\big]}{2}$$

$$+ \frac{\|\sigma^*(X)DV(X_t)\|^2}{2e^{\alpha t}} + \overline{V}(X_t) \le \alpha V(X_t), \quad a.s. \tag{A.2}$$

Then

$$\sup_{t\in[0,T]} \mathbb{E}\left[\exp\left(\frac{V(X_t)}{e^{\alpha t}} + \int_0^t \frac{\overline{V}(X_r)}{e^{\alpha r}}dr\right)\right] \le \mathbb{E}\left[e^{V(X_0)}\right]. \tag{A.3}$$

Proof Let $Y_t = \int_0^t \frac{\overline{V}(X_r)}{e^{\alpha r}} dr$. Applying the Itô formula to

$$Z(t, X_t, Y_t) := \exp\left(\frac{V(X_t)}{e^{\alpha t}} + Y_t\right),$$

we obtain

$$Z(t, X_t, Y_t) - e^{V(X_0)}$$

$$= \int_0^t e^{-\alpha r} Z(r, X_r, Y_r)(\overline{V}(X_r) - \alpha V(X_r)) dr$$

$$+ \int_0^t e^{-\alpha r} Z(r, X_r, Y_r) DV(X_r)\mu(X_r) dX_r$$

$$+ \frac{1}{2}\int_0^t e^{-\alpha r} Z(r, X_r, Y_r)\mathrm{Tr}\left[D^2 V(X_r)\sigma(X_r)\sigma^*(X_r)\right] dr$$

$$+ \frac{1}{2}\int_0^t e^{-2\alpha r} Z(r, X_r, Y_r)\left\|\sigma^*(X_r)DV(X_r)\right\|^2 dr.$$

Condition (A.2) implies that

$$\mathbb{E}\left[Z(t, X_t, Y_t)\right] - \mathbb{E}\left[e^{V(X_0)}\right]$$

$$= \mathbb{E}\left[\int_0^t Z(r, X_r, Y_r)\left(\frac{\overline{V}(X_r) - \alpha V(X_r)}{e^{\alpha r}} + \frac{DV(X_r)\mu(X_r)}{e^{\alpha r}}\right.\right.$$

$$\left.\left. + \frac{\mathrm{Tr}\left[D^2 V(X_r)\sigma(X_r)\sigma^*(X_r)\right]}{2e^{\alpha r}} + \frac{\|\sigma^*(X_r)DV(X_r)\|^2}{2e^{2\alpha r}}\right) dr\right] \leq 0.$$

This completes the proof of (A.3). □

A.4 Stochastic Multi-Symplectic Structure

Chapter 4 investigates the infinite-dimensional stochastic Hamiltonian system, which can be regarded as a stochastic evolution equation in time. When the spatial variable is also of interest, both the stochastic multi-symplectic Hamiltonian system (also called stochastic Hamiltonian partial differential equation) and stochastic multi-symplectic structure are involved.

Stochastic multi-symplectic Hamiltonian systems are initiated by Jiang et al. [146] in 2013. They are stochastic partial differential equations and formulated as

$$M dz + K z_x dt = \nabla S_1(z) dt + \nabla S_2(z) \circ dW(t), \tag{A.4}$$

where $z \in \mathbb{R}^n$, S_1, S_2 are smooth functions of z, and M, K are $n \times n$ skew-symmetric matrices. Here, we write z_x for the Jacobian matrix of z with respect to the spatial variable x, and $W(\cdot)$ is a **Q**-Wiener process. Concrete examples include stochastic Maxwell equations, stochastic nonlinear Schrödinger equations, nonlinear stochastic wave equations, and there may be more practical instances in this category.

Theorem A.4.1 (See [146]) *The stochastic multi-symplectic Hamiltonian system* (A.4) *preserves the stochastic multi-symplectic conservation law locally, i.e.,*

$$d\omega_2^{(1)}(t, x) + \partial_x \omega_2^{(2)}(t, x)dt = 0, \quad a.s.,$$

equivalently,

$$\int_{x_0}^{x_1} \omega_2^{(1)}(t_1, x)\, dx + \int_{t_0}^{t_1} \omega_2^{(2)}(t, x_1)\, dt$$

$$= \int_{x_0}^{x_1} \omega_2^{(1)}(t_0, x)\, dx + \int_{t_0}^{t_1} \omega_2^{(2)}(t, x_0)\, dt, \quad a.s., \tag{A.5}$$

where $\omega_2^{(1)} = \frac{1}{2}dz \wedge Mdz$ and $\omega_2^{(2)} = \frac{1}{2}dz \wedge Kdz$ are the differential 2-forms, and $(x_0, x_1) \times (t_0, t_1)$ *is the local definition domain of $z(x, t)$.*

From (A.5), it can be seen that the symplecticity changes locally and synchronously in time and space directions. An immediate question is what kind of numerical methods applied to discretizing the stochastic multi-symplectic Hamiltonian system could preserve the discrete form of the stochastic multi-symplectic conservation law. Now we start the discussion with a review of results on the stochastic multi-symplecticity and stochastic multi-symplectic methods for some specific and fundamental stochastic Hamiltonian partial differential equations one after the other.

Stochastic Schrödinger Equation Consider 1-dimensional stochastic nonlinear Schrödinger equation driven by multiplicative noise

$$idu = \Delta u dt + 2|u|^{2\sigma} u dt + \varepsilon u \circ dW(t), \tag{A.6}$$

where $\sigma \leq 2$, $\varepsilon > 0$, $u = u(x, t)$ with $t > 0$ and $x \in [0, 1]$ is $\mathbb{C}$-valued, and $W(\cdot)$ is an $\mathbf{L}^2([0, 1])$-valued **Q**-Wiener process. By setting $u = P + iQ$, we separate (A.6) into

$$\begin{cases} dP = \Delta Q dt + 2\left(P^2 + Q^2\right)^\sigma Q dt + \varepsilon Q \circ dW(t), \\ dQ = -\Delta P dt - 2\left(P^2 + Q^2\right)^\sigma P dt - \varepsilon P \circ dW(t). \end{cases}$$

Introducing two additional new variables $v = \frac{\partial P}{\partial x}$ and $w = \frac{\partial Q}{\partial x}$, and defining a state variable $z = (P, Q, v, w)^\top$, we arrive at the associated stochastic multi-symplectic Hamiltonian system

$$Mdz + Kz_x dt = \nabla S_1(z)dt + \nabla S_2(z) \circ dW(t), \tag{A.7}$$

where

$$M = \begin{bmatrix} 0 & -1 & 0 & 0 \\ 1 & 0 & 0 & 0 \\ 0 & 0 & 0 & 0 \\ 0 & 0 & 0 & 0 \end{bmatrix}, \quad K = \begin{bmatrix} 0 & 0 & -1 & 0 \\ 0 & 0 & 0 & -1 \\ 1 & 0 & 0 & 0 \\ 0 & 1 & 0 & 0 \end{bmatrix},$$

and

$$S_1(z) = \frac{1}{2}v^2 + \frac{1}{2}w^2 + \frac{1}{(\sigma+1)}\left(P^2 + Q^2\right)^{\sigma+1}, \quad S_2(z) = \frac{1}{2}\varepsilon\left(P^2 + Q^2\right).$$

Let $0 = x_0 < x_1 < \cdots < x_{M+1} = 1$ be the uniform partition of the interval $\mathcal{O}$ with the stepsize $\tau = \frac{1}{M+1}$ and take a partition $0 = t_0 \leq t_1 \leq \cdots \leq t_N = T$ with a uniform time-step size $h = t_{n+1} - t_n$, $n = 0, 1, \ldots, N-1$. Taking advantage of the midpoint method to discretize (A.7) in both temporal and spatial directions, a stochastic multi-symplectic method reads

$$M\left(\frac{z_{n+1}^{j+\frac{1}{2}} - z_n^{j+\frac{1}{2}}}{h}\right) + K\left(\frac{z_{n+\frac{1}{2}}^{j+1} - z_{n+\frac{1}{2}}^{j}}{\tau}\right) = \nabla S_1\left(z_{n+\frac{1}{2}}^{j+\frac{1}{2}}\right) + \nabla S_2\left(z_{n+\frac{1}{2}}^{j+\frac{1}{2}}\right)\Delta W_{n+\frac{1}{2}}^{j+\frac{1}{2}},$$

$$\tag{A.8}$$

where $z_{n+\frac{1}{2}}^{j} = \frac{z_{n+1}^{j}+z_n^{j}}{2}$, $z_n^{j+\frac{1}{2}} = \frac{z_n^{j+1}+z_n^{j}}{2}$, $z_{n+\frac{1}{2}}^{j+\frac{1}{2}} = \frac{z_n^{j}+z_n^{j+1}+z_{n+1}^{j}+z_{n+1}^{j+1}}{4}$, and $\Delta W_{n+\frac{1}{2}}^{j+\frac{1}{2}} = \frac{W(x_j,t_{n+1})-W(x_j,t_n)}{h}$.

Theorem A.4.2 (See [146]) *The full discretization (A.8) satisfies the discrete stochastic multi-symplectic conservation law:*

$$\left(\omega_{n+1}^{j+\frac{1}{2}} - \omega_n^{j+\frac{1}{2}}\right)\tau + \left(\kappa_{n+\frac{1}{2}}^{j+1} - \kappa_{n+\frac{1}{2}}^{j}\right)h = 0, \quad a.s.,$$

where $\omega_n^{j+\frac{1}{2}} = \frac{1}{2}dz_n^{j+\frac{1}{2}} \wedge Mdz_n^{j+\frac{1}{2}}$ *and* $\kappa_{n+\frac{1}{2}}^{j} = \frac{1}{2}dz_{n+\frac{1}{2}}^{j} \wedge Kdz_{n+\frac{1}{2}}^{j}$.

Proof Taking differential on both sides of (A.8) leads to

$$\tau M\left(dz_{n+1}^{j+\frac{1}{2}} - dz_{n}^{j+\frac{1}{2}}\right) + hK\left(dz_{n+\frac{1}{2}}^{j+1} - dz_{n+\frac{1}{2}}^{j}\right)$$
$$= \tau h\left(\nabla^2 S_1\left(z_{n+\frac{1}{2}}^{j+\frac{1}{2}}\right)dz_{n+\frac{1}{2}}^{j+\frac{1}{2}} + \Delta W_{n+\frac{1}{2}}^{j+\frac{1}{2}}\nabla^2 S_2\left(z_{n+\frac{1}{2}}^{j+\frac{1}{2}}\right)dz_{n+\frac{1}{2}}^{j+\frac{1}{2}}\right).$$

Making use of the symmetry of both $\nabla^2 S_1(z)$ and $\nabla^2 S_2(z)$, together with

$$dz_{n+\frac{1}{2}}^{j+\frac{1}{2}} := \frac{1}{2}\left(dz_{n+1}^{j+\frac{1}{2}} + dz_{n}^{j+\frac{1}{2}}\right) = \frac{1}{2}\left(dz_{n+\frac{1}{2}}^{j+1} + dz_{n+\frac{1}{2}}^{j}\right),$$

yields

$$\tau\left(dz_{n+1}^{j+\frac{1}{2}} \wedge M dz_{n+1}^{j+\frac{1}{2}} - dz_{n}^{j+\frac{1}{2}} \wedge M dz_{n}^{j+\frac{1}{2}}\right)$$
$$+ h\left(dz_{n+\frac{1}{2}}^{j+1} \wedge K dz_{n+\frac{1}{2}}^{j+1} - dz_{n+\frac{1}{2}}^{j} \wedge K dz_{n+\frac{1}{2}}^{j}\right) = 0,$$

which completes the proof.   □

Stochastic Maxwell Equation with Additive Noise The stochastic Maxwell equation driven by an additive noise has been taken into consideration in [50, 125], and possesses the following form

$$\begin{cases} \varepsilon d\mathbf{E} = \nabla \times \mathbf{H} dt - (\lambda_1, \lambda_1, \lambda_1)^\top dW(t), \\ \mu d\mathbf{H} = -\nabla \times \mathbf{E} dt + (\lambda_2, \lambda_2, \lambda_2)^\top dW(t), \end{cases} \tag{A.9}$$

where ε is the electric permittivity, μ is the magnetic permeability, $\lambda_1, \lambda_2 \in \mathbb{R}$, and $W(\cdot)$ is an $\mathbf{L}^2(\mathcal{O})$-valued $\mathbf{Q}$-Wiener process with $\mathcal{O} = [x_L, x_R] \times [y_L, y_R] \times [z_L, z_R] \subset \mathbb{R}^3$. Moreover, $\mathbf{E} = (\tilde{E}_1, \tilde{E}_2, \tilde{E}_3)^\top$ and $\mathbf{H} = (\tilde{H}_1, \tilde{H}_2, \tilde{H}_3)^\top$.

The formulation of the stochastic Hamiltonian partial differential equation of (A.9) is due to [50]. Let $u = \left(\tilde{H}_1, \tilde{H}_2, \tilde{H}_3, \tilde{E}_1, \tilde{E}_2, \tilde{E}_3\right)^\top$. Then (A.9) can be rewritten as

$$M du + K_1 u_x dt + K_2 u_y dt + K_3 u_z dt = \nabla S(u) \circ dW(t)$$

with $S(u) = -\lambda_2\left(\tilde{H}_1 + \tilde{H}_2 + \tilde{H}_3\right) - \lambda_1\left(\tilde{E}_1 + \tilde{E}_2 + \tilde{E}_3\right)$ and

$$M = \begin{bmatrix} 0 & -I_3 \\ I_3 & 0 \end{bmatrix}, \quad K_i = \begin{bmatrix} D_i & 0 \\ 0 & D_i \end{bmatrix}, \quad i = 1, 2, 3, \tag{A.10}$$

and

$$D_1 = \begin{bmatrix} 0 & 0 & 0 \\ 0 & 0 & -1 \\ 0 & 1 & 0 \end{bmatrix}, \quad D_2 = \begin{bmatrix} 0 & 0 & 1 \\ 0 & 0 & 0 \\ -1 & 0 & 0 \end{bmatrix}, \quad D_3 = \begin{bmatrix} 0 & -1 & 0 \\ 1 & 0 & 0 \\ 0 & 0 & 0 \end{bmatrix}.$$

In what follows, denote the mesh sizes along x, y and z directions by Δx, Δy and Δz, respectively, and let Δt be the time-step size. The temporal-spatial domain $[0, T] \times [x_L, x_R] \times [y_L, y_R] \times [z_L, z_R]$ is partitioned by parallel lines with $t_n = n\Delta t$ and $x_i = x_L + i\Delta x, y_j = y_L + j\Delta y, z_k = z_L + k\Delta z$ for $n = 0, 1, \ldots, N$ and $i = 0, 1, \ldots, I$, $j = 0, 1, \ldots, J$, and $k = 0, 1, \ldots, K$. The grid point function $u_n^{i,j,k}$ is the approximation of $Z(t, x, y, z)$ at the node (t_n, x_i, y_j, z_k), where $n = 0, 1, \ldots, N$ and $i = 0, 1, \ldots, I$, $j = 0, 1, \ldots, J$, and $k = 0, 1, \ldots, K$. We introduce the following difference operators $\delta_t, \bar{\delta}_t, \delta_x, \bar{\delta}_x$ defined by

$$\delta_t u_n^{i,j,k} = \frac{u_{n+1}^{i,j,k} - u_n^{i,j,k}}{\Delta t}, \quad \bar{\delta}_t u_n^{i,j,k} = \frac{u_{n+1}^{i,j,k} - u_{n-1}^{i,j,k}}{2\Delta t},$$

$$\delta_x u_n^{i,j,k} = \frac{u_{i+1,j,k}^n - u_n^{i,j,k}}{\Delta x}, \quad \bar{\delta}_x u_n^{i,j,k} = \frac{u_{i+1,j,k}^n - u_{i-1,j,k}^n}{2\Delta x}.$$

The definitions for the difference operators $\delta_y, \bar{\delta}_y, \delta_z, \bar{\delta}_z$ along y and z directions are analogous. Based on above notations, [50] proposes three stochastic multi-symplectic method for (A.9) as follows

- Scheme I:

$$\delta_t (E_1)_n^{i+\frac{1}{2},j+\frac{1}{2},k+\frac{1}{2}} = \delta_y (H_3)_{n+\frac{1}{2}}^{i+\frac{1}{2},j,k+\frac{1}{2}} - \delta_z (H_2)_{n+\frac{1}{2}}^{i+\frac{1}{2},j+\frac{1}{2},k} - \lambda_1 \Delta W_n^{i,j,k},$$

$$\delta_t (E_2)_n^{i+\frac{1}{2},j+\frac{1}{2},k+\frac{1}{2}} = \delta_z (H_1)_{n+\frac{1}{2}}^{i+\frac{1}{2},j+\frac{1}{2},k} - \delta_x (H_3)_{n+\frac{1}{2}}^{i,j+\frac{1}{2},k+\frac{1}{2}} - \lambda_1 \Delta W_n^{i,j,k},$$

$$\delta_t (E_3)_n^{i+\frac{1}{2},j+\frac{1}{2},k+\frac{1}{2}} = \delta_x (H_2)_{n+\frac{1}{2}}^{i,j+\frac{1}{2},k+\frac{1}{2}} - \delta_y (H_1)_{n+\frac{1}{2}}^{i+\frac{1}{2},j,k+\frac{1}{2}} - \lambda_1 \Delta W_n^{i,j,k},$$

$$\delta_t (H_1)_n^{i+\frac{1}{2},j+\frac{1}{2},k+\frac{1}{2}} = \delta_z (E_2)_{n+\frac{1}{2}}^{i+\frac{1}{2},j+\frac{1}{2},k} - \delta_y (E_3)_{n+\frac{1}{2}}^{i+\frac{1}{2},j,k+\frac{1}{2}} + \lambda_2 \Delta W_n^{i,j,k},$$

$$\delta_t (H_2)_n^{i+\frac{1}{2},j+\frac{1}{2},k+\frac{1}{2}} = \delta_x (E_3)_{n+\frac{1}{2}}^{i,j+\frac{1}{2},k+\frac{1}{2}} - \delta_z (E_1)_{n+\frac{1}{2}}^{i+\frac{1}{2},j+\frac{1}{2},k} + \lambda_2 \Delta W_n^{i,j,k},$$

$$\delta_t (H_3)_n^{i+\frac{1}{2},j+\frac{1}{2},k+\frac{1}{2}} = \delta_y (E_1)_{n+\frac{1}{2}}^{i+\frac{1}{2},j,k+\frac{1}{2}} - \delta_x (E_2)_{n+\frac{1}{2}}^{i,j+\frac{1}{2},k+\frac{1}{2}} + \lambda_2 \Delta W_n^{i,j,k},$$

where $\Delta W_n^{i,j,k} = \left(W_{n+1}^{i,j,k} - W_n^{i,j,k} \right)/\Delta t$. It can be verified that Scheme I preserves the discrete stochastic multi-symplectic conservation law

$$
\frac{\omega_{i+\frac{1}{2},j+\frac{1}{2},k+\frac{1}{2}}^{n+1} - \omega_{i+\frac{1}{2},j+\frac{1}{2},k+\frac{1}{2}}^{n}}{\Delta t} + \frac{(\kappa_1)_{i+1,j+\frac{1}{2},k+\frac{1}{2}}^{n+\frac{1}{2}} - (\kappa_1)_{i,j+\frac{1}{2},k+\frac{1}{2}}^{n+\frac{1}{2}}}{\Delta x}
$$

$$
+ \frac{(\kappa_2)_{i+\frac{1}{2},j+1,k+\frac{1}{2}}^{n+\frac{1}{2}} - (\kappa_2)_{i+\frac{1}{2},j,k+\frac{1}{2}}^{n+\frac{1}{2}}}{\Delta y} + \frac{(\kappa_3)_{i+\frac{1}{2},j+\frac{1}{2},k+1}^{n+\frac{1}{2}} - (\kappa_3)_{i+\frac{1}{2},j+\frac{1}{2},k}^{n+\frac{1}{2}}}{\Delta z} = 0, \quad a.s.,
$$

where

$$
\omega_{i,j,k}^{n} = \mathrm{d}u_{i,j,k}^{n} \wedge \mathrm{M}\mathrm{d}u_{i,j,k}^{n}, \quad \left(\kappa_p \right)_{i,j,k}^{n} = \mathrm{d}u_{i,j,k}^{n} \wedge \mathrm{K}_p \mathrm{d}u_{i,j,k}^{n}, \quad p = 1, 2, 3.
$$

- Scheme II:

$$
\bar{\delta}_t (E_1)_n^{i,j,k} = \bar{\delta}_y (H_3)_n^{i,j,k} - \bar{\delta}_z (H_2)_n^{i,j,k} - \lambda_1 \Delta \widetilde{W}_{n+1}^{i,j,k},
$$

$$
\bar{\delta}_t (E_2)_n^{i,j,k} = \bar{\delta}_z (H_1)_n^{i,j,k} - \bar{\delta}_x (H_3)_n^{i,j,k} - \lambda_1 \Delta \widetilde{W}_{n+1}^{i,j,k},
$$

$$
\bar{\delta}_t (E_3)_n^{i,j,k} = \bar{\delta}_x (H_2)_n^{i,j,k} - \bar{\delta}_y (H_1)_n^{i,j,k} - \lambda_1 \Delta \widetilde{W}_{n+1}^{i,j,k},
$$

$$
\bar{\delta}_t (H_1)_n^{i,j,k} = -\bar{\delta}_y (E_3)_n^{i,j,k} + \bar{\delta}_z (E_2)_n^{i,j,k} + \lambda_2 \Delta \widetilde{W}_{n+1}^{i,j,k},
$$

$$
\bar{\delta}_t (H_2)_n^{i,j,k} = -\bar{\delta}_z (E_1)_n^{i,j,k} + \bar{\delta}_x (E_3)_n^{i,j,k} + \lambda_2 \Delta \widetilde{W}_{n+1}^{i,j,k},
$$

$$
\bar{\delta}_t (H_3)_n^{i,j,k} = -\bar{\delta}_x (E_2)_n^{i,j,k} + \bar{\delta}_y (E_1)_n^{i,j,k} + \lambda_2 \Delta \widetilde{W}_{n+1}^{i,j,k}
$$

with $\Delta \widetilde{W}_{n+1}^{i,j,k} = \left(W_{n+1}^{i,j,k} - W_{n-1}^{i,j,k} \right)/(2\Delta t)$. This method possesses the discrete stochastic multi-symplectic conservation law

$$
\frac{\omega_{i,j,k}^{n+\frac{1}{2}} - \omega_{i,j,k}^{n-\frac{1}{2}}}{\Delta t} + \frac{(\kappa_1)_{i+\frac{1}{2},j,k}^{n} - (\kappa_1)_{i-\frac{1}{2},j,k}^{n}}{\Delta x}
$$

$$
+ \frac{(\kappa_2)_{i,j+\frac{1}{2},k}^{n} - (\kappa_2)_{i,j-\frac{1}{2},k}^{n}}{\Delta y} + \frac{(\kappa_3)_{i,j,k+\frac{1}{2}}^{n} - (\kappa_3)_{i,j,k-\frac{1}{2}}^{n}}{\Delta z} = 0, \quad a.s.,
$$

where

$$
\omega_{i,j,k}^{n+\frac{1}{2}} = \mathrm{d}u_{i,j,k}^{n} \wedge \mathrm{M}\mathrm{d}u_{i,j,k}^{n+1}, \quad (\kappa_1)_{i+\frac{1}{2},j,k}^{n} = \mathrm{d}u_{i,j,k}^{n} \wedge \mathrm{K}_1 \mathrm{d}u_{i+1,j,k}^{n}
$$

$$
(\kappa_2)_{i,j+\frac{1}{2},k}^{n} = \mathrm{d}u_{i,j,k}^{n} \wedge \mathrm{K}_2 \mathrm{d}u_{i,j+1,k}^{n}, \quad (\kappa_3)_{i,j,k+\frac{1}{2}}^{n} = \mathrm{d}u_{i,j,k}^{n} \wedge \mathrm{K}_3 \mathrm{d}u_{i,j,k+1}^{n}.
$$

• Scheme III:

$$\delta_t \left(E_1\right)_n^{i,j,k} = \bar{\delta}_y \left(H_3\right)_{n+\frac{1}{2}}^{i,j,k} - \bar{\delta}_z \left(H_2\right)_{n+\frac{1}{2}}^{i,j,k} - \lambda_1 \Delta W_n^{i,j,k},$$

$$\delta_t \left(E_2\right)_n^{i,j,k} = \bar{\delta}_z \left(H_1\right)_{n+\frac{1}{2}}^{i,j,k} - \bar{\delta}_x \left(H_3\right)_{n+\frac{1}{2}}^{i,j,k} - \lambda_1 \Delta W_n^{i,j,k},$$

$$\delta_t \left(E_3\right)_n^{i,j,k} = \bar{\delta}_x \left(H_2\right)_{n+\frac{1}{2}}^{i,j,k} - \bar{\delta}_y \left(H_1\right)_{n+\frac{1}{2}}^{i,j,k} - \lambda_1 \Delta W_n^{i,j,k},$$

$$\delta_t \left(H_1\right)_n^{i,j,k} = -\bar{\delta}_y \left(E_3\right)_{n+\frac{1}{2}}^{i,j,k} + \bar{\delta}_z \left(E_2\right)_{n+\frac{1}{2}}^{i,j,k} + \lambda_2 \Delta W_n^{i,j,k},$$

$$\delta_t \left(H_2\right)_n^{i,j,k} = -\bar{\delta}_z \left(E_1\right)_{n+\frac{1}{2}}^{i,j,k} + \bar{\delta}_x \left(E_3\right)_{n+\frac{1}{2}}^{i,j,k} + \lambda_2 \Delta W_n^{i,j,k},$$

$$\delta_t \left(H_3\right)_n^{i,j,k} = -\bar{\delta}_x \left(E_2\right)_{n+\frac{1}{2}}^{i,j,k} + \bar{\delta}_y \left(E_1\right)_{n+\frac{1}{2}}^{i,j,k} + \lambda_2 \Delta W_n^{i,j,k},$$

where $\Delta W_n^{i,j,k} = \left(W_{n+1}^{i,j,k} - W_n^{i,j,k}\right)/\Delta t$. Scheme III satisfies the discrete stochastic multi-symplectic conservation law

$$\frac{\omega_{i,j,k}^{n+1} - \omega_{i,j,k}^n}{\Delta t} + \frac{(\kappa_1)_{i+\frac{1}{2},j,k}^{n+\frac{1}{2}} - (\kappa_1)_{i-\frac{1}{2},j,k}^{n+\frac{1}{2}}}{\Delta x}$$

$$+ \frac{(\kappa_2)_{i,j+\frac{1}{2},k}^{n+\frac{1}{2}} - (\kappa_2)_{i,j-\frac{1}{2},k}^{n+\frac{1}{2}}}{\Delta y} + \frac{(\kappa_3)_{i,j,k+\frac{1}{2}}^{n+\frac{1}{2}} - (\kappa_3)_{i,j,k-\frac{1}{2}}^{n+\frac{1}{2}}}{\Delta z} = 0, \quad a.s.,$$

where

$$\omega_{i,j,k}^{n+1} = du_{i,j,k}^{n+1} \wedge M du_{i,j,k}^{n+1}, \quad (\kappa_1)_{i+\frac{1}{2},j,k}^{n+\frac{1}{2}} = du_{i,j,k}^{n+\frac{1}{2}} \wedge K_1 du_{i+1,j,k}^{n+\frac{1}{2}}$$

$$(\kappa_2)_{i,j+\frac{1}{2},k}^{n+\frac{1}{2}} = du_{i,j,k}^{n+\frac{1}{2}} \wedge K_2 du_{i,j+1,k}^{n+\frac{1}{2}}, \quad (\kappa_3)_{i,j,k+\frac{1}{2}}^{n+\frac{1}{2}} = du_{i,j,k}^{n+\frac{1}{2}} \wedge K_3 du_{i,j,k+1}^{n+\frac{1}{2}}.$$

Stochastic Maxwell Equation with Multiplicative Noise In [127], the authors study the following stochastic Maxwell equations driven by multiplicative noise

$$\begin{cases} \varepsilon d\mathbf{E} = \nabla \times \mathbf{H} dt - \lambda \mathbf{H} \circ dW(t), \\ \mu d\mathbf{H} = -\nabla \times \mathbf{E} dt + \lambda \mathbf{E} \circ dW(t), \end{cases} \tag{A.11}$$

with λ being the amplitude of the noise and $W(\cdot)$ being an $\mathbf{L}^2(\mathscr{O})$-valued $\mathbf{Q}$-Wiener process. Denoting $u = \left(\tilde{\mathbf{H}}_1, \tilde{\mathbf{H}}_2, \tilde{\mathbf{H}}_3, \tilde{\mathbf{E}}_1, \tilde{\mathbf{E}}_2, \tilde{\mathbf{E}}_3\right)^{\top}$, we derive the associated stochastic Hamiltonian partial differential equation

$$M du + K_1 u_x dt + K_2 u_y dt + K_3 u_z dt = \nabla S(u) \circ dW(t)$$

with $S(u) = \frac{\lambda}{2}\left(\left|\tilde{\mathbf{E}}_1\right|^2 + \left|\tilde{\mathbf{E}}_2\right|^2 + \left|\tilde{\mathbf{E}}_3\right|^2 + \left|\tilde{\mathbf{H}}_1\right|^2 + \left|\tilde{\mathbf{H}}_2\right|^2 + \left|\tilde{\mathbf{H}}_3\right|^2\right)$ and M, K$_i$, $i = 1, 2, 3$ being given by (A.10).

Based on the wavelet interpolation technique in space and the midpoint method in time, [127] presents a multi-symplectic method as follows

$$\left(\tilde{\mathbb{E}}_1\right)_{n+1} - \left(\tilde{\mathbb{E}}_1\right)_n = \Delta t \left(A_2 \left(\tilde{\mathbb{H}}_3\right)_{n+\frac{1}{2}} - A_3 \left(\tilde{\mathbb{H}}_2\right)_{n+\frac{1}{2}}\right) - \lambda \left(\tilde{\mathbb{H}}_1\right)_{n+\frac{1}{2}} \Delta_n W,$$

$$\left(\tilde{\mathbb{E}}_2\right)_{n+1} - \left(\tilde{\mathbb{E}}_2\right)_n = \Delta t \left(A_3 \left(\tilde{\mathbb{H}}_1\right)_{n+\frac{1}{2}} - A_1 \left(\tilde{\mathbb{H}}_3\right)_{n+\frac{1}{2}}\right) - \lambda \left(\tilde{\mathbb{H}}_2\right)_{n+\frac{1}{2}} \Delta_n W,$$

$$\left(\tilde{\mathbb{E}}_3\right)_{n+1} - \left(\tilde{\mathbb{E}}_3\right)_n = \Delta t \left(A_1 \left(\tilde{\mathbb{H}}_2\right)_{n+\frac{1}{2}} - A_2 \left(\tilde{\mathbb{H}}_1\right)_{n+\frac{1}{2}}\right) - \lambda \left(\tilde{\mathbb{H}}_3\right)_{n+\frac{1}{2}} \Delta_n W,$$

$$\left(\tilde{\mathbb{H}}_1\right)_{n+1} - \left(\tilde{\mathbb{H}}_1\right)_n = \Delta t \left(A_3 \left(\tilde{\mathbb{E}}_2\right)_{n+\frac{1}{2}} - A_2 \left(\tilde{\mathbb{E}}_3\right)_{n+\frac{1}{2}}\right) + \lambda \left(\tilde{\mathbb{E}}_1\right)_{n+\frac{1}{2}} \Delta_n W,$$

$$\left(\tilde{\mathbb{H}}_2\right)_{n+1} - \left(\tilde{\mathbb{H}}_2\right)_n = \Delta t \left(A_1 \left(\tilde{\mathbb{E}}_3\right)_{n+\frac{1}{2}} - A_3 \left(\tilde{\mathbb{E}}_1\right)_{n+\frac{1}{2}}\right) + \lambda \left(\tilde{\mathbb{E}}_2\right)_{n+\frac{1}{2}} \Delta_n W,$$

$$\left(\tilde{\mathbb{H}}_3\right)_{n+1} - \left(\tilde{\mathbb{H}}_3\right)_n = \Delta t \left(A_2 \left(\tilde{\mathbb{E}}_1\right)_{n+\frac{1}{2}} - A_1 \left(\tilde{\mathbb{E}}_2\right)_{n+\frac{1}{2}}\right) + \lambda \left(\tilde{\mathbb{E}}_3\right)_{n+\frac{1}{2}} \Delta_n W,$$

where

$$\left(\tilde{\mathbb{E}}_k\right)_n = \left((E_k)_n^{1,1,1}, (E_k)_n^{2,1,1}, (E_k)_n^{N_1,1,1}, \ldots, (E_k)_n^{1,N_2,1}, \ldots, (E_k)_n^{N_1,N_2,N_3}\right)^\top,$$

$$\left(\tilde{\mathbb{H}}_k\right)_n = \left((H_k)_n^{1,1,1}, (H_k)_n^{2,1,1}, (H_1)_n^{N_1,1,1}, \ldots, (H_1)_n^{1,N_2,1}, \ldots, (H_k)_n^{N_1,N_2,N_3}\right)^\top$$

for $k = 1, 2, 3$, and $A_1 = B^x \otimes I_{N_2} \otimes I_{N_3}$, $A_2 = I_{N_1} \otimes B^y \otimes I_{N_3}$ and $A_3 = I_{N_1} \otimes I_{N_2} \otimes B^z$ are skew-symmetric matrices corresponding to differential matrices B^x, B^y and B^z, respectively (see [271]). Moreover, $\Delta_n W = \mathbf{1} \otimes \left(\Delta_n W^1, \ldots, \Delta_n W^{N_1}\right)^\top$, $\mathbf{1} = (1, \ldots, 1)_{N_2 \times N_3}^\top$, and $\Delta_n W^i$ stands for an approximation of $\Delta_n W$ in the spatial direction with respect to x. This full discretization possesses the following discrete stochastic multi-symplectic conservation law

$$\frac{\omega_{i,j,k}^{n+1} - \omega_{i,j,k}^n}{\Delta t} + \sum_{i'=i-(\gamma-1)}^{i+(\gamma-1)} (B^x)_{i,i'} (\kappa_x)_{i',j,k}^{n+1/2}$$

$$+ \sum_{j'=j-(\gamma-1)}^{j+(\gamma-1)} (B^y)_{j,j'} (\kappa_y)_{i,j',k}^{n+1/2} + \sum_{k'=k-(\gamma-1)}^{k+(\gamma-1)} (B^z)_{k,k'} (\kappa_z)_{i,j,k'}^{n+1/2} = 0, \quad a.s.,$$

where γ is a positive even integer,

$$\omega_{i,j,k}^n = \frac{1}{2} du_{i,j,k}^n \wedge M du_{i,j,k}^n, \quad (\kappa_x)_{i',j,k}^{n+1/2} = du_{i,j,k}^{n+1/2} \wedge K_1 du_{i',j,k}^{n+1/2},$$

$$\left(\kappa_y\right)_{i,j',k}^{n+1/2} = du_{i,j,k}^{n+1/2} \wedge K_2 du_{i,j',k}^{n+1/2}, \quad (\kappa_z)_{i,j,k'}^{n+1/2} = du_{i,j,k}^{n+1/2} \wedge K_3 du_{i,j,k'}^{n+1/2}.$$

References

1. A. Abdulle, D. Cohen, G. Vilmart, K.C. Zygalakis, High weak order methods for stochastic differential equations based on modified equations. SIAM J. Sci. Comput. **34**(3), A1800–A1823 (2012)
2. A. Abdulle, G. Vilmart, K. C. Zygalakis, High order numerical approximation of the invariant measure of ergodic SDEs. SIAM J. Numer. Anal. **52**(4), 1600–1622 (2014)
3. A. Abdulle, G. Vilmart, K.C. Zygalakis, Long time accuracy of Lie-Trotter splitting methods for Langevin dynamics. SIAM J. Numer. Anal. **53**(1), 1–16 (2015)
4. S. Albeverio, A. Hilbert, E. Zehnder, Hamiltonian systems with a stochastic force: nonlinear versus linear, and a Girsanov formula. Stochastics Stochastics Rep. **39**(2–3), 159–188 (1992)
5. A. Alfonsi, Strong order one convergence of a drift implicit Euler scheme: application to the CIR process. Statist. Probab. Lett. **83**(2), 602–607 (2013)
6. D.F. Anderson, D.J. Higham, Y. Sun, Multilevel Monte Carlo for stochastic differential equations with small noise. SIAM J. Numer. Anal. **54**(2), 505–529 (2016)
7. L.C. Andrews, R.L. Phillips, *Laser Beam Propagation Through Random Media* (SPIE, Bellingham, 2005)
8. S. Anmarkrud, A. Kværnø, Order conditions for stochastic Runge–Kutta methods preserving quadratic invariants of Stratonovich SDEs. J. Comput. Appl. Math. **316**, 40–46 (2017)
9. C. Anton, Weak backward error analysis for stochastic Hamiltonian systems. BIT **59**(3), 613–646 (2019)
10. C. Anton, Explicit pseudo-symplectic methods based on generating functions for stochastic Hamiltonian systems. J. Comput. Appl. Math. **373**, 112433, 16 (2020)
11. R. Anton, D. Cohen, Exponential integrators for stochastic Schrödinger equations driven by Itô noise. J. Comput. Math. **36**(2), 276–309 (2018)
12. C. Anton, Y.S. Wong, J. Deng, Symplectic schemes for stochastic Hamiltonian systems preserving Hamiltonian functions. Int. J. Numer. Anal. Model. **11**(3), 427–451 (2014)
13. C. Anton, J. Deng, Y.S. Wong, Weak symplectic schemes for stochastic Hamiltonian equations. Electron. Trans. Numer. Anal. **43**, 1–20 (2014/2015)
14. R. Anton, D. Cohen, S. Larsson, X. Wang, Full discretization of semilinear stochastic wave equations driven by multiplicative noise. SIAM J. Numer. Anal. **54**(2), 1093–1119 (2016)
15. R. Anton, D. Cohen, L. Quer-Sardanyons, A fully discrete approximation of the one-dimensional stochastic heat equation. IMA J. Numer. Anal. **40**(1), 247–284 (2020)
16. L. Arnold, *Stochastic Differential Equations: Theory and Applications.* (Wiley-Interscience [John Wiley & Sons], New York-London-Sydney, 1974)
17. L. Arnold, I. Chueshov, G. Ochs, *Random Dynamical Systems Methods in Ship Stability: A Case Study* (Springer, Berlin, 2005)

© The Author(s), under exclusive license to Springer Nature Singapore Pte Ltd. 2022
J. Hong, L. Sun, *Symplectic Integration of Stochastic Hamiltonian Systems*,
Lecture Notes in Mathematics 2314, https://doi.org/10.1007/978-981-19-7670-4

18. G.B. Arous, Flots et séries de Taylor stochastiques. Probab. Theory Related Fields **81**(1), 29–77 (1989)
19. O. Bang, P.L. Christiansen, F. If, K. Ø. Rasmussen, Y.B. Gaididei, Temperature effects in a nonlinear model of monolayer scheibe aggregates. Phys. Rev. E **49**, 4627–4636 (1994)
20. F.G. Bass, Y.S. Kivshar, V.V. Konotop, G.M. Pritula, On stochastic dynamics of solitons in inhomogeneous optical fibers. Opt. Commun. **70**(4), 309–314 (1989)
21. F. Baudoin, *An Introduction to the Geometry of Stochastic Flows* (Imperial College Press, London, 2004)
22. P. Baxendale, Brownian motions in the diffeomorphism group. I. Compositio Math. **53**(1), 19–50 (1984)
23. C. Bayer, P. Friz, S. Riedel, J. Schoenmakers, From rough path estimates to multilevel Monte Carlo. SIAM J. Numer. Anal. **54**(3), 1449–1483 (2016)
24. D.A. Beard, T. Schlick, Inertial stochastic dynamics. I: Long-time-step methods for Langevin dynamics. J. Chem. Phys. **112**(17), 7313–7322 (2000)
25. S. Becker, A. Jentzen, Strong convergence rates for nonlinearity-truncated Euler-type approximations of stochastic Ginzburg-Landau equations. Stochastic Process. Appl. **129**(1), 28–69 (2019)
26. S. Becker, A. Jentzen, P.E. Kloeden, An exponential Wagner–Platen type scheme for SPDEs. SIAM J. Numer. Anal. **54**(4), 2389–2426 (2016)
27. L. Bertini, G. Giacomin, Stochastic Burgers and KPZ equations from particle systems. Commun. Math. Phys. **183**(3), 571–607 (1997)
28. L. Bertini, N. Cancrini, G. Jona-Lasinio, The stochastic Burgers equation. Commun. Math. Phys. **165**(2), 211–232 (1994)
29. J. Bismut, *Mécanique Aléatoire*. Lecture Notes in Mathematics, vol. 866 (Springer-Verlag, Berlin-New York, 1981)
30. D. Blackmore, A.K. Prykarpatsky, V.H. Samoylenko, *Nonlinear Dynamical Systems of Mathematical Physics: Spectral and Symplectic Integrability Analysis* (World Scientific, Hackensack, 2011)
31. S. Blanes, A. Iserles, Explicit adaptive symplectic integrators for solving Hamiltonian systems. Celestial Mech. Dynam. Astronom. **114**(3), 297–317 (2012)
32. S.G. Bobkov, F. Götze, Exponential integrability and transportation cost related to logarithmic Sobolev inequalities. J. Funct. Anal. **163**(1), 1–28 (1999)
33. L.L. Bonilla (ed.), *Inverse Problems and Imaging*. Lecture Notes in Mathematics, vol. 1943 (Springer-Verlag/ Fondazione C.I.M.E., Berlin/Florence, 2008)
34. N. Bou-Rabee, H. Owhadi, Boltzmann–Gibbs preserving Langevin integrators. arXiv:0712.4123v3
35. N. Bou-Rabee, H. Owhadi, Stochastic variational integrators. IMA J. Numer. Anal. **29**(2), 421–443 (2009)
36. C.E. Bréhier, J. Cui, J. Hong, Strong convergence rates of semi-discrete splitting approximations for stochastic Allen–Cahn equation. IMA J. Numer. Anal. **39**(4), 2096–2134 (2019)
37. P. Brenner, V. Thomée, On rational approximations of semigroups. SIAM J. Numer. Anal. **16**(4), 683–694 (1979)
38. L. Brugnano, F. Iavernaro, *Line Integral Methods for Conservative Problems*. Monographs and Research Notes in Mathematics (CRC Press, Boca Raton, 2016)
39. K. Burrage, P.M. Burrage, High strong order explicit Runge–Kutta methods for stochastic ordinary differential equations. Appl. Numer. Math. **22**(1–3), 81–101 (1996)
40. K. Burrage, P.M. Burrage, Order conditions of stochastic Runge–Kutta methods by B-series. SIAM J. Numer. Anal. **38**(5), 1626–1646 (2000)
41. K. Burrage, P.M. Burrage, Low rank Runge–Kutta methods, symplecticity and stochastic Hamiltonian problems with additive noise. J. Comput. Appl. Math. **236**(16), 3920–3930 (2012)
42. Y. Cao, J. Hong, Z. Liu, Approximating stochastic evolution equations with additive white and rough noises. SIAM J. Numer. Anal. **55**(4), 1958–1981 (2017)

43. Y. Cao, J. Hong, Z. Liu, Well-posedness and finite element approximations for elliptic SPDEs with Gaussian noises. Commun. Math. Res. **36**(2), 113–127 (2020)
44. C. Cardon-Weber, Cahn–Hilliard stochastic equation: existence of the solution and of its density. Bernoulli **7**(5), 777–816 (2001)
45. J.R. Chaudhuri, P. Chaudhury, S. Chattopadhyay, Harmonic oscillator in presence of nonequilibrium environment. J. Chem. Phys. **130**(23), 234109 (2009)
46. X. Chen, *Random Walk Intersections: Large Deviations and Related Topics*. Mathematical Surveys and Monographs, vol. 157 (American Mathematical Society, Providence, 2009)
47. C. Chen, J. Hong, Symplectic Runge–Kutta semidiscretization for stochastic Schrödinger equation. SIAM J. Numer. Anal. **54**(4), 2569–2593 (2016)
48. R. Chen, M. Tao, Data-driven prediction of general Hamiltonian dynamics via learning exactly-symplectic maps, in *Proceedings of the 38th International Conference on Machine Learning*, vol.139 (2021), pp. 1717–1727
49. C. Chen, D. Cohen, J. Hong, Conservative methods for stochastic differential equations with a conserved quantity. Int. J. Numer. Anal. Model. **13**(3), 435–456 (2016)
50. C. Chen, J. Hong, L. Zhang, Preservation of physical properties of stochastic Maxwell equations with additive noise via stochastic multi-symplectic methods. J. Comput. Phys. **306**, 500–519 (2016)
51. C. Chen, J. Hong, L. Ji, Mean-square convergence of a semidiscrete scheme for stochastic Maxwell equations. SIAM J. Numer. Anal. **57**(2), 728–750 (2019)
52. C. Chen, J. Hong, L. Ji, Runge–Kutta semidiscretizations for stochastic Maxwell equations with additive noise. SIAM J. Numer. Anal. **57**(2), 702–727 (2019)
53. C. Chen, J. Hong, D. Jin, Modified averaged vector field methods preserving multiple invariants for conservative stochastic differential equations. BIT **60**(4), 917–957 (2020)
54. C. Chen, J. Hong, D. Jin, L. Sun, Asymptotically-preserving large deviations principles by stochastic symplectic methods for a linear stochastic oscillator. SIAM J. Numer. Anal. **59**(1), 32–59 (2021)
55. L. Chen, S. Gan, X. Wang, First order strong convergence of an explicit scheme for the stochastic SIS epidemic model. J. Comput. Appl. Math. **392**, 113482 (2021)
56. C. Chen, J. Hong, D. Jin, L. Sun, Large deviations principles for symplectic discretizations of stochastic linear Schrödinger equation. Potential Anal. (2022). https://doi.org/10.1007/s11118-022-09990-z
57. C. Chen, J. Cui, J. Hong, D. Sheng, Convergence of density approximations for stochastic heat equation. arXiv:2007.12960.
58. C. Chen, T. Dang, J. Hong, Weak intermittency of stochastic heat equation under discretizations. J. Differ. Equ. **333**, 268–301 (2022)
59. C. Chen, T. Dang, J. Hong, T. Zhou, CLT for approximating ergodic limit of SPDEs via a full discretization. Stochastic Process. Appl. **157**, 1–41 (2023)
60. C. Chen, J. Hong, C. Huang, Stochastic modified equations for symplectic methods applied to rough Hamiltonian systems based on the Wong–Zakai approximation. arXiv:1907.02825v1.
61. P. Chow, Stochastic wave equations with polynomial nonlinearity. Ann. Appl. Probab. **12**(1), 361–381 (2002)
62. W.T. Coffey, Y.P. Kalmykov, *The Langevin Equation: With Applications to Stochastic Problems in Physics, Chemistry and Electrical Engineering*. World Scientific Series in Contemporary Chemical Physics, vol. 28 (World Scientific, Hackensack, 2017)
63. D. Cohen, On the numerical discretisation of stochastic oscillators. Math. Comput. Simulation **82**(8), 1478–1495 (2012)
64. D. Cohen, G. Dujardin, Energy-preserving integrators for stochastic Poisson systems. Commun. Math. Sci. **12**(8), 1523–1539 (2014)
65. D. Cohen, G. Dujardin, Exponential integrators for nonlinear Schrödinger equations with white noise dispersion. Stoch. Partial Differ. Equ. Anal. Comput. **5**(4), 592–613 (2017)
66. D. Cohen, L. Quer-Sardanyons, A fully discrete approximation of the one-dimensional stochastic wave equation. IMA J. Numer. Anal. **36**(1), 400–420 (2016)

67. D. Cohen, S. Larsson, M. Sigg, A trigonometric method for the linear stochastic wave equation. SIAM J. Numer. Anal. **51**(1), 204–222 (2013)

68. D. Cohen, J. Cui, J. Hong, L. Sun, Exponential integrators for stochastic Maxwell's equations driven by Itô noise. J. Comput. Phys. **410**, 109382, 21 (2020)

69. S. Cox, M. Hutzenthaler, A. Jentzen, Local Lipschitz continuity in the initial value and strong completeness for nonlinear stochastic differential equations. arXiv:1309.5595.

70. J. Cui, L. Dieci, H. Zhou, Time discretizations of Wasserstein–Hamiltonian flows. Math. Comp. **91**(335), 1019–1075 (2022)

71. J. Cui, J. Hong, Analysis of a splitting scheme for damped stochastic nonlinear Schrödinger equation with multiplicative noise. SIAM J. Numer. Anal. **56**(4), 2045–2069 (2018)

72. J. Cui, J. Hong, Strong and weak convergence rates of a spatial approximation for stochastic partial differential equation with one-sided Lipschitz coefficient. SIAM J. Numer. Anal. **57**(4), 1815–1841 (2019)

73. J. Cui, J. Hong, Absolute continuity and numerical approximation of stochastic Cahn–Hilliard equation with unbounded noise diffusion. J. Differ. Equ. **269**(11), 10143–10180 (2020)

74. J. Cui, J. Hong, Z. Liu, Strong convergence rate of finite difference approximations for stochastic cubic Schrödinger equations. J. Differ. Equ. **263**(7), 3687–3713 (2017)

75. J. Cui, J. Hong, Z. Liu, W. Zhou, Stochastic symplectic and multi-symplectic methods for nonlinear Schrödinger equation with white noise dispersion. J. Comput. Phys. **342**, 267–285 (2017)

76. J. Cui, J. Hong, Z. Liu, W. Zhou, Strong convergence rate of splitting schemes for stochastic nonlinear Schrödinger equations. J. Differ. Equ. **266**(9), 5625–5663 (2019)

77. J. Cui, J. Hong, L. Sun, On global existence and blow-up for damped stochastic nonlinear Schrödinger equation. Discrete Contin. Dyn. Syst. Ser. B **24**(12), 6837–6854 (2019)

78. J. Cui, J. Hong, L. Sun, Strong convergence of full discretization for stochastic Cahn-Hilliard equation driven by additive noise. SIAM J. Numer. Anal. **59**(6), 2866–2899 (2021)

79. J. Cui, J. Hong, L. Sun, Weak convergence and invariant measure of a full discretization for parabolic SPDEs with non-globally Lipschitz coefficients. Stochastic Process. Appl. **134**, 55–93 (2021)

80. J. Cui, S. Liu, H. Zhou, What is a stochastic Hamiltonian process on finite graph? An optimal transport answer. J. Differ. Equ. **305**, 428–457 (2021)

81. J. Cui, J. Hong, L. Ji, L. Sun, Energy-preserving exponential integrable numerical method for stochastic cubic wave equation with additive noise. arXiv:1909.00575.

82. J. Cui, J. Hong, D. Sheng, Density function of numerical solution of splitting AVF scheme for stochastic Langevin equation. Math. Comp. **91**(337), 2283–2333 (2022)

83. G. Da Prato, *An Introduction to Infinite-Dimensional Analysis*. Universitext (Springer-Verlag, Berlin, 2006)

84. G. Da Prato, A. Debussche, Stochastic Cahn–Hilliard equation. Nonlinear Anal. **26**(2), 241–263 (1996)

85. G. Da Prato, J. Zabczyk, *Stochastic Equations in Infinite Dimensions*. Encyclopedia of Mathematics and its Applications, vol. 152, 2nd edn. (Cambridge University Press, Cambridge, 2014)

86. R.C. Dalang, M. Sanz-Solé, Hölder-Sobolev regularity of the solution to the stochastic wave equation in dimension three. Mem. Amer. Math. Soc. **199**(931), vi+70 (2009)

87. R.C. Dalang, D. Khoshnevisan, C. Mueller, D. Nualart, Y. Xiao, *A Minicourse on Stochastic Partial Differential Equations*. Lecture Notes in Mathematics, vol. 1962 (Springer-Verlag, Berlin, 2009)

88. G. Darboux, Sur le problème de Pfaff. C. R. Acad. Sci. Paris **94**, 835–837 (1882)

89. A. de Bouard, A. Debussche, The stochastic nonlinear Schrödinger equation in H^1. Stochastic Anal. Appl. **21**(1), 97–126 (2003)

90. A. Debussche, E. Faou, Weak backward error analysis for SDEs. SIAM J. Numer. Anal. **50**(3), 1735–1752 (2012)

91. A. Dembo, O. Zeitouni, *Large Deviations Techniques and Applications*. Stochastic Modelling and Applied Probability, vol. 38 (Springer-Verlag, Berlin, 2010)

92. J. Deng, Strong backward error analysis for Euler–Maruyama method. Int. J. Numer. Anal. Model. **13**(1), 1–21 (2016)
93. J. Deng, C. Anton, Y.S. Wong, High-order symplectic schemes for stochastic Hamiltonian systems. Commun. Comput. Phys. **16**(1), 169–200 (2014)
94. A. Deya, A. Neuenkirch, S. Tindel, A Milstein-type scheme without Lévy area terms for SDEs driven by fractional Brownian motion. Ann. Inst. Henri Poincaré Probab. Stat. **48**(2), 518–550 (2012)
95. P. Dörsek, Semigroup splitting and cubature approximations for the stochastic Navier-Stokes equations. SIAM J. Numer. Anal. **50**(2), 729–746 (2012)
96. V. Duruisseaux, J. Schmitt, M. Leok, Adaptive Hamiltonian variational integrators and applications to symplectic accelerated optimization. SIAM J. Sci. Comput. **43**(4), A2949–A2980 (2021)
97. K.D. Elworthy, *Stochastic Differential Equations on Manifolds*. London Mathematical Society Lecture Note Series, vol. 70 (Cambridge University Press, Cambridge-New York, 1982)
98. P. Etter, Advanced applications for underwater acoustic modeling. Adv. Acoust. Vib. **2012**, 1–28 (2012)
99. S. Fang, D. Luo, Flow of homeomorphisms and stochastic transport equations. Stoch. Anal. Appl. **25**(5), 1079–1108 (2007)
100. K. Feng, M. Qin, *Symplectic Geometric Algorithms for Hamiltonian Systems* (Zhejiang Science and Technology Publishing House/Springer, Hangzhou/Heidelberg, 2010)
101. G. Ferré, H. Touchette, Adaptive sampling of large deviations. J. Stat. Phys. **172**(6), 1525–1544 (2018)
102. F. Flandoli, Dissipativity and invariant measures for stochastic Navier–Stokes equations. Nonlinear Differ. Equ. Appl. **1**(4), 403–423 (1994)
103. F. Flandoli, D. Dariusz, Martingale and stationary solutions for stochastic Navier–Stokes equations. Probab. Theory Related Fields **102**(3), 367–391 (1995)
104. P. Friz, S. Riedel, Convergence rates for the full Gaussian rough paths. Ann. Inst. Henri Poincaré Probab. Stat. **50**(1), 154–194 (2014)
105. P. Friz, N. Victoir, *Multidimensional Stochastic Processes as Rough Paths: Theory and Applications*. Cambridge Studies in Advanced Mathematics, vol. 120 (Cambridge University Press, Cambridge, 2010)
106. T. Fujiwara, H. Kunita, Stochastic differential equations of jump type and Lévy processes in diffeomorphisms group. J. Math. Kyoto Univ. **25**(1), 71–106 (1985)
107. A. Galka, T. Ozaki, H. Muhle, U. Stephani, M. Siniatchkin, A data-driven model of the generation of human eeg based on a spatially distributed stochastic wave equation. Cogn. Neurodyn. **2**(2), 101–113 (2008)
108. M.B. Giles, Multilevel Monte Carlo methods. Acta Numer. **24**, 259–328 (2015)
109. M.B. Giles, T. Nagapetyan, K. Ritter, Multilevel Monte Carlo approximation of distribution functions and densities. SIAM/ASA J. Uncertain. Quantif. **3**(1), 267–295 (2015)
110. D.T. Gillespie, The chemical Langevin equation. J. Chem. Phys. **113**(1), 297–306 (2000)
111. H. Goldstein, *Classical Mechanics*. Addison-Wesley Series in Physics (Addison-Wesley, Reading, 1980)
112. Y.N. Gornostyrev, M.I. Katsnelson, A.V. Trefilov, S.V. Tret'jakov, Stochastic approach to simulation of lattice vibrations in strongly anharmonic crystals: anomalous frequency dependence of the dynamic structure factor. Phys. Rev. B **54**, 3286–3294 (1996)
113. C. Graham, D. Talay, *Stochastic Simulation and Monte Carlo Methods: Mathematical Foundations of Stochastic Simulation*. Stochastic Modelling and Applied Probability, vol. 68 (Springer, Heidelberg, 2013)
114. J. Guckenheimer, P. Holmes, *Nonlinear Oscillations, Dynamical Systems, and Bifurcations of Vector Fields*. Applied Mathematical Sciences, vol. 42 (Springer-Verlag, New York, 1990)
115. I. Gyöngy, A note on Euler's approximations. Potential Anal. **8**(3), 205–216 (1998)
116. E. Hairer, C. Lubich, G. Wanner, *Geometric Numerical Integration: Structure-Preserving Algorithms for Ordinary Differential Equations*. Springer Series in Computational Mathematics, vol. 31, (Springer-Verlag, Berlin, 2002)

117. D.J. Higham, X. Mao, A.M. Stuart, Strong convergence of Euler-type methods for nonlinear stochastic differential equations. SIAM J. Numer. Anal. **40**(3), 1041–1063 (2002)
118. D.J. Higham, X. Mao, A.M. Stuart, Exponential mean-square stability of numerical solutions to stochastic differential equations. LMS J. Comput. Math. **6**, 297–313 (2003)
119. E. Hille, R.S. Phillips, *Functional Analysis and Semi-Groups*. American Mathematical Society Colloquium Publications, vol. XXXI (American Mathematical Society, Providence, 1974)
120. M. Hochbruck, T. Pažur, Implicit Runge–Kutta methods and discontinuous Galerkin discretizations for linear Maxwell's equations. SIAM J. Numer. Anal. **53**(1), 485–507 (2015)
121. M. Hochbruck, T. Jahnke, R. Schnaubelt, Convergence of an ADI splitting for Maxwell's equations. Numer. Math. **129**(3), 535–561 (2015)
122. D.D. Holm, T.M. Tyranowski, Stochastic discrete Hamiltonian variational integrators. BIT **58**(4), 1009–1048 (2018)
123. J. Hong, R. Scherer, L. Wang, Predictor-corrector methods for a linear stochastic oscillator with additive noise. Math. Comput. Model. **46**(5-6), 738–764 (2007)
124. J. Hong, S. Zhai, J. Zhang, Discrete gradient approach to stochastic differential equations with a conserved quantity. SIAM J. Numer. Anal. **49**(5), 2017–2038 (2011)
125. J. Hong, L. Ji, L. Zhang, A stochastic multi-symplectic scheme for stochastic Maxwell equations with additive noise. J. Comput. Phys. **268**, 255–268 (2014)
126. J. Hong, D. Xu, P. Wang, Preservation of quadratic invariants of stochastic differential equations via Runge–Kutta methods. Appl. Numer. Math. **87**, 38–52 (2015)
127. J. Hong, L. Ji, L. Zhang, J. Cai, An energy-conserving method for stochastic Maxwell equations with multiplicative noise. J. Comput. Phys. **351**, 216–229 (2017)
128. J. Hong, L. Sun, X. Wang, High order conformal symplectic and ergodic schemes for the stochastic Langevin equation via generating functions. SIAM J. Numer. Anal. **55**(6), 3006–3029 (2017)
129. J. Hong, X. Wang, L. Zhang, Numerical analysis on ergodic limit of approximations for stochastic NLS equation via multi-symplectic scheme. SIAM J. Numer. Anal. **55**(1), 305–327 (2017)
130. J. Hong, C. Huang, X. Wang, Symplectic Runge–Kutta methods for Hamiltonian systems driven by Gaussian rough paths. Appl. Numer. Math. **129**, 120–136 (2018)
131. J. Hong, X. Wang, *Invariant Measures for Stochastic Nonlinear Schrödinger Equations: Numerical Approximations and Symplectic Structures*. Lecture Notes in Mathematics, vol. 2251 (Springer, Singapore, 2019)
132. J. Hong, J. Ruan, L. Sun, L. Wang, Structure-preserving numerical methods for stochastic Poisson systems. Commun. Comput. Phys. **29**(3), 802–830 (2021)
133. J. Hong, L. Ji, X. Wang, J. Zhang, Positivity-preserving symplectic methods for the stochastic Lotka–Volterra predator-prey model. BIT **62**(2), 493–520 (2022)
134. J. Hong, L. Ji, X. Wang, Convergence in probability of an ergodic and conformal multi-symplectic numerical scheme for a damped stochastic NLS equation. arXiv:1611.08778.
135. L. Hornung, Strong solutions to a nonlinear stochastic maxwell equation with a retarded material law. J. Evol. Equ. **18**, 1427–1469 (2018)
136. Y. Hu, Semi-implicit Euler–Maruyama scheme for stiff stochastic equations, in *Stochastic Analysis and Related Topics, V (Silivri, 1994)*. Progress in Probability, vol. 38 (Birkhäuser Boston, Boston, 1996)
137. Y. Hu, G. Kallianpur, Exponential integrability and application to stochastic quantization. Appl. Math. Optim. **37**(3), 295–353 (1998)
138. Z. Huang, *Fundation of Stochastic Analysis* (Beijing Science Press, Beijing, 2001)
139. M. Hutzenthaler, A. Jentzen, On a perturbation theory and on strong convergence rates for stochastic ordinary and partial differential equations with nonglobally monotone coefficients. Ann. Probab. **48**(1), 53–93 (2020)
140. M. Hutzenthaler, A. Jentzen, P.E. Kloeden, Strong convergence of an explicit numerical method for SDEs with nonglobally Lipschitz continuous coefficients. Ann. Appl. Probab. **22**(4), 1611–1641 (2012)

141. M. Hutzenthaler, A. Jentzen, X. Wang, Exponential integrability properties of numerical approximation processes for nonlinear stochastic differential equations. Math. Comput. **87**(311), 1353–1413 (2018)
142. N. Ikeda, S. Watanabe, *Stochastic Differential Equations and Diffusion Processes*. North-Holland Mathematical Library, vol. 24 (North-Holland Publishing/Kodansha, Ltd., Amsterdam/Tokyo, 1989)
143. J.A. Izaguirre, D.P. Catarello, J.M. Wozniak, R.D. Skeel, Langevin stabilization of molecular dynamics. J. Chem. Phys. **114**(5), 2090–2098 (2001)
144. L. Jacobe de Naurois, A. Jentzen, T. Welti, Lower bounds for weak approximation errors for spatial spectral Galerkin approximations of stochastic wave equations, in *Stochastic Partial Differential Equations and Related Fields*. Springer Proceedings in Mathematics and Statistics, vol. 229 (Springer, Cham, 2018)
145. A. Jentzen, P.E. Kloeden, Overcoming the order barrier in the numerical approximation of stochastic partial differential equations with additive space-time noise. Proc. R. Soc. Lond. Ser. A Math. Phys. Eng. Sci. **465**(2102), 649–667 (2009)
146. S. Jiang, L. Wang, J. Hong, Stochastic multi-symplectic integrator for stochastic nonlinear Schrödinger equation. Commun. Comput. Phys. **14**(2), 393–411 (2013)
147. G. Katsiolides, E.H. Müller, R. Scheichl, T. Shardlow, M.B. Giles, D.J. Thomson, Multilevel Monte Carlo and improved timestepping methods in atmospheric dispersion modelling. J. Comput. Phys. **354**, 320–343 (2018)
148. R. Khasminskii, *Stochastic Stability of Differential Equations*. Stochastic Modelling and Applied Probability, vol. 66 (Springer, Heidelberg, 2012)
149. R. Kleiber, R. Hatzky, A. Könies, K. Kauffmann, P. Helander, An improved control-variate scheme for particle-in-cell simulations with collisions. Comput. Phys. Commun. **182**(4), 1005–1012 (2011)
150. A. Klenke, *Probability Theory: A Comprehensive Course*. Universitext. (Springer, Cham, 2020)
151. P. Kloeden, E. Platen, *Numerical Solution of Stochastic Differential Equations*. Applications of Mathematics (New York), vol. 23 (Springer-Verlag, Berlin, 1992)
152. V.N. Kolokoltsov, Stochastic Hamiltonian–Jacobi–Bellman equation and stochastic Hamiltonian systems. J. Dyn. Control Syst. **2**(3), 299–319 (1996)
153. Y. Komori, Weak second-order stochastic Runge–Kutta methods for non-commutative stochastic differential equations. J. Comput. Appl. Math. **206**(1), 158–173 (2007)
154. Y. Komori, K. Burrage, A stochastic exponential Euler scheme for simulation of stiff biochemical reaction systems. BIT **54**(4), 1067–1085 (2014)
155. Y. Komori, T. Mitsui, H. Sugiura, Rooted tree analysis of the order conditions of ROW-type scheme for stochastic differential equations. BIT **37**(1), 43–66 (1997)
156. Y. Komori, D. Cohen, K. Burrage, Weak second order explicit exponential Runge–Kutta methods for stochastic differential equations. SIAM J. Sci. Comp **39**(6), A2857–A2878 (2017)
157. M. Kopec, Weak backward error analysis for Langevin process. BIT **55**(4), 1057–1103 (2015)
158. M. Kopec, Weak backward error analysis for overdamped Langevin processes. IMA J. Numer. Anal. **35**(2), 583–614 (2015)
159. M. Kovács, S. Larsson, F. Saedpanah, Finite element approximation of the linear stochastic wave equation with additive noise. SIAM J. Numer. Anal. **48**(2), 408–427 (2010)
160. M. Kraus, T.M. Tyranowski, Variational integrators for stochastic dissipative Hamiltonian systems. IMA J. Numer. Anal. **41**(2), 1318–1367 (2021)
161. R. Kruse, *Strong and Weak Approximation of Semilinear Stochastic Evolution Equations*. Lecture Notes in Mathematics, vol. 2093 (Springer, Cham, 2014)
162. H. Kunita, *Lectures on Stochastic Flows and Applications*. Tata Institute of Fundamental Research Lectures on Mathematics and Physics, vol. 78 (Springer-Verlag, Berlin, 1986)
163. H. Kunita, *Stochastic Flows and Stochastic Differential Equations*. Cambridge Studies in Advanced Mathematics, vol. 24 (Cambridge University Press, Cambridge, 1990)
164. H. Kuo, *Introduction to Stochastic Integration*. Universitext (Springer, New York, 2006)

165. L. Kurt, T. Schäfer, Propagation of ultra-short solitons in stochastic Maxwell's equations. J. Math. Phys. **55**(1), 011503, 11 (2014)

166. A.M. Lacasta, J.M. Sancho, A.H. Romero, I.M. Sokolov, K. Lindenberg, From subdiffusion to superdiffusion of particles on solid surfaces. Phys. Rev. E **70**, 051104 (2004)

167. P. Lalanne, M.P. Jurek, Computation of the near-field pattern with the coupled-wave method for transverse magnetic polarization. J. Mod. Opt. **45**(7), 1357–1374 (1998)

168. P.S. Landa, Noise-induced transport of brownian particles with consideration for their mass. Phys. Rev. E **58**, 1325–1333 (1998)

169. J.A. Lázaro-Camí, J.P. Ortega, Stochastic Hamiltonian dynamical systems. Rep. Math. Phys. **61**(1), 65–122 (2008)

170. Y. Le Jan, Flots de diffusion dans $\mathbf{R}^d$. C. R. Acad. Sci. Paris Sér. I Math. **294**(21), 697–699 (1982)

171. M. Leok, J. Zhang, Discrete Hamiltonian variational integrators. IMA J. Numer. Anal. **31**(4), 1497–1532 (2011)

172. Q. Li, C. Tai, E. Weinan, Stochastic modified equations and adaptive stochastic gradient algorithms, in *Proceedings of the 34th International Conference on Machine Learning*, vol. 70 (2017), pp. 2101–2110

173. Q. Li, C. Tai, E. Weinan, Stochastic modified equations and dynamics of stochastic gradient algorithms I: mathematical foundations. J. Mach. Learn. Res. **20**, 40 (2019)

174. K.B. Liaskos, I.G. Stratis, A.N. Yannacopoulos, Stochastic integrodifferential equations in Hilbert spaces with applications in electromagnetics. J. Integral Equ. Appl. **22**(4), 559–590 (2010)

175. S. Lie, Zur theorie der transformationsgruppen. Christ. Forh. Aar. **5**, 553–557 (1888)

176. W. Liu, M. Röckner, *Stochastic Partial Differential Equations: An Introduction*. Universitext (Springer, Cham, 2015)

177. Z. Liu, Z. Qiao, Strong approximation of monotone stochastic partial differential equations driven by white noise. IMA J. Numer. Anal. **40**(2), 1074–1093 (2020)

178. G.J. Lord, A. Tambue, Stochastic exponential integrators for the finite element discretization of SPDEs for multiplicative and additive noise. IMA J. Numer. Anal. **33**(2), 515–543 (2013)

179. T. Lyons, Differential equations driven by rough signals. Rev. Mat. Iberoamericana **14**(2), 215–310 (1998)

180. Q. Ma, D. Ding, X. Ding, Symplectic conditions and stochastic generating functions of stochastic Runge–Kutta methods for stochastic Hamiltonian systems with multiplicative noise. Appl. Math. Comput. **219**(2), 635–643 (2012)

181. S.J.A. Malham, A. Wiese, Stochastic Lie group integrators. SIAM J. Sci. Comput. **30**(2), 597–617 (2008)

182. B.B. Mandelbrot, J.W. Van Ness, Fractional Brownian motions, fractional noises and applications. SIAM Rev. **10**, 422–437 (1968)

183. X. Mao, *Stochastic Differential Equations and Their Applications*. Horwood Publishing Series in Mathematics & Applications (Horwood Publishing Limited, Chichester, 1997)

184. X. Mao, F. Wei, T. Wiriyakraikul, Positivity preserving truncated Euler–Maruyama method for stochastic Lotka–Volterra competition model. J. Comput. Appl. Math. **394**, 113566 (2021)

185. L. Markus, A. Weerasinghe, Stochastic oscillators. J. Differ. Equ. **71**(2), 288–314 (1988)

186. J.E. Marsden, A. Weinstein, The Hamiltonian structure of the Maxwell-Vlasov equations. Phys. D **4**(3), 394–406 (1981/1982)

187. J.E. Marsden, M. West, Discrete mechanics and variational integrators. Acta Numer. **10**, 357–514 (2001)

188. J.C. Mattingly, A.M. Stuart, D.J. Higham, Ergodicity for SDEs and approximations: locally Lipschitz vector fields and degenerate noise. Stochastic Process. Appl. **101**(2), 185–232 (2002)

189. R.I. McLachlan, G.R.W. Quispel, Explicit geometric integration of polynomial vector fields. BIT **44**(3), 515–538 (2004)

190. S.P. Meyn, R.L. Tweedie, *Markov Chains and Stochastic Stability*. Communications and Control Engineering Series (Springer-Verlag London, London, 1993)

191. D. Mihalache, G.I. Stegeman, C.T. Seaton, E.M. Wright, R. Zanoni, A.D. Boardman, T. Twardowski, Exact dispersion relations for transverse magnetic polarized guided waves at a nonlinear interface. Opt. Lett. **12**(3), 187–189 (1987)
192. G.N. Milstein, Weak approximation of solutions of systems of stochastic differential equations. Teor. Veroyatnost. i Primenen. **30**(4), 706–721 (1985)
193. G.N. Milstein, *Numerical Integration of Stochastic Differential Equations*. Mathematics and its Applications, vol. 313 (Kluwer Academic Publishers Group, Dordrecht, 1995)
194. G.N. Milstein, M.V. Tretyakov, Quasi-symplectic methods for Langevin-type equations. IMA J. Numer. Anal. **23**(4), 593–626 (2003)
195. G.N. Milstein, M.V. Tretyakov, Numerical integration of stochastic differential equations with nonglobally Lipschitz coefficients. SIAM J. Numer. Anal. **43**(3), 1139–1154 (2005)
196. G.N. Milstein, M.V. Tretyakov, Monte Carlo methods for backward equations in nonlinear filtering. Adv. Appl. Probab. **41**(1), 63–100 (2009)
197. G.N. Milstein, M.V. Tretyakov, *Stochastic Numerics for Mathematical Physics*. Scientific Computation, 2nd edn. (Springer, Cham, 2021)
198. G.N. Milstein, Y.M. Repin, M.V. Tretyakov, Mean-square symplectic methods for Hamiltonian systems with multiplicative noise (2001)
199. G.N. Milstein, Y.M. Repin, M.V. Tretyakov, Numerical methods for stochastic systems preserving symplectic structure. SIAM J. Numer. Anal. **40**(4), 1583–1604 (2002)
200. G.N. Milstein, Y.M. Repin, M.V. Tretyakov, Symplectic integration of Hamiltonian systems with additive noise. SIAM J. Numer. Anal. **39**(6), 2066–2088 (2002)
201. T. Misawa, Conserved quantities and symmetry for stochastic dynamical systems. Phys. Lett. A **195**(3–4), 185–189 (1994)
202. T. Misawa, Energy conservative stochastic difference scheme for stochastic Hamilton dynamical systems. Jpn. J. Indust. Appl. Math. **17**(1), 119–128 (2000)
203. T. Misawa, Symplectic integrators to stochastic Hamiltonian dynamical systems derived from composition methods. Math. Probl. Eng. **2010**, Article ID 384937, 1–12 (2010)
204. P. Monk, *Finite Element Methods for Maxwell's Equations*. Numerical Mathematics and Scientific Computation (Oxford University Press, New York, 2003)
205. E.H. Müller, R. Scheichl, T. Shardlow, Improving multilevel Monte Carlo for stochastic differential equations with application to the Langevin equation. Proc. A. **471**(2176), 20140679 (2015)
206. N.J. Newton, Asymptotically efficient Runge–Kutta methods for a class of Itô and Stratonovich equations. SIAM J. Appl. Math. **51**(2), 542–567 (1991)
207. X. Niu, J. Cui, J. Hong, Z. Liu, Explicit pseudo-symplectic methods for stochastic Hamiltonian systems. BIT **58**(1), 163–178 (2018)
208. S. Ober-Blöbaum, Galerkin variational integrators and modified symplectic Runge–Kutta methods. IMA J. Numer. Anal. **37**(1), 375–406 (2017)
209. S. Ober-Blöbaum, N. Saake, Construction and analysis of higher order Galerkin variational integrators. Adv. Comput. Math. **41**(6), 955–986 (2015)
210. G.N. Ord, A stochastic model of Maxwell's equations in 1 + 1 dimensions. Int. J. Theor. Phys. **35**(2), 263–266 (1996)
211. E. Orsingher, Randomly forced vibrations of a string. Ann. Inst. H. Poincaré Sect. B (N.S.) **18**(4), 367–394 (1982)
212. E.S. Palamarchuk, An analytic study of the Ornstein-Uhlenbeck process with time-varying coefficients in the modeling of anomalous diffusions. Autom. Remote Control **79**(2), 289–299 (2018)
213. C. Park, D.J. Scheeres, Solutions of the optimal feedback control problem using Hamiltonian dynamics and generating functions, in *IEEE International Conference on Decision and Control*, vol. 2 (2003), pp. 1222–1227
214. C. Park, D.J. Scheeres, Solutions of optimal feedback control problems with general boundary conditions using Hamiltonian dynamics and generating functions, in *Proceedings of the 2004 American Control Conference*, vol. 1 (2004), pp. 679–684

215. G.A. Pavliotis, A.M. Stuart, K.C. Zygalakis, Calculating effective diffusiveness in the limit of vanishing molecular diffusion. J. Comput. Phys. **228**(4), 1030–1055 (2009)
216. S. Peng, Problem of eigenvalues of stochastic Hamiltonian systems with boundary conditions. Stochastic Process. Appl. **88**(2), 259–290 (2000)
217. S. Peszat, J. Zabczyk, Nonlinear stochastic wave and heat equations. Probab. Theory Related Fields **116**(3), 421–443 (2000)
218. E.P. Philip, Stochastic integration and differential equations. *Applications of Mathematics (New York)*. Stochastic Modelling and Applied Probability, vol. 21 (Springer-Verlag, Berlin, 2004)
219. R. Qi, X. Wang, An accelerated exponential time integrator for semi-linear stochastic strongly damped wave equation with additive noise. J. Math. Anal. Appl. **447**(2), 988–1008 (2017)
220. R. Qi, X. Wang, Error estimates of finite element method for semilinear stochastic strongly damped wave equation. IMA J. Numer. Anal. **39**(3), 1594–1626 (2019)
221. K. Rasmussen, Y.B. Gaididei, O. Bang, P. Christiansen, The influence of noise on critical collapse in the nonlinear Schrödinger equation. Phys. Lett. A **204**(2), 121–127 (1995)
222. R. Reigada, A.H. Romero, A. Sarmiento, K. Lindenberg, One-dimensional arrays of oscillators: energy localization in thermal equilibrium. J. Chem. Phys. **111**(4), 1373–1384 (1999)
223. J. Ren, X. Zhang, Stochastic flow for SDEs with non-Lipschitz coefficients. Bull. Sci. Math. **127**(8), 739–754 (2003)
224. M. Ripoll, M.H. Ernst, P. Español, Large scale and mesoscopic hydrodynamics for dissipative particle dynamics. J. Chem. Phys. **115**(15), 7271–7284 (2001)
225. C.M. Rohwer, F. Angeletti, H. Touchette, Convergence of large-deviation estimators. Phys. Rev. E **92**, 052104 (2015)
226. A. Rössler, Runge–Kutta methods for Stratonovich stochastic differential equation systems with commutative noise. J. Comput. Appl. Math. **164/165**, 613–627 (2004)
227. A. Rössler, Second order Runge–Kutta methods for Stratonovich stochastic differential equations. BIT **47**(3), 657–680 (2007)
228. R. Rudnicki, Long-time behaviour of a stochastic prey-predator model. Stochastic Process. Appl. **108**(1), 93–107 (2003)
229. S.M. Rytov, Y.A. Kravtsov, V.I. Tatarskiĭ, *Principles of Statistical Radiophysics*, vol. 3 (Springer-Verlag, Berlin, 1989)
230. B.L. Sawford, Turbulent relative dispersion. Annu. Rev. Fluid Mech. **33**, 289–317 (2001)
231. C. Scalone, Positivity preserving stochastic θ-methods for selected SDEs. Appl. Numer. Math. **172**, 351–358 (2022)
232. M. Schienbein, H. Gruler, Langevin equation, Fokker–Planck equation and cell migration. Bull. Math. Biol. **55**(3), 585–608 (1993)
233. M. Seeßelberg, H.P. Breuer, H. Mais, F. Petruccione, J. Honerkamp, Simulation of one-dimensional noisy Hamiltonian systems and their application to particle storage rings. Z. Phys. C Particles Fields **62**(1), 63–73 (1994)
234. M.J. Senosiain, A. Tocino, A review on numerical schemes for solving a linear stochastic oscillator. BIT **55**(2), 515–529 (2015)
235. T. Shardlow, Modified equations for stochastic differential equations. BIT **46**(1), 111–125 (2006)
236. C. Shi, Y. Xiao, C. Zhang. The convergence and MS stability of exponential Euler method for semilinear stochastic differential equations. Abstr. Appl. Anal. **2012**, Article ID 350407, 19 (2012)
237. B. Shi, S.S. Du, W. Su, M.I. Jordan, Acceleration via symplectic discretization of high-resolution differential equations, in *Advances in NIPS*, vol. 32 (2019)
238. R.D. Skeel, Integration schemes for molecular dynamics and related applications, in *The Graduate Student's Guide to Numerical Analysis '98 (Leicester)*. Springer Series in Computational Mathematics, vol. 26 (Springer, Berlin, 1999)
239. M. Song, X. Qian, T. Shen, S. Song, Stochastic conformal schemes for damped stochastic Klein-Gordon equation with additive noise. J. Comput. Phys. **411**, 109300, 20 (2020)

240. E. Sonnendrücker, A. Wacher, R. Hatzky, R. Kleiber, A split control variate scheme for PIC simulations with collisions. J. Comput. Phys. **295**, 402–419 (2015)
241. O.D. Street, D. Crisan, Semi-martingale driven variational principles. Proc. A. **477**(2247), 20200957 (2021)
242. A.H. Strømmen Melbø, D.J. Higham, Numerical simulation of a linear stochastic oscillator with additive noise. Appl. Numer. Math. **51**(1), 89–99 (2004)
243. D.W. Stroock, *Lectures on Topics in Stochastic Differential Equations*. Tata Institute of Fundamental Research Lectures on Mathematics and Physics, vol. 68 (Tata Institute of Fundamental Research/Springer-Verlag, Bombay/Berlin-New York, 1982)
244. J.B. Sturgeon, B.B. Laird, Symplectic algorithm for constant-pressure molecular dynamics using a Nosé–Poincaré thermostat. J. Chem. Phys. **112**(8), 3474–3482 (2000)
245. L. Sun, L. Wang, Stochastic symplectic methods based on the Padé approximations for linear stochastic Hamiltonian systems. J. Comput. Appl. Math. **311**, 439–456 (2017)
246. J. Sun, C. Shu, Y. Xing, Multi-symplectic discontinuous Galerkin methods for the stochastic Maxwell equations with additive noise. J. Comput. Phys. **461**, 111199 (2022)
247. A. Süß , M. Waurick, A solution theory for a general class of SPDEs. Stoch. Partial Differ. Equ. Anal. Comput. **5**(2), 278–318 (2017)
248. D. Talay, Stochastic Hamiltonian systems: exponential convergence to the invariant measure, and discretization by the implicit Euler scheme. Markov Process. Related Fields **8**(2), 163–198 (2002)
249. M. Tao, Explicit high-order symplectic integrators for charged particles in general electromagnetic fields. J. Comput. Phys. **327**, 245–251 (2016)
250. M. Thieullen, J.C. Zambrini, Probability and quantum symmetries. I: the theorem of Noether in Schrödinger's Euclidean quantum mechanics. Ann. Inst. H. Poincaré Phys. Théor. **67**(3), 297–338 (1997)
251. L.E. Thomas, Persistent energy flow for a stochastic wave equation model in nonequilibrium statistical mechanics. J. Math. Phys. **53**(9), 095208, 10 (2012)
252. D.J. Thomson, Criteria for the selection of stochastic models of particle trajectories in turbulent flows. J. Fluid Mech. **180**, 529–556 (1987)
253. A. Tocino, On preserving long-time features of a linear stochastic oscillator. BIT **47**(1), 189–196 (2007)
254. A. Tocino, J. Vigo-Aguiar, Weak second order conditions for stochastic Runge–Kutta methods. SIAM J. Sci. Comput. **24**(2), 507–523 (2002)
255. M.V. Tretyakov, Z. Zhang, A fundamental mean-square convergence theorem for SDEs with locally Lipschitz coefficients and its applications. SIAM J. Numer. Anal. **51**(6), 3135–3162 (2013)
256. T.M. Tyranowski, Stochastic variational principles for the collisional Vlasov–Maxwell and Vlasov–Poisson equations. Proc. A. **477**(2252), 20210167 (2021)
257. H. van Wyk, M. Gunzburger, J. Burkhardt, M. Stoyanov, Power-law noises over general spatial domains and on nonstandard meshes. SIAM/ASA J. Uncertain. Quantif. **3**(1), 296–319 (2015)
258. C. Viterbo, An introduction to symplectic topology. *Journées "Équations aux Dérivées Partielles" (Saint Jean de Monts, 1991)* (École Polytech, Palaiseau, 1991)
259. B. Vujanović, A variational principle for non-conservative dynamical systems. Z. Angew. Math. Mech. **55**(6), 321–331 (1975)
260. L. Wang, Variational integrators and generating functions for stochastic Hamiltonian systems. Ph.D. Thesis, Karlsruhe Institute of Technology, 2007
261. X. Wang, An exponential integrator scheme for time discretization of nonlinear stochastic wave equation. J. Sci. Comput. **64**(1), 234–263 (2015)
262. X. Wang, An efficient explicit full-discrete scheme for strong approximation of stochastic Allen-Cahn equation. Stochastic Process. Appl. **130**(10), 6271–6299 (2020)
263. L. Wang, J. Hong, Generating functions for stochastic symplectic methods. Discrete Contin. Dyn. Syst. **34**(3), 1211–1228 (2014)

264. L. Wang, J. Hong, L. Sun, Modified equations for weakly convergent stochastic symplectic schemes via their generating functions. BIT **56**(3), 1131–1162 (2016)
265. P. Wang, J. Hong, D. Xu, Construction of symplectic Runge–Kutta methods for stochastic Hamiltonian systems. Commun. Comput. Phys. **21**(1), 237–270 (2017)
266. H. Yang, J. Huang, First order strong convergence of positivity preserving logarithmic Euler–Maruyama method for the stochastic SIS epidemic model. Appl. Math. Lett. **121**, 107451 (2021)
267. Y. Yi, Y. Hu, J. Zhao, Positivity preserving logarithmic Euler–Maruyama type scheme for stochastic differential equations. Commun. Nonlinear Sci. Numer. Simul. **101**, 105895 (2021)
268. X. Zhang, Homeomorphic flows for multi-dimensional SDEs with non-Lipschitz coefficients. Stoch. Process. Appl. **115**(3), 435–448 (2005)
269. X. Zhang, Stochastic flows and Bismut formulas for stochastic Hamiltonian systems. Stoch. Process. Appl. **120**(10), 1929–1949 (2010)
270. Z. Zhang, G.E. Karniadakis, *Numerical Methods for Stochastic Partial Differential Equations with White Noise*. Applied Mathematical Sciences, vol. 196 (Springer, Cham, 2017)
271. L. Zhang, C. Chen, J. Hong, L. Ji, A review on stochastic multi-symplectic methods for stochastic Maxwell equations. Commun. Appl. Math. Comput. **1**(3), 467–501 (2019)
272. W. Zhou, L. Zhang, J. Hong, S. Song, Projection methods for stochastic differential equations with conserved quantities. BIT **56**(4), 1497–1518 (2016)
273. X. Zhou, H. Yuan, C.J. Li, Q. Sun, Stochastic modified equations for continuous limit of stochastic ADMM. arXiv:2003.03532
274. K.C. Zygalakis, On the existence and the applications of modified equations for stochastic differential equations. SIAM J. Sci. Comput. **33**(1), 102–130 (2011)

LECTURE NOTES IN MATHEMATICS 🐎 Springer

Editors in Chief: J.-M. Morel, B. Teissier;

Editorial Policy

1. Lecture Notes aim to report new developments in all areas of mathematics and their applications – quickly, informally and at a high level. Mathematical texts analysing new developments in modelling and numerical simulation are welcome.

   Manuscripts should be reasonably self-contained and rounded off. Thus they may, and often will, present not only results of the author but also related work by other people. They may be based on specialised lecture courses. Furthermore, the manuscripts should provide sufficient motivation, examples and applications. This clearly distinguishes Lecture Notes from journal articles or technical reports which normally are very concise. Articles intended for a journal but too long to be accepted by most journals, usually do not have this "lecture notes" character. For similar reasons it is unusual for doctoral theses to be accepted for the Lecture Notes series, though habilitation theses may be appropriate.

2. Besides monographs, multi-author manuscripts resulting from SUMMER SCHOOLS or similar INTENSIVE COURSES are welcome, provided their objective was held to present an active mathematical topic to an audience at the beginning or intermediate graduate level (a list of participants should be provided).

   The resulting manuscript should not be just a collection of course notes, but should require advance planning and coordination among the main lecturers. The subject matter should dictate the structure of the book. This structure should be motivated and explained in a scientific introduction, and the notation, references, index and formulation of results should be, if possible, unified by the editors. Each contribution should have an abstract and an introduction referring to the other contributions. In other words, more preparatory work must go into a multi-authored volume than simply assembling a disparate collection of papers, communicated at the event.

3. Manuscripts should be submitted either online at www.editorialmanager.com/lnm to Springer's mathematics editorial in Heidelberg, or electronically to one of the series editors. Authors should be aware that incomplete or insufficiently close-to-final manuscripts almost always result in longer refereeing times and nevertheless unclear referees' recommendations, making further refereeing of a final draft necessary. The strict minimum amount of material that will be considered should include a detailed outline describing the planned contents of each chapter, a bibliography and several sample chapters. Parallel submission of a manuscript to another publisher while under consideration for LNM is not acceptable and can lead to rejection.

4. In general, **monographs** will be sent out to at least 2 external referees for evaluation.

   A final decision to publish can be made only on the basis of the complete manuscript, however a refereeing process leading to a preliminary decision can be based on a pre-final or incomplete manuscript.

   Volume Editors of **multi-author works** are expected to arrange for the refereeing, to the usual scientific standards, of the individual contributions. If the resulting reports can be forwarded to the LNM Editorial Board, this is very helpful. If no reports are forwarded or if other questions remain unclear in respect of homogeneity etc, the series editors may wish to consult external referees for an overall evaluation of the volume.

5. Manuscripts should in general be submitted in English. Final manuscripts should contain at least 100 pages of mathematical text and should always include

   – a table of contents;
   – an informative introduction, with adequate motivation and perhaps some historical remarks: it should be accessible to a reader not intimately familiar with the topic treated;
   – a subject index: as a rule this is genuinely helpful for the reader.
   – For evaluation purposes, manuscripts should be submitted as pdf files.

6. Careful preparation of the manuscripts will help keep production time short besides ensuring satisfactory appearance of the finished book in print and online. After acceptance of the manuscript authors will be asked to prepare the final LaTeX source files (see LaTeX templates online: https://www.springer.com/gb/authors-editors/book-authors-editors/manuscriptpreparation/5636) plus the corresponding pdf- or zipped ps-file. The LaTeX source files are essential for producing the full-text online version of the book, see http://link.springer.com/bookseries/304 for the existing online volumes of LNM). The technical production of a Lecture Notes volume takes approximately 12 weeks. Additional instructions, if necessary, are available on request from lnm@springer.com.

7. Authors receive a total of 30 free copies of their volume and free access to their book on SpringerLink, but no royalties. They are entitled to a discount of 33.3 % on the price of Springer books purchased for their personal use, if ordering directly from Springer.

8. Commitment to publish is made by a *Publishing Agreement*; contributing authors of multiauthor books are requested to sign a *Consent to Publish form*. Springer-Verlag registers the copyright for each volume. Authors are free to reuse material contained in their LNM volumes in later publications: a brief written (or e-mail) request for formal permission is sufficient.

Addresses:
Professor Jean-Michel Morel, CMLA, École Normale Supérieure de Cachan, France
E-mail: moreljeanmichel@gmail.com

Professor Bernard Teissier, Equipe Géométrie et Dynamique,
Institut de Mathématiques de Jussieu – Paris Rive Gauche, Paris, France
E-mail: bernard.teissier@imj-prg.fr

Springer: Ute McCrory, Mathematics, Heidelberg, Germany,
E-mail: lnm@springer.com

Printed in the United States
by Baker & Taylor Publisher Services